STATISTICS-I

FOR BBA STUDENTS

AF632383

STATISTICS-I

FOR BBA STUDENTS

(As per WBUT New Syllabus)

Satyabrata Pal
Principal, Swami Vivekananda Institute of Management and Computer Science
Karbala Sonarpur, Kolkata
Formerly
Principal NSHM College of Management and Technology
Durgapur, W.B.
Dean, Post-Graduate Studies and Professor
Bidhan Chandra Krishi Viswavidyalaya, Mohanpur
Nadia, W.B.

Samarendra Nath Pal
Formerly, Scientist, Indian Jute Industries Research Association
Taratala, Kolkata

Subhabaha Pal
Consultant and Member of Academic Central
Sikkim Manipal University-DDE
&
Academic Coordinator (Management and Commerce)
Technology Enabled Learning Portal (SMU)
Formerly, Risk Management Quantification Specialist
Corporate Risk Management Department
Kuwait Petroleum Corporation, Kuwait

NEW AGE
WBUT

An Imprint of

NEW AGE INTERNATIONAL (P) LIMITED, PUBLISHERS
New Delhi • Bangalore • Chennai • Cochin • Guwahati • Hyderabad
Jalandhar • Kolkata • Lucknow • Mumbai • Ranchi
Visit us at **www.newagepublishers.com**

Copyright © 2010, New Age International (P) Ltd., Publishers
Published by New Age International (P) Ltd., Publishers
First Edition: 2010

All rights reserved.

No part of this book may be reproduced in any form, by photostat, microfilm, xerography, or any other means, or incorporated into any information retrieval system, electronic or mechanical, without the written permission of the copyright owner.

Branches:

- 36, Malikarjuna Temple Street, Opp. ICWA, Basavanagudi, **Bangalore**. ✆ (080) 26677815
- 26, Damodaran Street, T. Nagar, **Chennai**. ✆ (044) 24353401
- CC-39/1016, Carrier Station Road, Ernakulam South, **Cochin.** ✆ (0484) 4051303
- Hemsen Complex, Mohd. Shah Road, Paltan Bazar, Near Starline Hotel, **Guwahati**. ✆ (0361) 2543669
- No. 105, 1st Floor, Madhiray Kaveri Tower, 3-2-19, Azam Jahi Road, Nimboliadda, **Hyderabad**. ✆ (040) 24652456
- RDB Chambers (Formerly Lotus Cinema) 106A, 1st Floor, S.N. Banerjee Road, **Kolkata**. ✆ (033) 22275247
- 16-A, Jopling Road, **Lucknow**. ✆ (0522) 2209578
- 142C, Victor House, Ground Floor, N.M. Joshi Marg, Lower Parel, **Mumbai**. ✆ (022) 24927869
- 22, Golden House, Daryaganj, **New Delhi**. ✆ (011) 23262370, 23262368

ISBN : 978-81-224-2812-4

Rs. 150.00

C-09-10-4118

Printed in India at Ajit Printers, Delhi.
Typeset at Print & Publicity, Kolkata.

PUBLISHING FOR ONE WORLD
NEW AGE INTERNATIONAL (P) LIMITED, PUBLISHERS
4835/24, Ansari Road, Daryaganj, New Delhi-110002
Visit us at **www.newagepublishers.com**

Preface

Statistics –I for BBA students constitutes a distinct text and guide book covering the syllabus on Statistics (Paper) included under the BBA programme in West Bengal University of Science and Technology (WBUT) and also the syllabi on Statistics included under papers (Statistics) in ICWAI. ICAI, B.COM, M.COM, BBA and MBA examinations. The book is replete with ample (also typical) examples, solved problems from question papers of university and professional examinations and interesting exercises for follow-up enterprise by the students. The entire content is presented in simple, elegant, lucid and concise manner retaining the essential rigour. The organization of the book has taken care of the specificities in terms of the emphasis being given to address the depth in regard to the subject matter included under each of the above courses (stream-wise)–truly the intending students due to appear at such examinations will find the exposition offering a rich blend of theory along with their applications pertaining to the respective course for a deep and lasting understanding in respect of its content. In the process of composition of the book the authors have drawn upon their vast reserve of experience acquired over a long span of years of teaching imparted at different institutions towards dissemination of the academic matter embodied under the book.

A glimpse at the Chapter-titles in the book reveals the following sequence of Chapter arrangements.

Basis of Statistical Philosophy, Types of data and their Collection and Presentation, Frequency Distribution and its Representation, Characteristics of Frequency Distribution and their Measures, Statistical Treatment of Bivariate Data, Index Number, Time Series and Addendum.

It is expected that the book will be immensely useful to all categories of students, specially, for its student-friendly presentation and lucid delineation of the subject matter. The Index included at the end of the book is an wealth and it instantly helps the reader to easily get an access to the page corresponding to a term/definition of his interest.

Adequate care has been taken during the composition of the book, however, if any errors, crept during the process, are brought to the notice of the authors, the names of such persons will be duly acknowledged in the next edition of the book. The authors will be grateful to the readers who, in the near future, will communicate with the suggestions for improvement. It will be a matter of deep satisfaction if the communities of students to whom the book is addressed, find it useful.

Authors

STATISTICS -1 (WBUT SYLLABUS)

BBA-103 : STATISTICS - I

Course Content :

1. Introduction : Statistics as a subject; Functions, Importance and Limitations of Statistics; Planning and Execution of a statistical investigation; Census and sample investigation; Descriptive and Inferential statistics. [2L]

2. Collection, Editing and Presentation of Data : Primary data and secondary data; Methods of collection; Scrutiny of data. Presentation of data: textual and tabular presentations; Construction of a table and the different components of a table. Diagrammatic representation of data: Line diagrams, Bar diagrams, Pie charts and divided-bar diagrams. [3L]

3. Frequency Distributions : Attribute and variable; Frequency distribution of an attribute; Discrete and continuous variables; Frequency distributions of discrete and continuous variables; Bivariate and Multivariate Frequency Distributions. Diagrammatic representation of a frequency distribution: case of an attribute; case of a discrete variable: column diagram, frequency polygon and step diagram; case of a continuous variable: histogram and ogive. [4L]

4. Measures of Central Tendency : Definition and utility; Characteristics of a good average; Different measures of average; Arithmetic Mean; Median; Other positional measures - quartiles, deciles, percentiles; Mode; Relation between Mean, Median and Mode; Geometric and Harmonic Mean. Choice of a suitable measure of central tendency. [6L]

5. Measures of dispersion : Meaning and objective of dispersion; Characteristics of a good measure of dispersion; Different measures of dispersion - Range, Quartile deviation, Mean deviation, Mean Absolute deviation, Standard deviation; Comparison of the different measures of dispersion. Measures of relative dispersion - Coefficient of Variation. Combined mean and standard deviation. Chebyschev's Theorem. Graphical measure of dispersion - Lorenz curve; Gini coefficient. [7L]

6. Moments, Skewness and Kurtosis : Moments; Coefficients based on moments; Sheppard's correction; Skewness; Measures of skewness; Kurtosis and its measures. [3L]

7. Correlation and Regression : Analysis of Bivariate data. Correlation Analysis -Meaning of correlation; Scatter Diagram; Karl Pearson's coefficient of linear correlation; Calculation of the correlation coefficient from grouped data; Properties of the correlation coefficient; Advantages and limitations of the coefficient of correlation; Idea of rank correlation; Spearman's rank correlation coefficient. Regression Analysis - Two lines of regression; Some important results relating to regression lines; Correlation Coefficient and the two Regression Coefficients; Coefficient of determination; Concept of ultiple regression.[7L]

8. Index Numbers : Definition, characteristic and uses of index numbers; Methods of constructing price and quantity indices (simple and aggregate); Value index; Cbmparison of Laspeyres' and Paasche's Index Numbers; Tests of adequacy; Chain-base index numbers; Base shifting, splicing and deflating; Consumer Price Index Numbers; Problems in the construction of index numbers. [4L]

9. Analysis of Time Series : Objective of time series analysis; Causes of variations in time series data; Components of a time series; Decomposition - Additive and Multiplicative Models; Determination of trend - Moving averages method and method of least squares; Measurement of secular trend; Seasonal variations; Measurement of cyclical variations; Measurement of random variations. [4L]

Contents

STATISTICAL METHOD(S)

INDEX NUMBER & TIME SERIES

CHAPTER 1

BASIS OF STATISTICAL PHILOSOPHY

1.1. Introducing Statistics

The act of recording numerical facts and figures followed as a sequel to human discovery of counting procedure. The formal way of recording such numerical facts, however, emerged much later. It is developing to the present day when computer is used extensively to keep record of such information in terms of numerical data. As examples of numerical facts the following may be cited: the number of television sets of different makes produced each year, the daily prices of agricultural commodities in different markets, the export and import figures over the last fifteen years, the amount spent on Defence as provided in the annual budgets over the last ten years, the gene-bank keeping the record of varied information on different species. Prior to the advent of twentieth century such numerical facts and figures, which are, in fact, numerical expressions of information (also called information) are popularly called Statistics. Indeed, each of the above cited examples is a result of a particular kind of human activity, and as such these information are also called Data. A single numerical figure/observation is called a Datum in this sense, information, data and statistics are synonymous. From the above it is seen that data are the outcomes of an activity, and, indeed, every activity, be it past or present, performed on earth is an eternal source of data though the more important ones find their places in human record, the rest being lost for ever. It is interesting to note the various types of data that can be generated through human or natural activities. As human activities are diverse encompassing multitudes of areas, let us take an account of the daily activities performed by a human being—a student, a labourer, an employee in an office, an administrator, a politician, a businessman or a man engaged in any other profession. Any individual, irrespective of his profession, commences his daily routine by taking a tooth brush with a spread of paste on it. The particular information being generated from his brushing and washing activities relate to the quality of such activities, and can be recorded in terms of "either good or bad", and form a kind of data. Similarly, the number of cigarettes smoked per day, the mode of the journey conducted, the distance travelled to reach the place of work, the mode of spending the leisure time, the time taken to complete the dinner, are all examples of daily routine activities, and the data that emanate from such activities are either qualitative or quantitative (see Sec. 2.3) in nature.

The foregoing paragraph discusses what statistics (or data) are and from where these are generated. This meaning of the word "Statistics" is used in the plural sense and is the most familiar to us. However, apart from this usage of the term, the word "Statistics" is the name of a subject, and in this sense it refers to the singular meaning of the word. Statistics, as a subject, made its formal beginning in the early years of twentieth century, and by the end of it we observe that its growth is not only enormous but remarkable. It has established its potential use in almost all branches of science, technology and management. The following paragraph deals with the definition of the subject "Statistics."

The definition of the subject "Statistics" is embedded in the process of generation of data from activities of various kinds. In fact, the outcomes of such activities (or experiments or investigations) represent data, and such collected data are later on scanned and analysed and finally summarised conclusions are drawn out of these. The body of scientific concepts and methods applied to all the above four stages(namely, collection, tabulation, analysis & interpretation) constitute the art and science of the subject "Statistics". Though the above definition of the subject "Statistics" is a general one, yet this subject deals primarily with situations (experiments or investigations) where the outcomes can not be predicted with certainty and there exists variation among the observations (outcomes) of the experiment. To understand the above the following experiments (or investigations) are mentioned: (a) In order to find out the percentage of students promoted to class IX in high schools in West Bengal let an investigation be made to collect data on percentage of promotees in 20 high schools all over West Bengal. The conclusion of such an investigation may not be fully deterministic (with respect to the true value) owing to incompleteness of the data(as only 20 schools are considered). Uncertainties of such kind prevail in almost all types of scientific investigations designed either to verify the truth of a proposition or to form an idea of the underlying law governing the phenomenon under study. (b) In the second type of experiments, viz. agricultural experiments, it is observed that even if all sorts of controls are exerted during experiment, repeated observations on yields from plants of a particular species (sown at the same time) differ. This inherent variation is another aspect of concern with real-life investigations. In such cases statistical tools are used extensively at every stage of the inquiry to elicit the truth. Thus, Statistics(the subject) is also defined as a collection of scientific concepts and methods applied during the course of collection, tabulation, analysis and interpretation of data in persence of uncertainty and variation, in the context of an investigation.

Precisely, Statistics strives to discover truth in the midst of uncertainty and diversity. In this sense, Statistics is not like Physics or Chemistry or Biology. In either of the above sciences a law is formulated based on repeated observations on a hypothesis (framed on a phenomenon) with subsequent confirmation (by taking further observations). In the course of such repeated experimentation the law is ultimately established via some hypothesis. Laws make up the foundation and help in the development of any science. Statistcs, however, is not based on laws, rather it formulates concepts and methods by which the truth of a hypothesis can be verified on a set of observations, and in the real world where uncertainties prevail over almost all scientific investigations, the formulated laws (or hypothesis) are bound to be non-deterministic (or probabilistic) and under such situations, Statisics, the science of uncertainty, models order in prevailing disorder.

Modern statistical theory and methodology are applied in analysis and interpretation of data emanating from different disciplines of science, namely, agriculture, biotechnology, engineering, physics, biochemistry, philosophy, economics, hydrology, social, environmental and medical sciences.

1.2. Historical Development of Statistics

1.2.1 Western Scenario

Though the western part of the globe witnessed the initial activities related to the building up of the foundation on the subject Statistics in the twentieth century only, the practice of recording data is fairly ancient and its origin can be traced back to the age, when even the act of counting was not perfected. Rulers of the ancient kingdoms all over the world employed accountants who collected detailed data about people and resources of their states. A very early record of statistics reveals that at the time(3000 B.C.) of Royal Mace (an Egyptian King); 1,20,000 prisoners, 4,00,000 oxen and 10,22,000 goats were captured in a war. Census records are perhaps the most ancient documents collected in a planned manner. Evidence of census records, collected during the reign of Hisa Dynasty (2000 B.C.) in China, are available. The fourth book of the old Testament contains references to early censuses conducted about 1500 B.C. and instruction to Moses to conduct a census of the fighting men of Israel. The Latin word '*Censere*' is the origin of the word 'Census', its meaning in Latin language being '*to tan*'. Censuses in ancient Babylonia, China and Egypt were conducted as early as 300 B.C. Roman census was conducted during the period of the king Servious Tullius (534-578 B.C.). An extended census was taken by Caesar Augustus in 5 B.C. Occasional statistical surveys were conducted on land, land use, land owners, tenants, servants and livestock during the Mediaeval Ages. The data on mortality started appearing in the early years of the sixteenth century in London. Spanish conquerors conducted the earliest census in America. The system of decennial census commenced in 1790 in U.S.

The term 'statistics' has its roots in the Latin words 'status' (meaning state) and '*Statista*' (meaning statesman). A comparative description of 158 states is presented in Aristotle's "*Polity*". Italian and German scholars developed the art of presentation of comparative description into a subject called *Statistics,* the word, however, was coined by the German Scholar Gottfried Achenwall around the middle of the eighteenth century. Sir John Sinclair presented statistical account of Scotland in a series of volumes published between 1791 and 1799 and used the term Statistics. The use of the term Statistics found place in Encyclopedia Britannica in 1797. Statistics was then regarded as the science of statecraft—its function was to be the eyes and ears of the government. Statistics appears to have been recognised as a science when it was included as a section by the British association for the Advancement of Science. The Royal Statistical Society was founded in 1834. By then, Statistics was considered as "facts relating to men, which are capable of being expressed in numbers, which sufficiently multiplied, to indicate general laws". The Central Statistical Bureau was established in France in 1800, the first one in the world. During 1830–1855, statistical societies were founded in some more countries and statistical offices were set up for the purpose of "procuring, arranging and publishing facts with a view to illustrating the conditions and prosperity of the society". During the years 1853-1876, international congresses (about 10 in number) were held in Europe (participated by different countries) to agree upon the concepts and definitions, and uniform methods of collection of data related to the determination of factors responsible for growth, to have a comparative idea on the growth of different countries, to exchange experiences and to set up common standards. The first congress was held in Brussels in 1853 with the participation of 156 delegates representing 26 countries. The proposal of establishing International Statistical Society was accepted in the diamond jobilee celebration of Statistical Society of London in 1885 and to that effect an organisation called "International Statistical Institute" was set up in 1885, its permanent office (publication) was, however, established much later at the Hague in 1913.

The discussion on the development of Statistics requires understanding of the concept of uncertainty and randomness which has been considered a baffling issue over a long period of time. From the time of Aristotle philosophers recognised the role of uncertainty (absence of order) in human life and considered it incomprehensible. Uncertainty prevails in all human activities, and decisions taken under uncertain conditions become more meaningful if quantification of uncertainty inherent in the process (activity) can be made. The theory of probability which originated more than 300 years ago, offers an useful tool for quantifying the uncertainty or studying chance. The concept of randomness implies the uncertainty or studying chance. The concept of randomness implies that the uncertainty (in other words chaos or entropy) is maximum. The optimum way to deal with an uncertain phenomenon is to invoke the mechanism of probability to discover the hidden laws of chance. The possible alternatives under such situations are found, and probabilities of their occurrences are recorded as measures of uncertainty. The random phenomenon of drawing of numbers in a lottery reveals the existence of some order in the sequence of numbers so produced in the long run. The probability theory provides a tool to determine the order. Thus chance deals with order in presence of disorder. On the other hand, when the process is deterministic (i.e. in order) a part of it may exhibit randomness (disorder) locally, and this phenomenon is known as chaos. Thus, chaos deals with local disorder in presence of global order. In the following paragraph history of development of probability is discussed.

It can be said that the publication of the book 'Ars Conjectandi' by Jacob Bernoulli marked the beginning of the theory of probability (as the limiting form of relative frequency or frequency theory) in 1713. Thereafter many attempts were made to formulate the foundation model of the empirical realm of relative frequencies, the most successful work in this direction is due to A. N. Kolmogorov (1933). DeMoiver derived the normal distribution, the most important probability law, in 1733, Galileo (1564-1642) was aware of the errors in observations. Laplace established the Normal curve as the appropriate probability distribution of errors in 1783 and later Gauss (1777-1855) studied the probability law of errors in measurements, and proposed an optimum way of combining the observations to estimate unknown magnitudes. Adolph Quetelet (1869) drew the bar diagram of a large group of people and observed its resemblance with the normal curve. This led him to use the concept of probability to describe social and biological phenomena. This work could be regarded as the precursor to modern statistics with probabilistic foundation. Gregor Mendel (1870) formulated his laws of heredity through simple chance mechanisms like rolling a die, Boltzman (1866) gave a statistical interpretation to one of the most fundamental propositions of theroetical physics, the second law of thermodynamics, Galton used the concept of correlation and regression for the first time on a data set relating to the heights of fathers and sons.

The foundation of Statistics was laid in Great Britain with the pioneering works of Sir R. A. Fisher in experimental design and also in other areas at Rothamsted Experimental Station (during the period 1919 to 1933) and also with the works of Karl Pearson at the University College, London since 1890. Pearson and his associates were involved in the application of statistical methods to a wide variety of problems in evolution, heredity and related fields. Such works, however, were based on probabilistic considerations. The distinction between population and sample was first clearly mentioned in the works of Fisher. The formulation of the definition of random sampling may be attributed to John Venn and C.F. Pierce.

It is generally held that the two branches of statistics, namely, descriptive statistics and theoretical statistics have distinct methodologies and through the works of Fisher and Pearson this distinction emerged almost with definiteness. In the former, the object is to summarise a data set or to make comparison among different data sets with the help of statistical measures of location, dispersion, moments of higher orders, correlation and regression, and also exposition of salient features through graphs, diagrams and two dimensional charts without any reference to the underlying stochastic mechanism (or probability distribution). This kind of statistical analysis is referred to as descriptive data analysis. In theoretical statistics (the foundation of which was laid by R.A. Fisher in 1922), the object is again summarisation of data but with reference to a specific family (or model) of underlying probability distributions. The summary of descriptive statistics depends on the specific stochastic model, and their distributions are used to specifiy margins of uncertainty in inference about the unknown parameter involved in the model. Such methodology is referred to as Inferential Data Analysis. Karl Pearson tried to bring the two branches closer and used the power of descriptive measures (moments and histograms) towards identification of the underlying family of distributions. For this purpose, he invented the celebrated test criterion, the chi-square statistics, for testing the hypothesis that a given set of data arose from a specified family of distributions. Karl Pearson also invented a variety of probability distributions being distinguishable by means of four moments. The first proper formulation of the problem of point estimation is due to Laplace. Later, Gauss presented a somewhat more satisfactory formulation. The familiar theory of least squares resulted from the work of Gauss who minimised the Quadratic Loss Function for estimation purpose. The criteria to be possessed by " a good estimate" was given by Fisher. Fisher proposed the method of maximum likelihood for point estimation and introduced the concept of Fiducial Interval. Pearson advocated the method of moments. Neyman, in the early thirties, proposed the concept of Interval Estimation. Fisher developed exact small sample tests for a variety of hypotheses under normality assumption and advocated their use with the help of tables of certain critical values, usually 5 percent and 1 percent quantiles of test criterion.

Contemporary literature shows numerous contributions in the area of test of significance using exact sampling theory, the contributors were Gosset(Student), Hotelling, Bose, Roy & Wilks among other. Fisher introduced the technique of Analysis of Variance in 1923 and also the method of data collection by adopting a systematic procedure called 'Design of Experiments'. W.A. Shewart (1931) introduced Statistical Quality control methods wherein control charts were used for detecting change in a production process. J. Neyman and E.S. Pearson presented a systematic approach to the theory of inference through a series of papers (collected paper, 1928). A. Wald introduced the sequential methods for application in sampling inspection and also presented the Decision Theoretic Approach to the problem on inference. Almost contemporarily, the theory of Multivariate Analysis was introduced by Fisher and this area was further enriched by a contribution made by Hotelling, Mahalanobis, S. N. Roy and C.R. Rao among others. Pitman (1935) introduced non-parametric methods. Such methods use test criterion whose distributions are independent of the underlying stochastic model for the data. Since then the area has developed remarkably till the recent times. The development of sample survey was made in the forties and this area has also seen a vast growth in the mean time. The two major areas of Statistics, namely descriptive statistics and theoretical statistics do not consider any mechanism by which defects contained in the data can be eliminated. In fact, methods for understanding the defects and special features of data were not given the importance, and there was a need to select the suitable stochastic model or a class a models for the analysis of data. Tukey (1962, 1977) and Mosteller & Tukey (1968) introduced the technique of Exploratory Data Analysis to extract all available information (being not merely confined to answering specific questions). It is regarded as a landmark in the history to developemnt of Statistics. Great advances have been made in the theory of Outliers, Residual analysis, Regression theory and Spatial analysis and distribution. The advent of computers has greatly assisted the selection of appropriate stochastic models for a data situation and to substantiate the findings with simulation results. In can be said that the future of the development of statistics lies in more collaborative activities of statisticians with scientists belonging to other disciplines of science.

In U.S.A., H.L. Rietz instituted courses in Statistics with an option in Mathematics at the University of lowa in 1918, though courses in Statistical Methods were offered as early as 1915. G.W. Snedecor and A.E. Brandt established the Statistical Laboratory in 1927 under Mathematics Department at Iowa. Samuel Wilks, Allen Craig received their Ph. D. Degrees (under guidance of Rietz) from the University of Iowa. W. G. Cochran joined the Statistical Laboratory in 1939. Soon after the development of Statistics Department at Iowa, a programme in Statistics was organised at the University of Michigan. H.C. Carver was the teacher of Statistics there and in 1930 he founded and then published Annals of Mathematical Statistics (in collaboration with American Statistical Association), until it was taken over by the Institute of Mathematical Statistics in 1938 (with S.S. Wilks as the first editor). H.L. Rietz was the first President of the Institute of Mathematical Statistics (1935-36). A. L. Bowley (1913) made, perhaps, the initial serious efforts at applying Statistical theory to guide the design and interpretation of results from sample surveys. Outstanding contributions in survey methodology were made thereafter by J. Neyman around the middle of the fourth decade. Early developments in Medical Statistics and related fields were made by Raymond Pearl and Lowell T. Reed at Johns Hopkins University. The Department of Bio-Statistics was estabilished in 1918 and it was the First Department in the United States with the word Statistics in its title.

1.2.2 Indian Scenario

P.C. Mahalanobis in his work "The Foundations of Statistics", has mentioned that qualitative statistical thinking has its origin in the works of Jaina Philosopher, Bhadrabahu of the fourth century. *Arthasastra* of Kautilya (written between 321-300 BC) bears ample evidence of the existence of a highly organised system of administrative statistics in ancient India. *Ain-e-Akbari* of Abul Fazal(written betwen 1596-1597) contains detailed information on the extent of resources, social condition, population, industry and wealth of the empire of Akbar. Population census in India as is conducted even today, started in 1871-72 during the British rule. A large scale survey over the Eastern India was conducted by East India Company in the years 1807-1815.

The pioneering role and diverse activities of P.C. Mahalanobis marked the foundation of Statistics as a discipline in India. In fact, the second and third decades of the twentieth century were the formative years of his statistical thought (he was familiar with works of Fisher and Pearson on statistical methods) during which period he gathered round him a band of young, talented and devoted scientists and undertook statistical studies at the Statistical Laboratory, at Presidency College, which led to the foundation of the Indian Statistical Institute in 1931. The year 1933 marked the beginning of a famous journal, *Sankhyā*. Thus Mahalanobis, a Professor of Physics, is rightly regarded as the "Father" of the discipline of Statistics in India. Mahalanobis and his associates were absorbed in making significant and path breaking contributions to the discipline of Statistics since the establishment of the Institute. This is amply revealed in the volumes of *Sankhyā* and other scientific journals of that time. Realising the need of Statistics as a separate and fully developed discipline of study, he organised a full-fledged post-graduate course in Statistics in Calcutta University in 1941. The statistics department of Presidency College introduced undergraduate honours teaching in the subject for the first time in the country, in 1944.

In 1947, Prof. Mahabanobis developed the Indian Statistical Institute, under government patronage, into a world renowend centre for advanced study and reasearch in Statistics. By then he realised the need for better organisation of the official statistical system in independent India and played an important role in the establishment of Central Statistical Unit in 1949 under Government of India, for carrying out such activities. Such activities, however, culminated in setting up of the Central Statistical Organisation and the Department of Statistics under Government of India. Again, it was through his efforts that the National Sample Survey (renamed as National Sample Survey Organisation in 1970) was established in 1950 for the collection of socio-economic data through sample surveys on continuing basis. It was, in fact, the first large scale multi-purpose survey. Mahalanobis's interaction with Indian Industrialists was also very effective in the sense that he was able to convince them of the utility of application of statistical methods in their efforts to improvement of the quality of their products owing to its impact almost all manufacturing or trading concerns have separate statistical wings.

The two decades after 1950 witnessed a wealth of research contributions penned through the eminent statisticians built up under the aegis of the Department of Statistics, Calcutta University, Indian Statistical Institute, Bombay University and Poona University. The Government of India introduced Indian Statistical Service in the later part of the seventh decade of this country.

This section on the history of development of statistics in Indian context will remain incomplete if the following few outstanding personalities are not mentioned in particular. The statistical school built up by Professor Mahalanobis was enriched with the association of R. C. Bose, S. N. Roy, C. R. Rao, M. C. Chakrabarti, H. K. Nandi and others for their outstanding contributions to teaching and research in statistics and even today India-born scientists are recognised in the United States and elsewhere in the world especially for their commitment and contribution towards promotional activities in the area of research in this discipline.

Apart from the development of Statistics in different Universities and Institutions, the following chronological events contributed towards the development of statistics in totality in India.

1895 : Establishment of Statistical Bureau in Calcutta.

1905 : Establishment of Department of Commercial Intelligenece and Statistics in Calcutta.

1932 : Creation of the Imperial (now) Indian Council of Agricultural Research with a statistical wing.

1933 : Starting of *Sankhyā*, the Indian Journal of Statistics.

1951 : Establishment of Central Statistical Organisation and State Staistical Bureaus.

1952 : Starting of Calcutta Statistical Association Bulletin from the Department of Statistics, Calcutta University.

1.3. Application of Statistics in Differenent Disciplines of Science

Survival of human race through ages has been made possible only by advancement of techniques needed at different stages of development of cultivation of principal crops consumed by mankind. Towards this process of development, Biology, Physics, Chemistry, Engineering etc have lent their supports till the present day when statistical methods are used extensively to obtain estimates of crop production in advance, to judge the best variety from among a number of competitive varieties belonging to a particular crop, to predict the amount of damage of crops owing to insect/ pest attacks, to quantify the yield advantage from intercrops, and to solve many other vital problems related to agricultural science in totality. Similarly, the procedures of estimation of, (i) the livestock population of a state/country at any period, (ii) the milk production in a region, (iii) the population of fish in a poand, (iv) the population of lions and birds in a forest. (v) the population of trees in a forest, (vi) the amount of timber that can be made available through replacement of old trees by new plants, are based on statistical methods, the related disciplines being Animal science in cases (i) & (ii), Fishery in case (iii), Wild life in case (iv), and Forestry in (v) & (vi). The study of population dynamics, be it human, animal or wild life, requires knowledge of suitable statistical methods.

In the realm of fundamental sciences, viz. Physics and Chemistry, statistical techniques have been in use for a long time. Though the basic laws of Physics were assumed to be deterministic, a fundamental change took place when the basic laws themselves were expressed in probabilistic terms specially at micro level of the behaviour of fundamental particles. The two areas in Physics which require statistical considerations for explanation are Mechanics and Thermodynamics. The introduction of statistical ideas in Physics started with the need to deal with the errors in astronomical measurements. Biochemical analysis has to take into account stochastic aspects and as such statistical methods play dominant role in such analysis for comprehension of the underlying causes and effects more precisely.

Statistical Quality Control (S. Q. C.) methods have been applied rigorously for maintenance of quality of products since the early part of the twentieth century. A decision on the variation (whether tolerable or not) in the products as produced through the application of scientific methods developed under different engineering sciences has been made possible at a specified level of exactitude with the help of S.Q.C. methods. Another area of Engineering which has undergone a remarkable growth in the last few decades is Hydroglogy, and it can be shown that significant development in this field has been due to the application of various statistical methods. The processes which link rainfall with riverflows/ stream discharges are essentially deterministic, but the boundary conditions being complicated a purely deterministic model does not provide the answer to simulation of this hydrologic process. Stochastic models deal with such simulation more precisely.

Modern Geology and Mining Engineering require the technical aspect of spatial distribution/ analysis which is a newly emerged area under the discipline of Statistics. Sir R.A. Fisher mentioned that even in 1833, statistical thinking had worked behind the discovery of geological time scale by Charter Lyett.

Social scientists adored the advent of different statistical techniques as and when they were evolved since these techniques offered valuable props in the understanding of the effects, in the quantitative terms, due to the existence of various factors in social system.

In detective work, statistics help in analysing bits and pieces of information, which individually may appear to be unrelated or even inconsistent to see an underlying pattern.

Data analysis under medical sciences has offered a challenging aspect of research to the statisticians and towards this process a new area of study called "Medical Statistics" has emerged. The principles of Design of Experiments are used in screening of drugs and in clinical trials. Statistical methods have now been made available to the estimation of HIV infected cases and projection of AIDS incidence.

Techniques under Quantitative Genetics are virtually statistical as laws of heredity and evolution are not deterministic and chance mechanism governs these important biological events. Thus laws of probability are widely used to explain such phenomena.

The statistical science has been found very useful in economics, commerce and managerial sciences owing to its power of analysing the effects of the involved factors under uncertainty or variation which usually prevail in managerial decisions, and in economic and commercial activities.

Modern economics uses regression theory, stochastic processes and many other statistical tools to describe and successfully explain the interplay of causes and effects governing the economic system. It is now fairly impossible to become a good economist without having a solid background of statistical methods. Of late, mathematical and statistical metohds have acquired wide applicability in analysis of stocks and shares. The collection of important statistical tools used in Economics is embodied under the caption "Econometrics", a branch developed under the joint discipline of statistics and economics. Of late, statistical methods are applied to forecast election results also.

Ecology is a branch of science concerned with the inter relationship of organisms and their environment. As has already been mentioned that biological processes operating in an environment are not fully deterministic, and therefore, statistical methods are applied to ensure better representations of the laws governing such phenomena.

In the quantitative analysis on polymerisation, application of the theory of probability and some related probability distributions are essential. To determine the mechanism of polyreactions it is imperative to determine the stochastic behaviour of molar concentration in the polymeric substance. In such situations the statistical methods are predominantly useful.

The spectre of pollution is looming large over our environment. Airborne particles, chemical (industrial) exhausts and waste products, automobile exhausts, thining of ozone layer etc. are adding significantly to the grave risk to human life on the planet. The assessment of this risk is a challenge to the scientific community. In this quest for assessment of risk owing to pollution bio-statistical principles play a fundamental role. Regression diagnostic versus ecological perspectives, measurement errors, generalised linear models, bio-equivalence and accelarated life testing are important basic statistical issues in this regard.

For the Government of a country, Statistics is the means by which it can make short and long range plans to achieve specified economic and social goals. Nowadays the Government collects vast amount of socio-economic, agricultural and demographic data through its various agencies and advanced statistical methods are applied for processing and analysing the data, and viable decisions are taken in regard to formulation of public policies related to administration.

Forecasting techniques based on sound statistical principles are extremely useful in meteorological predictions. Indeed, the development of statistical models in this area, has paved the way for making reliable weather forecasts.

Though the subject statistics made its formal beginning in the early part of 20th century but by the end of it we find that its remarkable growth has not only enriched its own content but has also established its potential application in almost all other branches of science.

1.4. Limitations and Misuses of Statistics

As a branch of science, Statistics has its own limitations. The following discussion relates to the limitations and misuses of this subject.

At the very outset it can be mentioned that the following facts on Statistics and Statistical Knowledge are held true : (i) Statistics is both science and art, (ii) Statistics is a part of abstract knowledge that enriches the knowledge of other branches of sciences or builds up a systematic knowledge in any field of developmental activity. As the subject has been developed on the basis of objective evidences and subsequent objective analysis, some limitations, misinterpretations and misuses are possible. These are presented below :

1. Statistical statements do not throw light on the characteristics of individual experimental units generally, (limitation).
2. Statistical statement are liable to be distrusted, (limitation).

Illustration : Suppose an agricultural bulletin has published that a particular variety of cow produces 18-20 kg of milk per day. Knowing the information a farmer procures the particular variety of cow and observes that the production of milk is only 12 kg per day during its lactation period. Obviously, the farmer begins to distrust the published statistics (obtained by use of statistical method). Such a distrust can be generated owing to the incompleteness in regard to presentation/understanding of statistical knowledge. The following considerations lead to apparent distrust on the soundness of statistical findings.

(a) The published information on milk yield may have been reported without its verification through planned and repeated experimentation.

(b) The average result may have been reported by random omission of observations (selected by using personal judgment). In such cases, however, the selection procedure should be governed by the application of the theory of *outliers* and not by personal judgment.

(c) The reporting of the average value only (without any mention of the confidence interval) keeps the user in dark about the inherent variability existing in the milk yield of cows (even in case of a particular breed)

(d) That the minimum procurement number of cows is also an important criterion to achieve the mentioned average amount of milk is also to be understood by the publishing agency and the user of such data; and hence the minimum procurement number of cows is to be mentioned along with the average milk yield.

3. Statistics are liable to be misunderstood (misuse).

Illustration : The following statements using the methods of Statistics may be examined.

(i) A man of height 5.5 feet can easily cross a river having its average depth 5 feet.

(ii) The daily steel production of a factory and the daily number of road accidents in the locality are highly correlated (since the correlation coefficient between these two variables as obtained from the data collected from the locality is 0.85).

The above two statements may appear to be idiotic to an educated person and hence he may mock at the findings obtained by application of statistical tools. In fact, such statements may arise, (i) due to lack of proper knowledge and experience in using tools or (ii) owing to the use of a particular statistical tool in every decision making problems. In fact, the statement (i) should be written as :

"A man of height 5.5 feet can easily cross a river having maximum depth 5 feet". Essentially, the appropriate statistical tool in this case should be the extreme (maximum)value and not the mean value.

In the second example, the variables considered do not have even causal ralationship and hence in such cases it is nonsensical to calculate correlation coefficient between the two variables. Most statistical analyses use certain basic assumptions. There may be data situations wherein basic assumptions are not satisfied. Thus diagnostic checks must be employed on the data set to determine the validity of basic assumptions, and if these assumptions are not met, remedial measures should be adopted on such data before vaild statistical conclusions are drawn.

1.5. Some useful Notations

A variable, such as height, weight, blood pressure, number of petals in a flower, is generally denoted by a capital letter of English alphabet (X, Y, Z etc) and the corresponding values of the variables are denoted generally by small letters x, y, z etc.

(a) Use of the single subscript : Whenever a set of observations is available, the observations are recorded serially as first, second, third and so on. Thus each observation bears a corresponding serial number we use a subscript i to the variable. The above can be understood with the following illustration.

Let the *height (in inches) of 10 students be observed as 50, 49, 52, 60, 65, 63, 58, 66, 64, 61. We use the subscript* i to identify the student, and therefore the corresponding observation (height) can be represented by x_i, $i = 1, 2, \ldots\ldots 10$.

Thus $x_1 = 50$ inch, $x_4 = 60$ inch, $x_8 = 66$ inch etc.

In the general set up, a set of observations on the variable *X* (say, height) can be represented by $x_1, x_2 \ldots . x_n$ or x_i, $i = 1, 2, \ldots, n$. Thus x_i is the observation corresponding to the i-th student (also x_i is called the i-th observation in the set).

The sum of *n* observations, $x_1 + x_2 + \ldots + x_n$, can be abbreviated by using the Greek capital letter sigma (Σ), and the sum is denoted by variable subscript and its range of values, in the text.

The sum, $\sum_{i=1}^{n} x_i$ is read as 'Sigma x_i, i from 1 to *n*'.

Some simple results on the use of Σ notation on single subscript are :

1. $x_i = c,\ i = 1, 2, \ldots\ldots n;$ then $\sum_{i=1}^{n} x_i = nc$

2. $\sum_{i=1}^{n} cx_i = c\sum_{i=1}^{n} x_i$, c being a constant.

3. $\sum_{i=1}^{n} (x_i \pm c) = \sum_{i=1}^{n} x_i \pm nc$

4. $\sum_{i=1}^{n} x_i = \sum_{\substack{i=1 \\ (x_i<m)}}^{k} x_i + \sum_{\substack{i=k+1 \\ (x_i \geq m)}}^{n} x_i$, $1 \leq k < n$ with k number of x_i values are less than m.

5. $\sum_{i=1}^{n} (x_i \pm y_i) = \sum_{i=1}^{n} x_i \pm \sum_{i=1}^{n} y_i$

(b) *Use of double subscripts* : Let us consider the height – observations of 15 students from 3 different classes, 5 students being taken from each class. The different classes can be taken in three rows (horizontal classification) and students of a particular class will be considered in a row serially. The representation of the above 15 observations will require the use of two subscripts, *i* and *j*, *i* representing the class identity of the students and *j* representing the serial number of the student belonging to the particular class.

The height-observations are presented in the following way.

Class (*i*) \ Serial No. (*j*)	1	2	3	4	5
1	$x_{11} = 39$	$x_{12} = 40$	$x_{13} = 42$	$x_{14} = 38$	$x_{15} = 50$
2	$x_{21} = 49$	$x_{22} = 52$	$x_{23} = 47$	$x_{24} = 45$	$x_{25} = 58$
3	$x_{31} = 50$	$x_{32} = 49$	$x_{33} = 60$	$x_{34} = 57$	$x_{35} = 52$

The observations corresponding to the (*i, j*)-th cell have been represented by x_{ij} $i = 1, 2, 3$; $j = 1, 2, 3, 4, 5$. If i = 2, j = 3, then the observation corresponding to (2, 3)-th cell is x_{23} (= 47). In fact, the subscript *i* denotes the class number and the subscript *j* denotes the serial number corresponding to the particular class. In the general set up when there are *m* - rows and *n*-columns, the *m* × *n* observations can be written in the following two-way table (row × column table).

Row (*i*) \ Column (*j*)	1	2	3	...	*j*	...	*n*	*Row-total*
1	x_{11}	x_{12}	x_{13}	...	x_{1j}	...	x_{1n}	x_{10}
2	x_{21}	x_{22}	x_{23}	...	x_{2j}	...	x_{2n}	x_{20}
3	x_{31}	x_{32}	x_{33}	...	x_{3j}	...	x_{3n}	x_{30}
⋱	⋱	⋱	⋱	⋱	⋱	⋱		
i	x_{i1}	x_{i2}	x_{i3}		x_{ij}		x_{in}	x_{i0}
⋮								
m	x_{m1}	x_{m2}	x_{m3}		x_{mj}		x_{mn}	x_{m0}
Column-total (x_{oj})	x_{01}	x_{02}	x_{03}		x_{0j}		x_{0n}	x_{00} = Grand total

The row-totals, column-totals and grand-totals can be summarized in the following way.

Sum of the observations in the i- th row

$$= x_{i1} + x_{i2} + ... + x_{in}$$

$$= \sum_{j=1}^{n} x_{ij} = x_{io}, \; i = 1, 2, ..., m$$

Sum of the observations in the j -th column

$$= x_{1j} + x_{2j} + ... + x_{mj}$$

$$= \sum_{i=1}^{n} x_{ij} = x_{oi}, \; j = 1, 2, ...n \, .$$

Sum of the $m \times n$ observations

$$= (x_{11} + x_{12} + ... + x_{1m}) + (x_{21} + x_{22} + .. + x_{2m}) ... + ... (x_{m1} + x_{m2} + ... + x_{mn})$$

$$= (x_{10} + x_{20} + ... + x_{m0}) = (x_{01} + x_{02} + ... + x_{0n}) = \sum_{i=1}^{m} x_{i0} = \sum_{i=1}^{m} \left(\sum_{j=1}^{n} x_{ij} \right)$$

$$= \sum_{j=1}^{n} \left(\sum_{i=1}^{m} x_{ij} \right) = \sum_{j=1}^{n} x_{0j}$$

The product of n observations, $x_1, x_2, ... x_n$, is denoted by $\prod_{i=1}^{n} x_i$ or $\prod_{1}^{n} x_i$ or $\prod x_i$,

$\prod_{i=1}^{n} x_i$ is read as '*Capital pie* x_i, i *from* 1 *to* n'.

Some results on the use of $\prod$ notation.

1. $\prod_{i=1}^{n} (cx_i) = c^n \prod_{i=1}^{n} x_i$, c being a constant.

2. $\prod_{i=1}^{n} x_i y_i = \left(\prod_{i=1}^{n} x_i \right) \left(\prod_{i=1}^{n} y_i \right)$

3. $\prod_{i=1}^{n} \frac{x_i}{y_i} = \frac{\prod_{i=1}^{n} x_i}{\prod_{i=1}^{n} y_i}$.

4. $\prod_{i} \prod_{j} x_{ij} = \prod_{i} \left(\prod_{j} x_{ij} \right) = \prod_{j} \left(\prod_{i} x_{ij} \right)$

CHAPTER 2

TYPES OF DATA AND THEIR COLLECTION AND PRESENTATION

2.1. Introduction

The meaning of data is factual information, in fact, data are the basis to specify an objective or to elicit information or to arrive at a particular conclusion with reference to an investigation. Scientists, sociologists and philosophers have been engaged in collection of data for scientific and social inquiries / investigations pertaining to everyday life. In early days, scientific inquires were related to exact types of findings, e.g., to determine the composition of air, water, blood; to evaluate the distance between two planets and to locate their paths of movement etc. To produce quality items such as sculptures, ornaments, idols, they exploited the resources—man, material and environment. But now-a-days most of the scientific inquires are related to obtaining the best output with the help of best (optimum) inputs intending to maintain sustainable growth in future. For example, to determine what fertiliser mixture will give the optimum yield, or to determine what amount as well as type of resin, temperature, time and pressure will produce the optimum usable reinforced components, we cannot follow the approach similar to that adopted for collection of information for the determination of composition of light or for the determination of focal length of a lens. Trial and error approach generally takes a lot of time to get the result. In such cases of uncertain environment realisation of statisticians' comprehension of data is essential for the best use of time, money and the basic resources. So proper knowledge of the characteristics of data is essential for those involved in scientific investigations or inquires related to varied aspects of everyday life.

2.2. Primary and Secondary Data

For the subsequent process of an inquiry data can be collected either from the actual field of investigation, *i.e.*, primary source or from published reports, *viz.*, newspapers, journals, books, which are, in fact, secondary sources. Data which originate from the primary source of investigation are called 'primary data' and data collected from the secondary sources are called 'secondary data'. As an illustration, a Government department publishes bulletins on agricultural and industrial productions each year. The data published in the bulletins are the secondary data, and the data actually collected in the areas relating to agriculture and industries (i.e. real-life data) that help to produce the ultimate information (data) in the bulletin are the primary data. There are some *merits and demerits* of both primary data and secondary data. A choice between primary data and secondary data generally depends on the nature and scope of the inquiry, financial position of a investigator, time available and accuracy needed. Though the primary data are more accurate, directly refer to the objective and are more informative, yet they are costlier and more time consuming to acquire than the secondary data. Though the latter may not be adequate, may be so extensive as to require judgment in isolating relevant portion from redundant portion, may not be very reliable and may not exactly fit in the present project of investigation, the advangage is that it requires less money and less time to acquire it.

2.3. Qualitative and Quantitative Data

The nature of data, either primary or secondary, depends also on the types of information obtained from the respondents (individuals or units). Respondents are those on which data are collected. These may be human beings or animals or plots or land or samples of blood or the samples of urine or pieces of cloth, etc. Let a producer produce aluminium plates. If the data obtained from the plates are recorded in the sequence of 'good or bad', 'defective or non-defective', then such types of data are called *qualitative data',* and if the data are recorded in the form of dimensions (involving units of measurement) of the plates, then the data so obtained are called the *quantitative data*. Data on qualitative characters, like, religion, sex, colour, occupation, marital status etc., are called qualitative data. So 'yes or no', 'true or false', 'defective or non-defective' and different codes may be regarded as different forms of the qualitative data. Data arising from some measurable characters, observed on individuals (units), such as, height, age, weight, blood pressure of human beings, number of petals in flowers, number of peas in pea-pods, scores on intelligence tests, are called quantitative data.

2.4. Frequency Type data and Non-frequency Type Data

Whatever be the data, either primary or secondary, can be placed on the Real Number line and can also be classified quantitatively. Such data are called *frequency type data*. Both qualitative and quantitative data can be treated as the *frequency type data.* Such types of data are called frequency type since against a particular value/ range of values, against a particular scoring/ coding, it is possible to have one or more individuals (units). As an illustration, let us take the weights (in kg) of 10 students of class IX as 37,41.5, 42.2, 37.9, 45.1, 38.7, 44.3,43.9, 39.49, 40.7. We can place and classify the data on the real line as below :

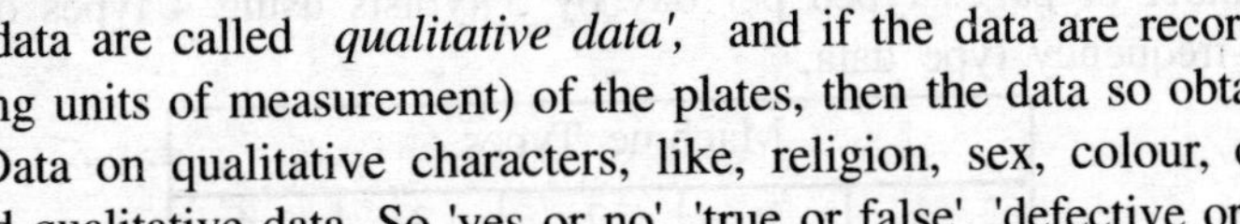

Classified real line : (35 – 38) (38 – 41) (41 – 44) (44 – 47)
(weight group)

Frequency : (2) (3) (3) (2)

Similarly, we can also place the data (on number of peas in pods) 3, 4, 4, 3, 2, 4, 5, 4, 3, 5, and 6 on the real line as below.

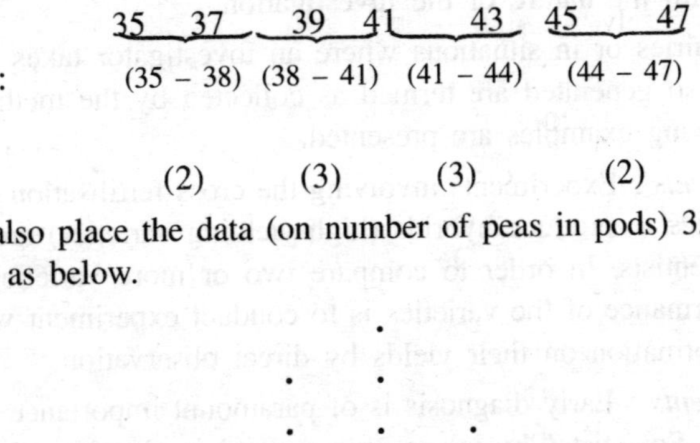

Discrete value : 2 3 4 5 6

Frequency : 1 3 4 2 1

It should be noted that if the data set contains just three observations, 2, 3 & 4, representing the number of peas in 3 pea-pods only, then these data are also frequency type data.

Qualitative data can also be placed on the real line if the qualities are coded numerically. If the data are observed either with respect to places or with respect to years/periods or with respect to some of the economic, social sectors etc., then these data are called *non-frequency type data*. Here each observation is included maintaining its identity in the series. Data on yearly production of steel from 1980 to 1989, data on yield of paddy in 1990 of different countries, like, Bangladesh, Pakistan, India, China etc., are examples of non-frequency type data. Thus time series data or historical data and spatial series data constitute non-frequency type data. The data presented in Tables, 2.1, 2.2, 2.5, 2.6, are the examples of non-frequency type data. The following data representing the number of pages typed per day by 3 typists using 4 types of machines, is also an example of non-frequency type data.

		Machine Types			
		1	2	3	4
Typists	1	31	26	35	29
	2	25	32	34	21
	3	30	35	36	33

2.5. Collection of data

Data Collection is an important, methodical and scientific area of activity for conduction of a current investigation or to facilitate any future inquires. The process of investigating any phenomenon involves the collection of data either from a primary source or from a secondary source. Collection of data from the primary source may be done,

(i) by taking direct observations, or,

(ii) by interviewing with question schedule, or,

(iii) by mailing the questionnaire,

in accordance with the nature of the investigation.

In scientific inquiries or in situations where an investigator takes observations on different units himself, the data so generated are termed as collected by the method of direct observation. Contextually, the following examples are presented.

Genetic Experiments : Experiments involving the cross-fertilisation of different genetic types of plants /animals species to produce hybrids (high-yielding varieties) are of prime interest to the agricultural/ animal scientists. In order to compare two or more varieties, the only way to learn about the relative performance of the varieties is to couduct experiment with them at a number of locations to collect information on their yields by direct observation.

Clinical Experiments : Early diagnosis is of paramount importance for successful treatments of many fatal diseases. Scientists/Physicians are engaged in development of effective diagnostic process by adoption of which the patients can check themselves. To establish the merits of a newly developed process, data should be collected directly from the field trial of the new process on a large number of persons who should also have undergone the conventional check up for comparison.

For the following investigations data may be collected either by interviewing the respondents with *question schedule* or by mailing a *questionnaire*. Sociologists make a distinction between questionnaire and question schedule as the former could be used for getting information by post while the latter for personal interview of the respondents. Questionnaire or question schedule is a list of questions organised in such a fashion that investigator's objective is fulfilled scientifically, economically and with minimum harrasment to the respondents. Questions in the question schedule should be clear, to the point, elegant, smooth, not repetitive and short answer type. The investigators (self or appointed) should be trained on the objective and also on the content of the inquiry (or investigation). In fact, the questions to be included in a question schedule or questionnaire should be simple, short, unambiguous, explicit, relevant, practical, objective and precise as far as possible. Questions on personal affairs and those affecting sentiments of the respondents should be excluded. Generally, a draft questionnaire is prepared at the very outset and information are taken on the basis of it (i.e., by way of pilot survey). The shortcomings and defects in the draft schedule are then rectified to prepare the final questionnaire. A model questionnaire is given in Annexure.

Socioeconomic and Sociological Surveys : In the interdisciplinary areas of sociology, economics and political science, studies are undertaken on such aspects as the economic wellbeing of the different ethinc groups, expenditure patterns at different income levels, farm/cattle management, effectiveness of a training procedure and that of a teaching aid, census, etc. For such kinds of survey we use two terms, 'sample' and 'population' or statistical population which is the aggregate of units/persons and which forms the domain of a study. 'Sample' means any representative part of a population. For any study, data should be collected only from a respresentative sample which preserves all characteristics of the population. The sample is selected by statistical methods.

Socio ecomomic surveys are performed by interviewing personally or by mailing questionnaires to a representative group of persons. In "mailing the questionnaire" method, printed questionnaire are despatched (with return postage pre-paid) to selected persons or agencies connected with the objective of the inquiry along with a request for supplying the information in respect of the underlying objective. The question schedule should be self-instructive in respect of the definitions of various terms and also of the definite way of filling up the questionnaire. Though this method of collection of data is less costly, less time consuming and covers a large geographical area, yet this method suffers from a few defects. In fact, the data obtained from the replies are less reliable and are generally poor in respect of the response criterion (in such cases, the questionnaires are to be sent for the second-time to reduce the number of non-respondents).

Investigators may depend upon the secondary sources of data in full or in part according to the nature of case study, provided the available data are reliable, i.e., they are published in official publications of the Government or Semi-Government or a reputed private agency.

2.6. Processing of data

Processed data are like processed foods which are consumed according to the tastes and habits of the consumers. Processed data are essential for the present and the future investigators.

In a primary or initial stage of an inquiry, data are usually contained in schedules or in questionnaries or in books/journals/periodicals, and are not presented in an easily assimilable from. To bring out the salient features of the data collected through a questionnaire (which, infact, is the first step towards processing) it is necessary to classify and tabulate the collected information, and in case when the data are collected from a secondary source, a rearrangement of these into new groups is necessary to have a fresh look on the collected information. The steps towards processing of data are discussed in the following :

(a) Scrutiny : As a first step towards data processing, the data collected through question

schedules or questionnaires are scrutinized for completeness, consistency, accuracy and homogeneity. For complenteness any important question should not remain unanswered and to check the consistency of the data one may apply ordinary common sense, whereas for accuracy special knowledge on units (both experimental and mathematical) is required, By homogeneity we mean that the data on a particular question can be understood from the same sense, e.g. for the question on income these should not be the case like—data against some experimental units (respondents or interviewers) are mentioned on weekly basis, data against some other experimental units are mentioned on monthly basis and so on, nonuniformly. The scrutineer should also be aware of the different types of positive human errors that may arise at the time of recording the data. If we scrutinize the following data on heights (in cm.) of 10 Indians : 171, 163, 142, 165, 184, 138, 156, 614, 169, 160; we see that the 8th observation, 614, is wrongly copied. It can be actually 164 or 146. In all cases, such rectifications may not be possible and hence deletion of the suspected observations is advisable.

(b) Classification and Tabulation : The first step in tabulating scrutinized data is to work out a detailed system of classification. Classification of logically connected data is a fundamental operation in respect of scientific analysis. Classification demands uniform and precise definition of categories. Classification makes handling of large volume of data easy. The data can broadly be classified, (i) area-wise, (ii) year-wise, (iii) quality-wise, or (iv) quantity-wise.

After classification, the data are sorted or counted according to the various categories in the classification and put in the form of a table.

(i) Area-wise classification : If the area-wise classification is chosen for tabulation of data, the data is grouped in terms of geographical divisions as countries, cities, states, etc. The data on imports of jute and jute goods by EEC can be classified area-wise as in following table.

Table 2.1 —Imports of Jute and Jute goods by EEC (year 1991)

Countries	Imports ('000 M Tons)
U.K.	50.1
France	10.0
FRG	23.9
Belgium	107.8
Netherlands	22.7
Italy	17.4
Spain	8.3
Portugal	0.8
Greece	3.2

Source : Eurostat / ITC- Brussels, 1991

(ii) Year-wise classification : For certain kinds of statistical data units of time such as a day, week, month or year (or even a second, minute or hour) may be used as a basis of classification. If year is taken as a basis of classification for the export of Jute goods from India, we have the following table.

Table 2.2 — Exports of Jute Goods to U.S.A. from India

Year	1980	1981	1982	1983	1984	1985
Quantity (000 MTons)	121.07	57.02	65.03	37.05	30.00	22.06
Value (Rs. Crores)	83.26	44.44	39.51	21.96	20.85	26.19

Source : Indian Jute, June 1992.

(iii) Quality -wise classification : Data arising from the attributes like, religion, sex, literacy, employment etc. are classified under the different groups of the attribute under study. For example, in order to study the secular nature of an organisation, the primary task may be to find out how many persons of that organisation belong to different religious groups for the stated objective. If Hindu, Muslim and Christian are the only three religious groups, the collected data on the religious attachment of the members of the organisation under study may be classified as follows :

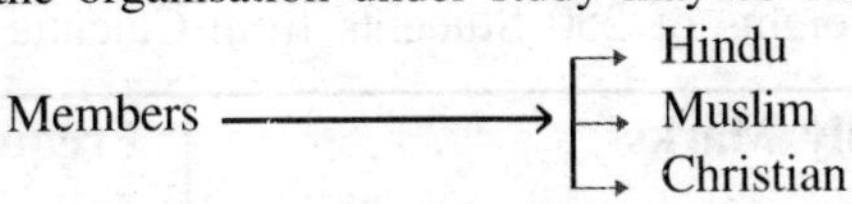

For a somewhat complex study on sex-wise literacy and employment in two major religious groups in a particular population, each religious group should be divided in dichotomous classification like male and female, literate and illiterate, employed and unemployed, etc. Qualitywise classification for the above study is shown below :

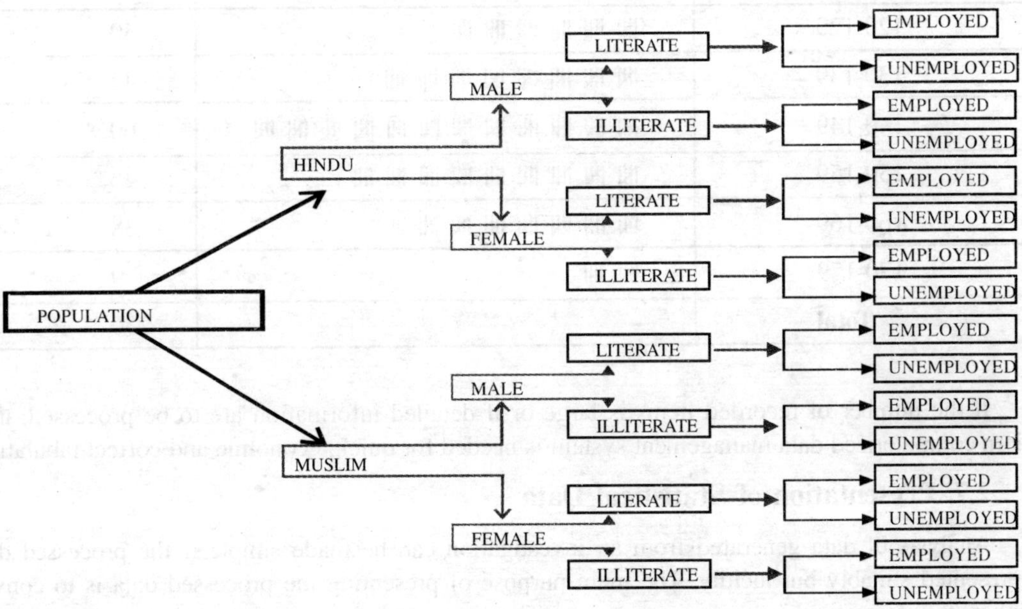

Similarly, we can classify a population quality-wise accommodating many more attributes in a befitting manner according to the objective of our study.

(iv) Quantity-wise classification : When the data are recorded by direct measurement or by counting on a character or individuals, quantitative classification is necessary for further analysis. For example, the data on height (recorded by measurement) or a group of individuals or the data on the number of peas (recorded by counting) in pea-pods are classified under quantity-wise classification. Since the data, recorded by measurement, are any values on the measurement scale, those are generally classified in continuous intervals. Data, recorded by counting, are generally classified by the single values since they are recorded by isolated values on the scale.

The frequency distribution (which will be discussed later) is an excellent illustration of this type of classification. For example, we record below the weights (lbs) of 250 post graduate students of the University of Calcutta as 91, 109, 139, 112, 110,....., 170, 179 (vide question 12 Exercise 2). The common method of classification is to enter each record in the appropriate partition (perferably of equal width and placed adjacently) on the weight scale by putting a tally mark for tabulating each record.

Table 2.3— Classified Distribution of Weights of 250 Students from Calcutta University

Classified Interval of Weights (1bs.)	Tally Marks	Frequency
90-99	𝍸 II	7
100-109	𝍸 IIII	9
110-119	𝍸 IIII	9
120-129	𝍸 𝍸 𝍸 𝍸 𝍸 𝍸	30
130-139	𝍸 𝍸 𝍸 𝍸 𝍸 𝍸 𝍸 𝍸 II	42
140-149	𝍸 𝍸 𝍸 𝍸 𝍸 𝍸 𝍸 𝍸 𝍸 𝍸 𝍸 𝍸	60
150-159	𝍸 𝍸 𝍸 𝍸 𝍸 𝍸 𝍸 𝍸 𝍸 I	46
160-169	𝍸 𝍸 𝍸 𝍸 𝍸 𝍸 𝍸 III	38
170-179	𝍸 IIII	9
Total		250

If the number of recorded items is large or if detailed information are to be processed, then modern computerized data management system is needed for quick, economic and correct tabulation.

2.7. Presentation of Statistical Data

Analysis of data generated from an investigation can be made simple if the processed data are presented suitably but lucidly. The main purpose of presenting the processed data is to convey its inherent information in a precise manner. The processed data can be presented convincingly through, (i) textual presentation, (ii) statistical tables, (iii) diagrams or charts, and (iv) graphs. These types of presentation can be found on the business pages of newspapers, in the annual reports of the companies, Government departments, scientific and other institutions, on the display boards of hospitals and trade centres such as Bank, LIC, UTI. There is no hard and fast rule for unique presentation of data, but the act of presentation depends upon individual's talent and artistic concept. In the following, some typical types of presentation of statistical data are illustrated.

(i)Textual Presentation : By this method the information is presented through a text by one or more paragraphs. This mode is generally used by official agencies and Government departments. An example of the textual presentation is given in the following paragraphs extracted from Tea Auction Report in 'The Statesman, May 25, 1993.

"Calcutta Auction Sale No. 21 as held on May 24 with a good general demand for all varieties. A total of 57,916 packages were offered against 51,139 packages in Sale No. 21 of 1992.

There were 31,592 packages of CTC teas on offer. The pieces obtained were generally firm on rates with occasional easier tendency by up to Re. 1 per Kg., especially for fanning grades. The CIS buyers operated up to Rs. 52 per kg. The domestic market lent good support, especially the western Indian buyers, who were strong on good liquoring varieties. The major blenders were active with selective inquiry from Polish buyers. Assam sold between Rs. 51 and Rs. 59 per kg. While Dooars ranged between Rs. 49 and Rs. 56 per kg. A line of Borasapari BOP sold at Rs. 71 per kg. which is the top price for a CTC tea this week.

There were 3945 packages of Darjeeling teas on offer which met the strong general demand".

Textual presentation should be brief, precise, objective and unambiguous. For a large mass of data or if the data contain a large number of comparisons, textual presentation is disavantageous. *Firstly*, it is ineffective owing to its presentation in narrative form. *Secondly*, the reader has to go through the entire text to know the information which he needs, and also the reader cannot grasp the whole of the information easily. *Thirdly*, all the information given in the text may not necessarily be put into a table. The main advantage of textual presentation in that any stray as well as important information can be incorporated in the text.

In most of the real life examples of the textual presentation there is little scope to think that it is a preceding stage of tabular presentation of statistical data. In fact, the textual presentation is resorted to in those situations where difficulty is experienced in tabular presentation of the information.

(ii) Statistical Tables : Statistical table is a systematic, compact and non-repetitive representation of quantitative information relating to both qualitative and quantitative characters. A table may be simple or complex according to the nature and the purpose of the investigation. Sometimes a table may serve as a report, such as reports on industrial and agricultural production, reports on population and vital statistics etc., which are prepared for reference in future. There are some other types of tables which may be used for further derivation and analysis. A fine example of the latter type of tables is frequency distribution table. Whatever be the type of table, the generally accepted parts and principles in the construction of statistical tables are as follows :

1. Every table should be identified by a *number* for its reference. In a report or dissertation the number, whether in Arabic or Roman numerals, should be allotted serially all-through or chapterwise.
2. Every table must have a suitable *title* by which we can get a precise idea about the contents of the table without going through its text. Title should not be too short or too long. It should be written in such a fashion so that the place, time, subject (unit),

quantitative and qualitative characters involved in the data can easily be identified. For example, in the title. "AGE AND SEX-WISE DISTRIBUTION OF EMPLOYEES OF INDIAN BANKS IN THE YEAR 1990", the place, time, subject, quantitative and qualitative characters are India, year 1990, employee, age, and sex respectively.

3. The *caption or column heading* and the *stub or row heading* of a table should be clearly and briefly (without abbreviation) outlined at the top and on the side respectively for qualitative and / or quantitative characters. The place & time identifications can be used as the caption and stub, respectively.

4. The numerical information regarding row-headings and column-headings are contained in the *body* of the table. This is the most important part of the table.

5. Row headings and column-headings should be brief. If there exists any ambigutiy or confusion regarding the definitions or the formulae, clarify these as briefly as possible in the footnote of the table using the symbols *, †, ~§ etc. If there exists any outlier (it need not be used for calculations) in data, it can be mentioned in the footnote of the statistical table.

6. At the right bottom of the table the *source* of data should be indicated. This part of a table is usually needed when data have been procured from a secondary source.

A typical format of a table is given below.

Table number : ←——Title——→

	←——CAPTION——→				
	Column Numbers				
	(1)	(2)	(3) = (2) ÷ (1)		(*n*)
STUB (1)					
(2)					
(3) = (1) ÷ (2)					
Row Numbers					
(*m*)					

7. Sometimes row headings and / or column headings are numbered numerically or alphabetically to facilitate the presentation of the rule of construction of the rows or columns whose elements have been derived from the elements of some other rows or columns.

Illustration 2.1. In an anthropometric study, summarised data using three percentile points (5%, 50%, 95%) and arithmetic mean on measurements of human body (both male and female) dimensions, such as, vertical reach, stature, eye-to-floor, elbow-to-floor, arm reach from wall, chest circumference, waist circumference, hip circumference and foot length, have to be presented in the form of a table. Given below is a blank table prepared to accommodate the above mentioned information.

Table 2.4—Percentiles and Mean Values on values on selected Human Body Mesurements.

Body Dimensions (in inches)	**Male**			**Female**			**Arithmetic Mean**	
	5%	**50%**	**95%**	**5%**	**50%**	**95%**	**Male**	**Female**
Vertical Reach	a*	b†	c§	..	..	..	..	..
Stature	..	..	..	..	..	..	..	..
Eye-to-floor	..	..	..	..	..	..	..	..
Elbow-to-floor	..	..	..	..	..	..	..	..
Arm reach from wall	..	..	..	..	..	..	..	..
Chest Circumference	..	..	..	..	..	..	..	..
Waist Circumference	..	..	..	..	..	..	..	..
Hip Circumference	..	..	..	..	..	..	..	..
Foot length	..	..	..	..	..	..	..	..

Source : An Anthoropometric Survey

Footnotes : For vertical reach

* 5% of male population having the vertical reach below '*a*' inch

† 50% of male population having the vertical reach below '*b*' inch

§ 95% of male population having the vertical reach below '*c*' inch

(iii) Diagrams or Charts : Diagrams are artistic presentations of some systematically related information. Diagrams are prepared with the help of lines, blocks, circles, pictures etc, on graph paper or on plain papers which are suitably graduated. Diagrams can be of two -or three-dimensional types, each type can again be of single or multiple nature.

(a) Line diagram : When data are recorded against time, i.e., week, month, year etc,. presentation of such data (i.e., time-series data) in a 'line diagram' can be useful for further analysis. Definition of line diagram can be given as follows :

Line diagram is a diagrammatic representation of a set of data recorded against time where the points, obtained by plotting the data along y-axis against the corresponding time along x-axis, are joined consecutively by line segments.

At first, we draw two perpendicular axes taking the time scale along the horizontal axis and the numerical scale covering the variable values along the y-axis. Then, we plot the given observations against the respective time points. Finally, we join the plotted points by segments of straight lines. Now let us consider the data of Table 2.2, which show the changing pattern of exports of jute goods (in value) during 1980-85. The data are presented conveniently in Figure 2.1.

Sometimes two or more line diagrams may be drawn under the common perpendicular axes to form comparative analysis of two or more time series data. Such a composite drawing of single line diagrams for each of the given comparable variables is known as 'multiple line diagram'. In multiple line diagram different identification marks (e.g., colours, broken lines, geometrical figure plots etc.) are necessary to distinguish between the line diagrams. Table 2.5 shows the exports and

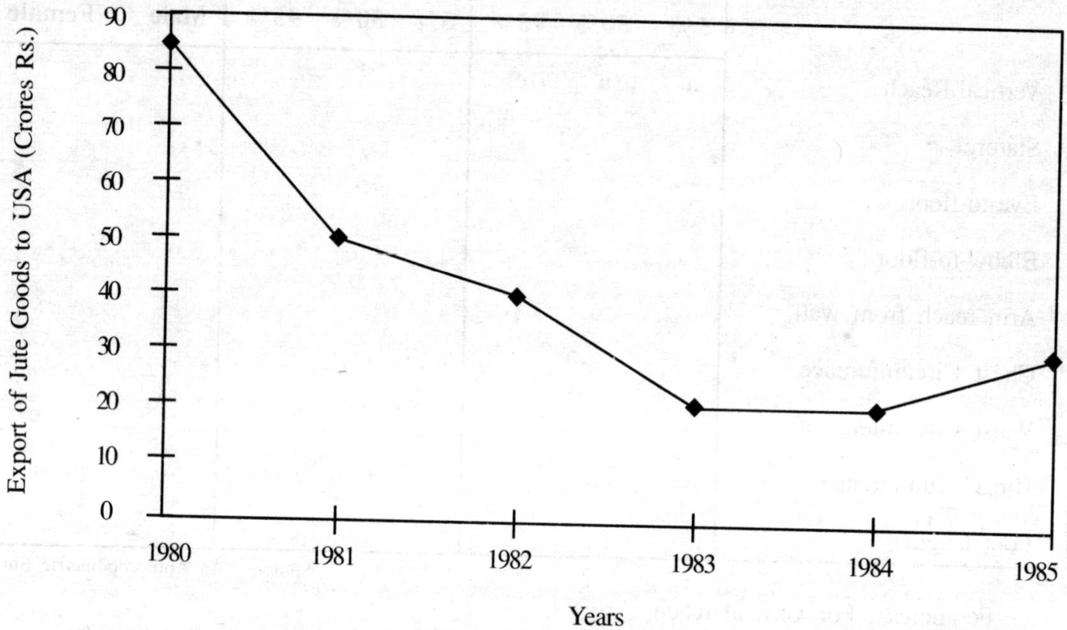

Fig. 2.1. Line Diagram

imports of India from 1998 to 1992. The lucid way to represent such data is to use a multiple line diagram. It is shown in Figure 2.2.

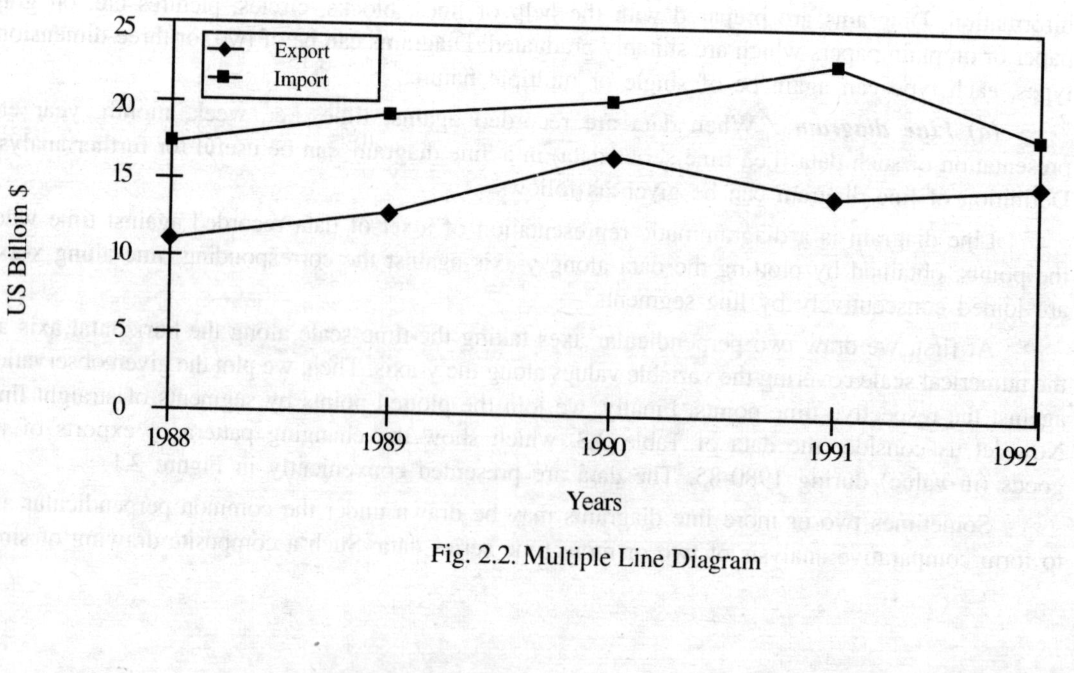

Fig. 2.2. Multiple Line Diagram

Table 2.5—Exports and Imports of India 1988-1992

Year	**1988**	**1989**	**1990**	**1991**	**1992**
Exports (US $ Billion)	11.8	13.5	16.5	17.9	17.8
Imports (US $ Billion)	17.2	19.0	20.9	23.3	19.5

Source : Economic Survey. 1992-93. Ministry of Finance.

The space between the Exports and Imports lines gives us an idea of the trade gap or the balance of payments in different years.

(b) Ratio chart or Semi-logarithmic chart : Generally diagrams are drawn on a graph paper on natural scale. Sometimes the representation of data with respect to time on a natural scale may not convey the facts on growth or decay, especially for the comparison of two or more time series data. In such cases the ratio chart may be more meaningful. In fact, the ratio chart serves the purpose of presentation of data when we are interested in relative change rather than absolute change in data over successive time periods. Here we have to use ratio scale, i.e., scaling based on G.P. series for plotting the given observations. Examples of natural scale and ratio scale are presented below.

Natural Scale : 0 1 2 3 4 5 6 7 8

Ratio Scale of Base 2 :1 2 4 8 16 32 64 128 256

(becomes a natural scale if we take log with base 2)

Ratio Scale of base 3 :1 3 9 27 81 243 729 2187 6561

(becomes a natural scale if we take log with base 3)

Generally, ratio scale of base 10 is used since logarithmic values with respect to base 10 are readily available. Use of ratio chart is advisable to represent a single series of data which increase or decrease in a constant ratio either exactly or approximately and also to represent two or more series of comparable data (which may or may not grow in a constant ratio) which start from significantly different orgins. In case of a data set wherein the observations do not grow in a constant ratio, the ratio chart will provide an useful means of representation. Table 2.6 shows the profits of an industrial unit and that of the whole industry from 1950 to 1956.

Table 2.6— Profit of an industrial unit and the industry as a whole

Year	1950	1951	1952	1953	1954	1955	1956
Industrial unit (.0000Rs)	1.50	2.00	2.18	2.69	3.50	6.00	14.00
Industry (.000000 Rs)	3.2	4.0	5.4	6.8	8.0	11.0	19.00

Source : I.C.W.A.Jan, '69

Such data can be meaningfully presented by a ratio chart (Figure 2.3) by plotting logarithms of given variable values on the natural vertical scale and corresponding time (year) on the horizontal scale. For the construction of ratio chart the following calculations have been made.

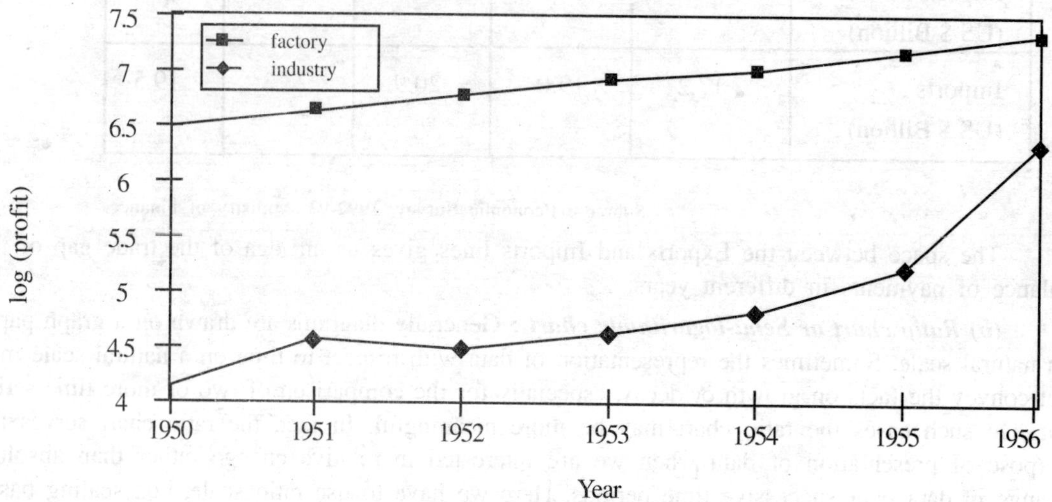

Fig. 2.3. Profit Profile (over years) Using Ratio Scale

Year	1950	1951	1952	1953	1954	1955	1956
log (Unit Profits)	4.18	4.30	4.26	4.43	4.54	4.78	5.97
log (Industry Profits)	6.50	6.60	6.73	6.83	6.90	7.04	7.28

(c) Band chart or Component part chart or Area chart : When data are given with respect to a number of components, e.g., production of paddy, wheat and other crops, each component being observed over the same time period and where the total value of the components at that time is as also important for the purpose of presentation, a compact way to represent such types of data diagrammatically is to use 'band chart'. A band chart is essentially a multiple line diagram where instead of composite drawing of line diagrams for individual components, we draw a number of line diagrams, one for each component and superimposed over another. More specifically, the data for the components have to be plotted by adding the data against the components successively. Let us have to plot three components A, B, and C (e.g., rice, wheat and pulses) with observations x, y and z. respectively against a particular time point(say, a year or a month) etc. To construct a band chart, we plot x for A, x + y for B and x + y + z for C component against the relevent time points. Such plotting task is to be done for all the given successive time points. Next join the plotted points coresponding to each component by line segments. Now the area between the undulated lines for the first component A and the horizontal lines (used for the time scale), the area between the line for component A and that for component B and the area between the line for B and that for the third component C are used for comparison and analysis purposes. Table 2.7 shows the production of cereals in India form 1962 to 1967. Such observations can be conveniently presented in a band chart (Figure 2.4)

Table 2.7—Production of cereals of India from 1962 to 1967 (In Million Ton)

Year	Rice	Wheat	Pulses	Other Cereals
1962	30	10	10	14
1963	32	11	8	18
1964	33	8.5	11.5	20
1965	35	12	11	20
1966	36	10	10	22
1967	38	11	9	23

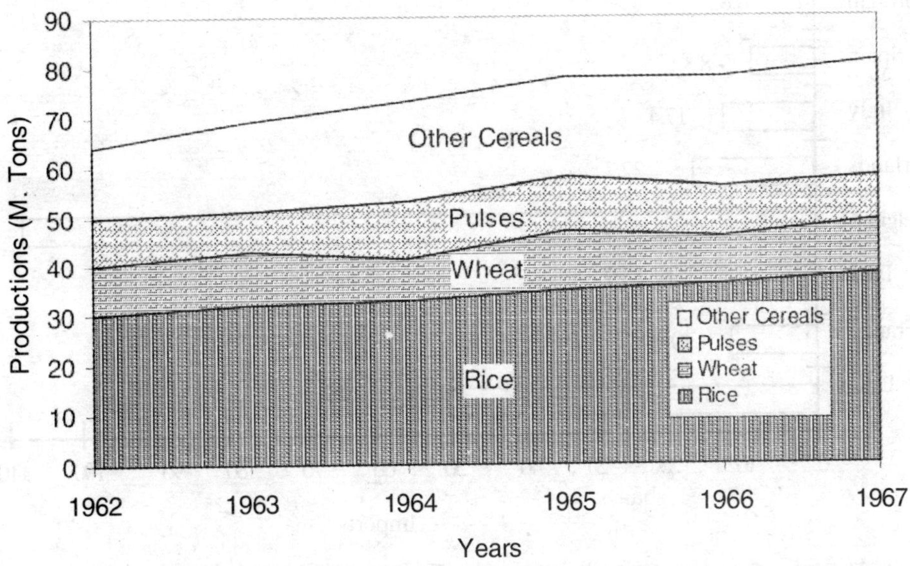

Fig. 2.4 Band Chart

(d) Bar diagrams : Bar diagram is the most common type of diagram used for depicting data. Bars of equal widths should be erected on a fixed base line, the height of a bar representing a particular observation with respect to a year or a place or a category. The detailed procedure for construction of bar diagrams is as follows :

Two perpendicular axes are drawn taking the variable values along the vertical axis, and years or other categories along the horizontal axis. The width of the bars should be equal and neither too wide nor too narrow with respect to the outline of the paper on which it is drawn. Gaps between the bars should be equal and their widths should not be wider than those of the bars, and also should not be narrower than half of the width of the bar in each case. It is desirable to write the respective observations at the top of each bar so that a reader can know the exact value without going through the vertical scale. Horizontal bars are customarily taken to represent the area-wise classified data.

Bar diagrams are of the following types.

I. Single bar diagram, II. Multiple bar diagram, III. Deviation bar diagram, IV. Component part bar diagram and V. Percentage bar diagram.

The methods of construction and usable situations for the above mentioned diagrams are illustrated as follows :

I. Single bar diagram

A single bar diagram is used to respresent data on one variable only. The observations on exports, imports, sales, production etc., at different times or places or objects may be shown by single bar diagram. The figures of Table 2.1 show the imports of jute and jute goods by EEC

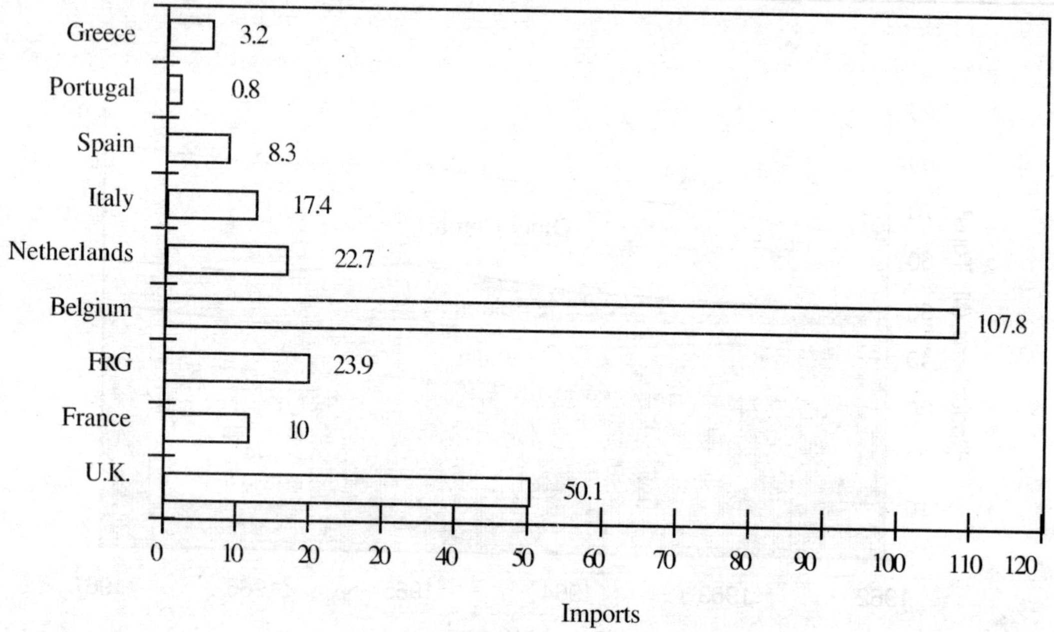

Fig. 2.5. Horizontal Bar Digram

countries from India in 1991. The tabulated data are presented by a single horizontal bar diagram in Figure 2.5. The data on the export values of Table 2.2 are presented in Figure 2.6 by vertical bar diagram. It is customary to repesent the data over space by horizontal bar diagram and the data over time by vertical bar diagram.

II. Multiple bar diagram

In a 'multiple bar diagram' we represent two or more series of data on comparable variables, e.g., exports and imports; demand, supply and price; in a poll the number of persons who contested and the number of persons who got elected belonging to major political parties etc., on the same graph paper. We draw contiguous bars for the given variables for each of the observed times or places or other categories. The bars for each variable should be distinct. Thus to represent the data

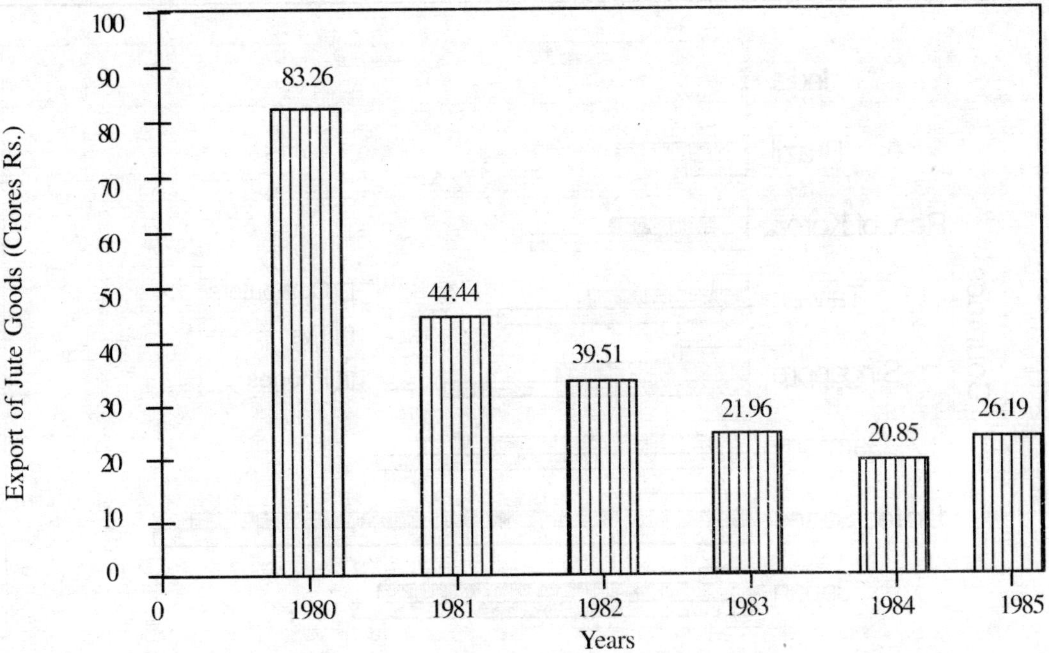

Fig. 2.6. Single Bar Diagram

of Table 2.8 on the use of information technology, the number of phones, TVs and computers per 1000 people in different developed and developing countries including India, we may use multiple bar diagram. (Figure 2.7).

Table 2.8— Use of Information Technology

Countries	Information Media		
	Phones	**TVs**	**Computers**
Japan	640	600	83
United States	960	900	243
Germany	599	482	87
Singapore	490	310	107
Taiwan	405	260	56
Rep. of Korea	350	220	25
Brazil	100	210	6
India	9	9	0.7

Source : World Bank Policy Research Bulletin, Mar-Apr,. 1993

III. Deviation bar diagram

Sometimes compilation or tabulation of data with respect to the net quantities of a variable may lead to both positive and negative values over the given sequence of times or places or other categories, e.g., data on net profits, on net growth rates, on net gains, on inflation rates etc. In

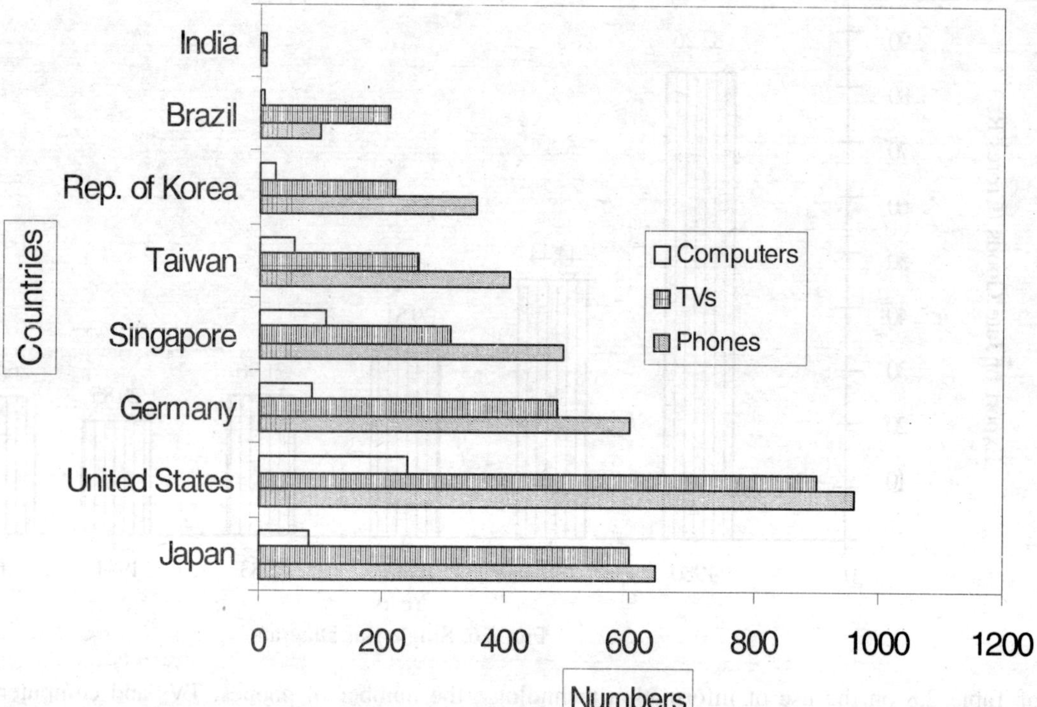

Fig. 2.7 Multiple Bar Diagram

Figure 2.8, the bars represent those net quantities. Such bars can have positive or negative values. The bars above the base line represent positive values and inverted bars represent the negative values. This diagram is known as 'deviation bar diagram'. The data of Table 2.9 on inflation in rural India can be represented conveniently by deviation bar diagram (Figure 2.8)

Table 2.9 Inflation Rate for Agricultural Labours in Rural India

Provinces	**Inflation***
Kerala	14.3
J & K	13.4
Punjab, Delhi Haryana, H.P.	7.7
Assam, Manipur, Meghalaya, Tripura	3.1
West Bengal	2.9
Orissa	2.3

(Table 2.9 continued)

Provinces	Inflation*
Bihar	2.2
Uttar Pradesh	0.6
Tamil Nadu	0.5
All India	–2.0
Karnataka	–5.6
Rajasthan	–6.3
Madhya Pradesh	–6.3
Andhra Pradesh	–7.9
Gujarat	–9.1
Maharastra	–18.5

Source : Labour Bureau, Shimla, Ministry of Labour

* Based on consumer price index for agricultural labours for July 1993, base year 1960-'61 = 100

IV. Component part bar diagram or Divided bar diagram

A component part bar diagram looks like a single bar diagram wherein each bar is divided into a number of components representing the data for the respective variables. The height of each bar represents the total of observations (which is also a variable and can be used for analysis) on the variables under study. To distinguish among the different components, it is needful to use different shades of colours. The data of Table 2.7 can be represented by component part bar diagram as shown in Figure 2.9.

V. Percentage bar diagram

When data on a variable are given category-wise e.g., data on Five Year Plan, on Family Budget Inquiry etc., we can represent such data by a 'percentage bar diagram' since the percentages are expected to be more representative values for different categories rather than the absolute values. When such diagrams are constructed, the length of the bar is taken as 100 and the width is taken arbitrarily. To compare the two or more sets of percentage figures (e.g., to compare First Five Year Plan to Fifth Five Year Plan) by percentage bar diagrams, the widths are made proportional to the respective aggregates. If a bar with certain width is chosen, then its length is divided into a number of sections by drawing lines parallel to the base in such a way that the area of each section represents the percentage of the corresponding category. The lines are drawn at the cumulative percentage lengths of the bar for the successive categories. It is similar to the plotting of observation in a band chart or in a component part bar diagram. The data of Table 2.10 shows the distributions of operating profits of India-Ashahi Glass Company for the year 1990-91 and 1991-92. The data have been represented conveniently by percentage bar diagram in the Figure 2.10 for comparison and analysis.

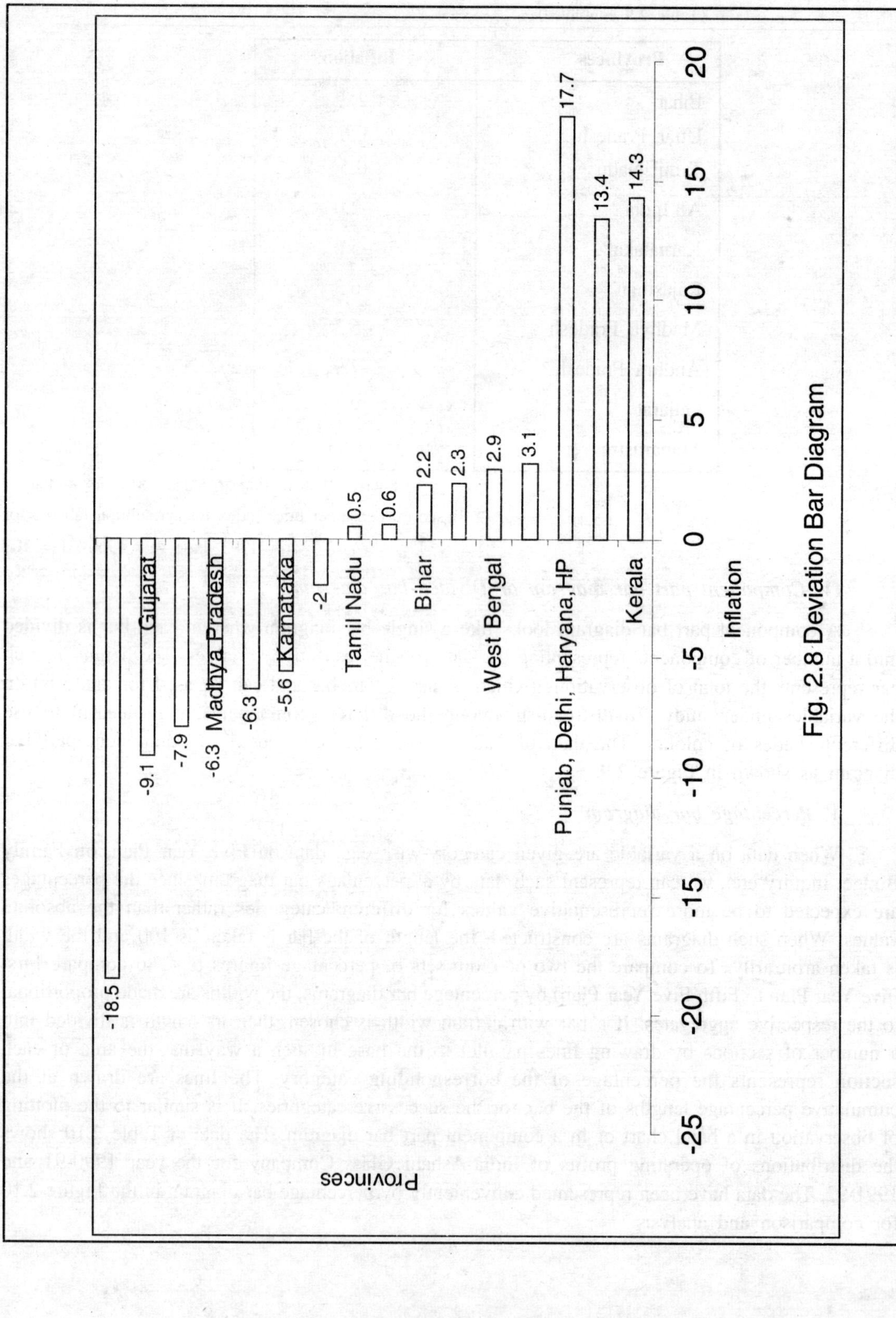

Fig. 2.8 Deviation Bar Diagram

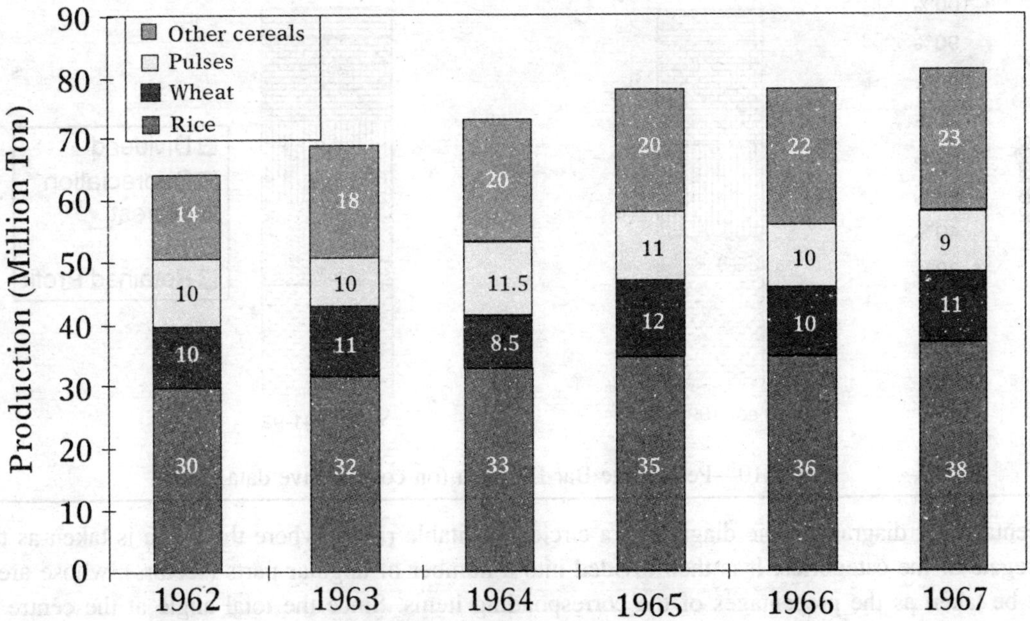

Fig. 2.9—Component Part Bar Diagram

Table 2.10—Distribution of Operating Profits (In corores Rs) of Indo-Ashahi Glass Co.

	1990-91	1991-92	Percentage Values*	
			1990-91	1991-92
Retained Profit	15.80	20.24	29.48(106⁰)	27.27(102⁰)
Tax	5.69	15.08	10.62(38⁰)	21.06(76ᵒ)
Interest	13.03	15.33	24.31(88⁰)	21.41(77⁰)
Depreciation	14.53	15.60	27.11(98⁰)	21.79(78⁰)
Dividend	4.54	5.35	8.48(30⁰)	7.47(27⁰)
Total	53.59	71.60	100.00(360⁰)	100.00(360⁰)

Source : The Economic Times, 2nd Nov., 1992

* Percentage values for the construction of Percentage bar diagram

(.) : Corresponding angles obtained by multiplying 3.6 to the percentage values for the construction of Pie diagram.

Here the widths of the two bars are taken as 1.8 and 2.4 respectively to make ajdustments (by taking proportionals) for varying total operating profits for the years 1990-91 and 1991-92 (as the ratio of the total operating profits = 1 : 1.33 = 1.8 : 2.4

(e) Pie diagram

Pie diagram is used for the same purpose as that for the represenatation of data by a

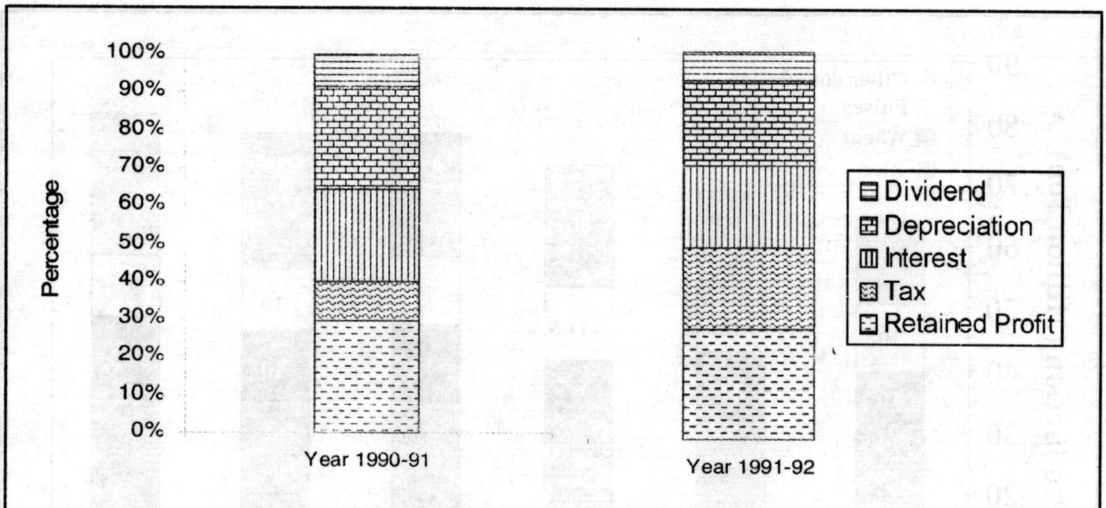

Fig. 2.10—Percentage Bar Diagram (on comparative data)

percentage bar diagram. A pie diagram is a circle of suitable radius where the circle is taken as the aggregate of the categories. It is then divided into a number of angular parts (sectors), whose areas may be taken as the precentages of the corresponding items. Since the total angle at the centre of a circle is 360^0, it is clear that for any particular item/category, the angle should be 3.6 (= 360/100) times the corresponding percentage. The data in Table 2.10 have been represented by two comparable pie diagrams in Figure 2.11. Evidently πr^2 =aggregate, since the ratio of square roots of total operating profits = 1 : 1.16 or 2.5 : 2.9, the radius of Pie diagram to represent data of 1990-91, is taken as 2.5 cm and that of the other is taken as 2.9cm.

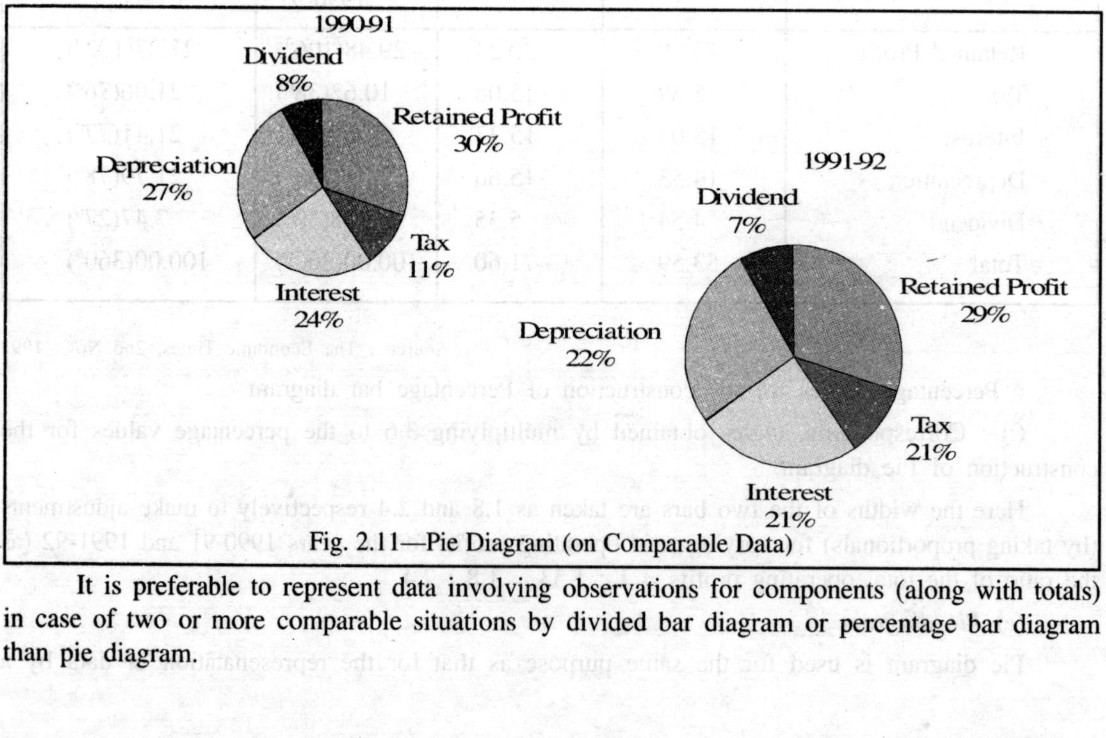

Fig. 2.11—Pie Diagram (on Comparable Data)

It is preferable to represent data involving observations for components (along with totals) in case of two or more comparable situations by divided bar diagram or percentage bar diagram than pie diagram.

(f) Pictorial diagram or Pictogram

As mentioned at the outset, there does not exist a unique set of hard and fast rules to present statistical data. We generally adopt a diagram which is relatively vivid, attractive and easily comprehensible. Presentation of data by pictures, constructed by unit idols or by unit squares, cubes, circles etc., is common in daily newspapers. The usefulness of such diagrams is no less than other diagrams. The following steps are taken for the construction of pictograms :

(i) Each of the unit symbols should represent certain number (numerical value) and a proportional part of it should represent the same proportion of the number.

(ii) The unit symbols should be of same size and shape, be self-explanatory, and the fraction of the symbol should be numerically understandable. If the data are concerned with the number of graduates, the symbol should be an outline of a man in convocation robe. If the data are concerned with TV sales the symbol should be the outline of a TV set.

(iii) The diagram should be as clear as possible. Many facts should not be presented at a time.

The pictogram possesses the following demerits :

(i) The pictorial diagrams give an overall picture without giving minute details existing in data.

(ii) There may be many facts (comparative or isolated) which cannot be shown pictorially.

The data of Table 2.11 shows the production of cars of a factory in a year from January to December. Such data can be presented by a pictogram as in Figure 2.12.

Table 2.11—Monthly Production of Cars in a year

Month	Jan	Feb.	Mar	Apr	May	June	July	Aug	Sept	Oct	Nov	Dec
Number of Cars	5720	4930	6110	5930	6040	4610	3060	4700	5605	3275	6850	6130

(g) Statistical maps

Presentation of data on maps is a very important tool to show industrial, agricultural and social variations from one geographical area to another. A map is used not only for graphic presentation of facts but it is indispensable in locating social problems and in drawing hidden facts also. We may use statistical maps when we are required to show diagrammatically the variation in family composition in different districts of a state, the average number of persons who come to a city every day from nearby towns, the variation in rainfall in different parts of a country etc. Remote sensing data on weather or on agricultural crops through statistical maps are very much convincing.

Some of the ways to construct statistical maps are discussed below :

(a) Geometrical Figures (dots, triangles, circles etc.) can be used to show area- wise density and location of the variable under study. Each geometrical unit figure should represent a certain magnitude.

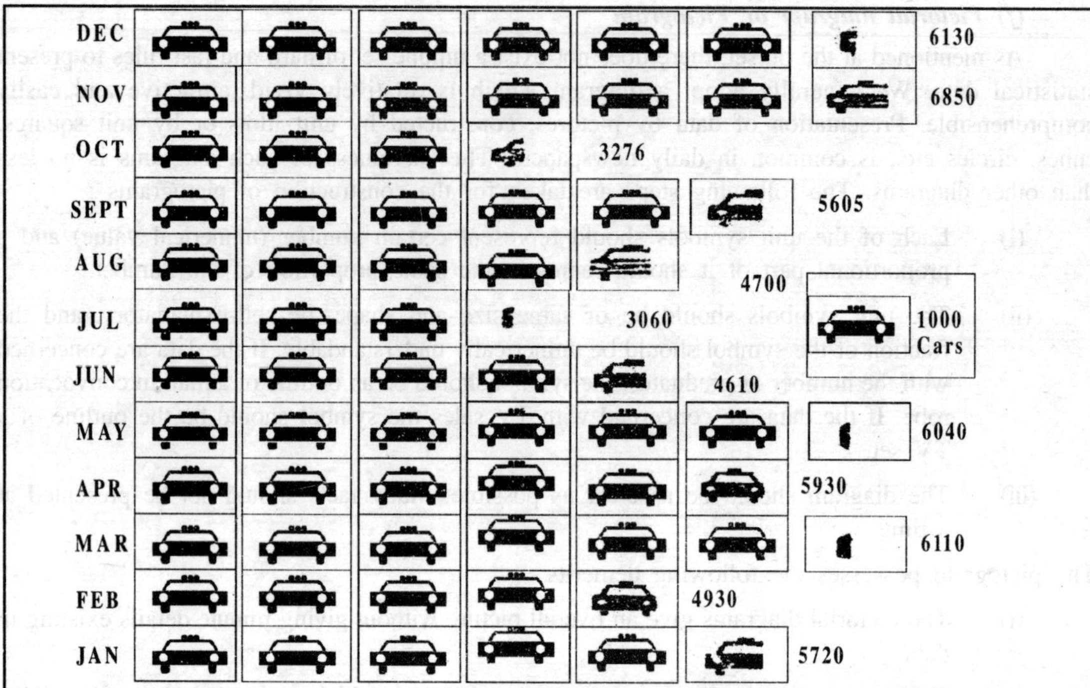

Fig. 2.12—Pictorial Diagram showing number of cars produced monthly by a company in a year

(*b*) Areawise data, related to a character, for a country can be represented by a bar diagram or a pie diagram, wherein each datum corresponding to an area is represented by a suitable bar (or any other suitable diagram) depicted on the related geographical area shown on the map.

(*c*) Areawise data variation relating to a particular character can be shown on a map by shaded contours. The shading can be done by geometrical designs or by light to deep colours.

The figures given in the table 2.12 are based on the Statistics supplied by National AIDS Control Organisation (NACO) and the information have been presented in the Statistical map (Fig. 2.13) in which the status of HIV infection in India as on December 31, 1994 has been reflected. From the map we can also understand how fast is the AIDS virus spreading its tentacles in some states of India, namely, Tamilnadu, Maharashtra, Kerala, Manipur and Delhi.

Table 2.12—Number of HIV infected people in the different states of India

State/Territory	Number of HIV inflected people	State/Territory	Number of HIV infected people
Andhra Pradesh	143	Maharashtra	5428
Assam	6	Manipur	2758
Bihar	3	Mizoram	72
Delhi	978	Nagalanad	112
Goa	357	Orissa	33
Gujrat	513	Pandichery	1009

Haryana	120	Punjab	165
Himachal Pradesh	13	Rajasthan	43
Jammu & Kashmir	10	Tamilnadu	2766
Karnataka	1569	Uttar Pradesh	475
Kerala	180	West Bengal	251
Madhya Pradesh	64		

Source : NACO

Exercises and Class Projects

2.1. What important things have you to keep in mind while preparing a question-naire?

2.2. What are the differences and /or similarities between data, objective evidence and information?

2.3. Distinguish between Primary data and Secondary data; Qualitative data and Quantitative data; Frequency type data and Non-frequency type data; Scientific inquiry and social inquiry.

2.4. Discuss briefly the statistical procedures of investigations relating to a scientific inquiry and a sociological inquiry.

2.5. Explain briefly : Line chart, Ratio chart, Bar chart, Pie chart and Percentage bar chart and mention how are these useful in presentation of data.

2.6. Represent the following data by appropriate diagrams and comment.

Table A — Year-wise Production of Wheat

Year :	1984	1985	1986	1987	1988	1989	1990
Production (000 tons) :	37	38	37	40	41	45	50

(ICWA. June '92)

Table B — Profit and Loss of a Company
(in ten days of a month)

Year :	1	2	3	4	5	6	7	8	9	10
Profit (in Rs.) :	400	800	600	900	200	200	—	—	800	1000
Loss (in Rs.) :	—	—	—	—	—	—	400	100	—	—

(B.U.B. Com 1963)

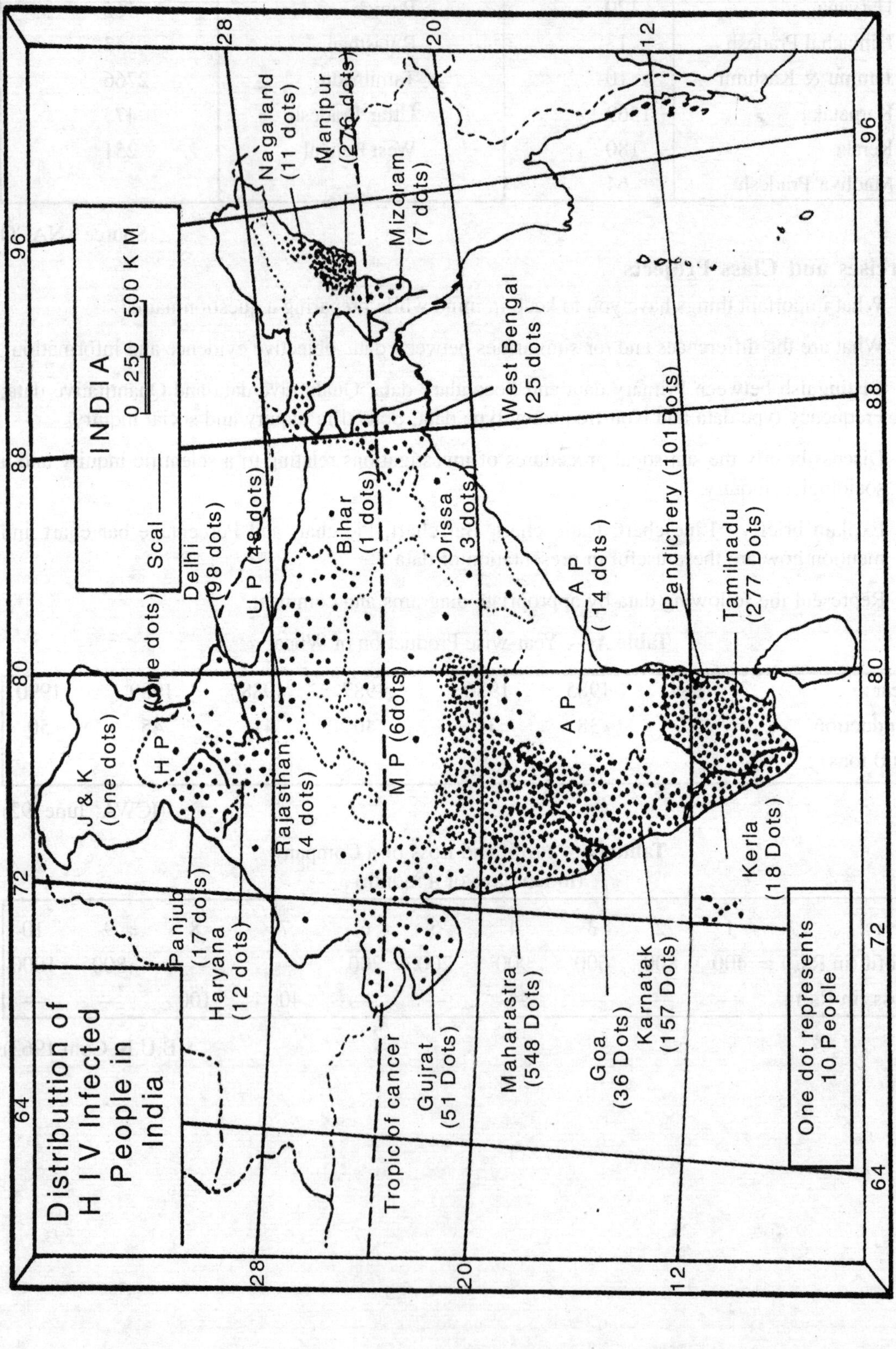

Fig. 2.13 Statistical map showing the status of HIV infection in India till 31st Dec. 1994

Table C — Irrigated Area in India, 1954-55

Source of Irrigation.	Irrigated Area (000 Acres)
Canals	22,224
(i) Government	19,356
(ii) Private	2,868
Ponds	9,889
Tube-wells	16,562
Others	5,5,87

(C.U.B. Com 1967)

Table D — Production (in million ton) of Wheat, Rice and Sugar for 1962-66 in India

Year	1962	1963	1964	1965	1966
Wheat	10	12.3	13.0	12.0	14
Rice	8	10.0	10.5	11.0	13
Sugar	7	7.5	9.0	8.5	8

Table E — Five Year Plan outlay of two states in a year

Sectors	Province A (in crores Rs.)	Province B (in crore Rs.)
Agriculture	8,000	9,500
Industry and Mines	7,000	6,500
Irrigation and Power	4,000	7,000
Transports	5,500	5,000
Miscellaneous	2,500	3,000

2.7. In a certain data, following four main characteristics with their sub-characteristics have been presented :

Main-characteristics	Sub-characteristics
Locality	Urban, Rural
Religion	Hindu, Non-Hindu
Sex	Male, Female
Age	0-30, 30-60, 60 and over

Prepare a blank table.

2.8. In a group of 1000 wage earners the monthly wages of 4 percent are below Rs. 60 and those of 15 percent are under Rs. 62.50. 15 percent earn Rs. 95 or more, and 5 percent get Rs. 100 or more. Present the above textual information in a table.

2.9. Complete the following table relating to the suitable diagrammatic representation of each of the data sets given in tables of Chapter 2.

Table F — Suitability Chart

Table Number	Diagrams to represent Most suitable	May be suitable	Not at all suitable
Table 2.1	Horizontal Bar Diagram	Pictorial Diagram	Other Diagrams
Table 2.2	—	—	—

2.10. The following table gives the area under wheat with irrigation facilities and yield rate of the same in India during the period 1950-55. Represent the data under the same co-ordinate axes on the graph paper.

Year	Irrigated Area ('000 acres)	Yield Rate (lbs per acre)
1950	7618	584
1951	8407	592
1952	8506	582
1953	9121	681
1954	9601	670
1955	9810	717

(Hints : Draw multiple-line diagram taking two vertical axes at both the ends of the horizontal axis for the two sets of data).

2.11. The following table gives the total value of products and elements of production costs in a country.

Year	Cost of raw Materials ('00 Rs)	Labour Costs ('00 Rs)	Overhead Cost & Profit ('00 Rs)	Total Value ('00 Rs)
1973	347	110	148	605
1975	359	107	160	626
1977	348	108	166	622
1979	382	116	202	700
1981	217	72	122	411

Depict the facts graphically in respect of absolute and relative changes in the different components over the years. Depict the same by drawing the bars as well.

2.12. The weights (in lbs.) of 250 Post-graduate students are presented in the following table :

91,	109,	139,	112,	110	165,	108,	117,	121,	166,
111,	172,	127,	159,	160,	171,	113,	118,	177,	178,
120,	128,	114,	141,	134,	143,	115,	173,	122,	116,

128,	130,	153,	135,	164,	167,	147,	174,	175,	145,
130,	145,	146,	138,	154,	138,	134,	149,	139,	147,
126,	162,	153,	135,	148,	146,	169,	139,	138,	148,
129,	148,	143,	148,	151,	144,	152,	157,	168,	151,
123,	163,	158,	154,	137,	133,	125,	105,	132,	161,
100,	149,	140,	159,	156,	160,	165,	167,	155,	152,
166,	154,	163,	176,	161,	143,	146,	120,	123,	168,
121,	158,	137,	135,	157,	162,	164,	146,	148,	169,
167,	122,	101,	149,	144,	141,	169,	164,	147,	108,
167,	155,	156,	127,	154,	145,	144,	148,	161,	166,
143,	104,	124,	145,	131,	136,	135,	104,	133,	133,
125,	132,	136,	149,	143,	165,	168,	161,	141,	131,
126,	133,	141,	151,	164,	141,	146,	132,	129,	103,
142,	159,	166,	152,	141,	146,	158,	127,	151,	142,
124,	145,	162,	167,	144,	147,	123,	163,	160,	161,
92,	150,	140,	142,	156,	126,	98,	144,	153,	150,
140,	131,	134,	151,	141,	148,	123,	127,	132,	135,
93,	148,	143,	141,	126,	133,	135,	155,	153,	154,
124,	137,	130,	141,	157,	152,	165,	150,	151,	131,
124,	148,	146,	135,	133,	156,	154,	95,	126,	132,
140,	142,	95,	143,	154,	145,	144,	140,	152,	141,
94,	127,	131,	133,	153,	155,	155,	151,	170,	179,

Present the data in the form of a statistical table.

Class projects :

2.1. Collect the diagrammatic presentations published in the daily newspapers like The Statesman, The Economic times, The Business Standard etc.

2.2. Prepare a questionnaire for a study on fuel consumption in your locality.

2.3. Present the data on monthly expenditure of your family on food, fuel, electricity, clothing, transport, education, etc. for the past five years and prepare a report.

CHAPTER 3

FREQUENCY DISTRIBUTION AND ITS REPRESENTATION

3.1. Attribute and variable

Any kind of activity, whatever be its form, generates information or data. Such data are processed and ultimately summarised into a compact but elegant form as mentioned earlier. An important point to note in this connection is that the observations emanating from any activity or statistical inquiry relate to measurement of characters on units (or individuals). The term "unit" in Statistics has wide applicability. For example, in measuring diameter of screws, each screw becomes a unit; in case of production of milk yield of cows. each cow becomes a unit; in counting the number of petals of flowers, each flower becomes a unit.

Qualitative characters of data are called attributes whereas quantitative characters of data are called variables. Religion, sex, colour, some human behaviours, dresses etc. are attributes and number of petals in flowers, height/weight/age of a man, number of persons arriving per minute in front of a sales counter are examples of a variable.

The attribute under study cannot be measured in general; one can only check whether it (the character) is absent or present in respondents (or units or individuals). For example, if the attribute under study is religion, we can only find out which persons belong to what religious groups. On the other hand, quantitative characteristics of units, since they (the characters) can be measured on a numerical scale, are called variables. In the study on variables the data are numerical to start with, whereas in the study of attributes numerical figures have to be generated secondarily. For example, when record of the weights of individuals or of the number of nodes in a plant is availble, we find numerical data at first, while when a record of the quality of electrical switches in the sequence of good or bad as "*b g b b b g g g g g b b g b b g g g b g*" is available, we can't get numerical data at first but at the next step numerical data will appear if the above data are presented as : number of good articles (switches) is 11 and number of bad articles (switches) is 9 out of 20 observations on the quality of the switches.

3.2. Discrete and Continuous Variables

A study of data obtained from an investigation of quantitative characters reveals that these characters are of two types — *discrete and continuous*. In the first type the concerned variable can take only a limited (finite) number of isolated values or countably infinite number of values in a given range. Such variables are called discrete variables. Examples of discrete variables are, number of defective articles in a lot, number of spots on leaves, number of letters per page in a book, number of letters in a word (word length) etc. In the second type, the variable can conceivably take any of infinite number of values in a given range. Such a variable is called a continuous variable. The height or weight of a plant, amount of rainfall per day, amount or radiation per day, diameter of a screw are examples of a continuous variable. It can be noted that a continuous variable takes uncountably infinite number of values. In case of continuous variables, limitation of measuring instrument may induce apparent discreteness on the values of a continuous variable, *e.g.*, height of an individual may be recorded in feet and inches only by a standard instrument whereas a more precise instrument may record the height correct to a hundredth part of an inch. So apparent discreteness is due to nonexactness of recording device and measuring instrument only. The distinction between a discrete variable and a continuous variable is to be carefully understood as their statistical treatments differ in some cases.

3.3. Construction of Frequency Distribution Tables

When we collect enormous raw data from question schedules, we cannot easily comprehend the characteristics of the whole (population or aggregate) from which the data have been collected as a part (sample). A frequency distribution table is grouping of raw data in numbers over the different values (single or group) on the scale according to the range of variation of the quantitative character under study. Frequency distribution helps in summarisation of data in a manageable form thus enabling further processing. From the frequency distribution table, we know what values (single or group) of the variable appear with what frequency (number of times). Data on a qualitative character can also be presented in the form of a frequency distribution table.

3.3.1. Frequency Distribution of An Attribute

An investigation on the awareness of the pilots agitation during january 1993 in Indian Airlines conducted by interview of 1000 inhabitants of Bombay, Calcutta, Delhi and Madras generated data in the sequence "*YYNYNNNYNY*....." where, Y = YES, N = NO. The data can be put into a systematic form after counting the occurrence of Y and N as in the following table.

Table 3.1 — Result of Survey on Pilots agitation in Indian Airlines

State of knowledge (about the agitation)	Frequency (number of people)	Relative frequency
Yes	463	0.463
No	537	0.537
Total	1000	1.000

Relative frequency = Frequency/total frequency

The above table presents the data on the attribute—state of knowledge, under one classification with two states. The state of knowledge can also be presented in a table involving two classifications, one of the classifications having more than two classes. For getting such a two-way classified data let a survey be undertaken in which the respondents ('respondent" is more appropriate term than the term 'unit' in social surveys) be asked to comment on a question on the possibility of winning of a particular political party in an election. Suppose a group of 5000 persons in a locality, of which 2391 are women, is selected. The replies are obtained in the sequence, "*YNYYINYYIYNIYYIN* ...." where Y = yes, N = no, I = indifferent, i.e., no comment. The data can be summarised as shown in the table below.

Table 3.2— Sexwise Opinion Poll for a Particular Political Party

State of knowledge	Sex		
	Male	Female	Total
Yes	1232	1057	2289
No	1127	991	2118
Indifferent	250	343	593
Total	2609	2391	5000

Entries in the table represent frequencies. Actually the above data represent the two-way frequency distribution of two attributes.

3.3.2. Frequency Distribution of a variable

The following two tables present data on two types of variables, namely, discrete and continuous variables. In Table 3.3, the number of letters (discrete variable) used in first 100 words in an editorial on the occasion of Netaji's birth anniversary in a daily newspaper are given. In table 3.4, the results emanating from a 'funnel experiment' conducted 90 times are given. The description of the experiment is as follwos. A funnel of 1 foot height is taken and is set inclined to the ground from a stand. An iron ball of diameter 1 cm is allowed to fall from a glass channel inclined to the mouth of the funnel. The iron ball moves spirally and ultimately comes out of the funnel. The time (in seconds) taken by the ball to pass through the funnel is recorded. The time (a continuous variable) in the above mentioned 90 experiments is presented in Table 3.4.

Table 3.3 — Word Lengths of 100 Words

8,	3,	3,	7,	10,	3,	5,	11,	2,	12,	6,	7,
4,	2,	8,	9,	13,	4,	3,	11,	9,	9,	5,	2,
7,	10,	7,	3,	12,	2,	6,	4,	9,	4,	3,	9,
2,	5,	4,	5,	3,	4,	2,	5,	10,	9,	3,	6,
2,	6,	2,	9,	2,	4,	6,	3,	1,	6,	2,	1,
7,	10,	6,	3,	8,	3,	5,	6,	1,	8,	5,	1,
9,	10,	4,	8,	8,	7,	2,	6,	3,	5,	3,	2,
3,	4,	2,	3,	7,	4,	4,	3,	5,	2,	8,	3,
8,	7,	3,	3.								

Table 3.4 — Results (time taken in seconds) of Funnel Experiment (90 Times)

11.0	11.3	14.4	14.3	12.7	11.7	19.8	13.6	13.0
12.7	12.4	17.3	15.1	11.2	14.2	10.3	15.8	14.6
14.0	12.1	16.2	13.1	15.5	13.3	10.4	14.8	12.0
16.0	17.0	10.9	14.9	18.6	19.5	15.1	18.2	14.2
12.9	17.4	13.8	18.7	14.1	17.7	11.5	15.2	10.7

(**Table 3.4** continued)

16.5	13.6	16.7	14.1	15.5	12.2	16.9	16.7	13.4
14.5	13.9	12.5	12.6	17.5	15.7	15.3	13.7	15.9
11.9	15.2	15.6	13.2	18.1	14.0	14.7	13.8	16.2
17.6	12.8	19.2	14.6	14.6	13.5	11.6	14.2	17.8
16.8	14.7	18.7	13.5	14.7	14.3	16.6	15.6	14.8

The data presented in the above two tables as such do not expose their significance to the reader readily. It must be summarised. The systematic procedure developed to condense voluminous data for ready comprehension of significance by means of tabular presentation leads to the constructions of a frequency distribution table. In a frequency distribution table the total frequency is distributed over a number of classes (or values) and each frequency corresponds either to a class in case of an attribute or to a value (or a class containing an infinite number of values) in case of a variable, the classes being mutually exclusive in nature. The marginal sacrifice of information resulting through condensation does not pose a serious drawback owing to the fact that the statistical science exposes features related to the aggregate of individuals and not to each specific individual belonging to the aggregate. The systematic procedure adopted for summarisation of data in case of the above two tables is described below.

(a) Discrete Case *(Table 3.3)*

On examining the whole set of observations, the largest and the smallest values are noted as 13 and 1, respectively. In a table, the numbers 1, 2, 3,........, 13 (i.e., the word-lengths) are written in the first column (in other words each class is supposed to contain only one variable value; the total number of classes in the first column being 13). Next the observations of the Table 3.3 are read one by one from the beginning and a stroke (or tally mark) is put against each tabulated figure serially in the second column. The strokes are put in a group of five to facilitate counting of the frequencies. The third column shows the frequencies (or absolute frequencies) corresponding to a different word-lengths. The ratio of the frequency to the total frequency corresponding to a particular word-length is called its relative frequency and when the same relative frequency is multiplied by 100, the per cent frequency is obtained. The relative frequency corresponding to a specific word-length gives the proportion of words having the specified number of letters. In case of frequency distribution of a discrcte variable two other types of frequencies, namely, cumulative frequencies less than type and cumulative frequencies greater than type are of particular importance. Cumulative frequency (less than type) corresponding to a particular word-length gives the total number of words which have number of letters equal to the specified number of letters or less, and is thus the sum of the frequencies starting from the top upto the frequency corresponding to the sepcified word-length. Similarly, cumulative frequency (greater than type) corresponding to a particular word-length gives the total number of words which have number of letters equal to the specified number of letters or more and is thus the sum of the frequencies starting from the frequency corresponding to the specified word length upto the bottom most frequency (comresponding to maximum word length). Cumulative relative frequencies and cumulative percentage frequencies can also be defined. Relevant calculation in regard to diffrent types of frequencies on the data on word-length are shown in Table 3.5.

Table 3.5 — Frequency distribution of word-lengths of 100 words.

Number of letters per Word	Tally marks	Frequency		Cumulative frequency		Percent cumulative frequency	
		Absolute	Relative	Less than type	Greater than type	l.t.t	g.t.t.
1	IIII	4	0.04	4	100	0.04	1.00
2	卌 卌 IIII	14	0.14	18	96	0.18	0.96
3	卌 卌 卌 卌	19	0.19	37	82	0.37	0.82
4	卌 卌	11	0.11	48	63	0.48	0.63
5	卌 IIII	9	0.9	57	52	0.57	0.52
6	卌 IIII	9	0.9	66	43	0.66	0.43
7	卌 IIII	8	0.8	74	34	0.74	0.34
8	卌 IIII	8	0.8	82	26	0.82	0.26
9	卌 III	8	0.8	90	18	0.90	0.18
10	卌	5	0.5	95	10	0.95	0.10
11	II	2	0.2	97	5	0.97	0.05
12	II	2	0.2	99	3	0.99	0.03
13	I	1	0.1	100	1	1.00	0.01
Total		**100**	**1.00**				

***(b) Continuous Case** (Table 3.4)*

In Table 3.4 the variable (time in seconds is continous in nature and by definition it may take infinite number of possible values and as such it will be improper to calculate frequency for each value (time point) of the variable (each different value being regarded as a class) since then the number of classes will be very large and the very purpose of condensation will be lost. Thus, it is pertinent to partition the total range of variate values = (maximum time – minimum time) into some intervals. Each class corresponds, of course, to an interval. The construction of classes is guided by the following principles :

(1) The classes should be exhaustive and each value must fall in any one of the classes.

(2) The classes should be non-overlapping. None of the values can be included in two classes.

(3) The number of classes should be neither too large nor too small. While too many classes induce spurious patterns in the frequencies, too few classes obscure the true pattern of the observed frequency distribution. As such neither is desirable. As a working rule the number classes should lie between 10 and 20 for a total frequency of 1000 or more. For choosing the number of classes the follwoing formulae have been suggested by statisticians.

Generally, statisticians choose the number of classes by common sense, but several attempts have been made to suggest an objective rule for selecting the number of classes (for the construction of a histogram of equal class width). Strages suggested that the number of classes *(k)* should be approximately equal to $1 + 3.3 \log n$. According to Scott the value of *k* is equal to $R\sqrt{n}/3.49\, s$, and according to Davies and Goldsmith the value of k is equal to $\sqrt{n}$, where *R* is range and *s* is standard deviation of the *n* given observations.

(4) The classes should be of equal width. Unequal class widths are not generally suitable for application of some statistical methods. Class frequencies wherein classes are of unequal widths, are not comparable.

(5) The mid values of the classes are called class marks. While forming classes it should be borne in mind that ideally there should exist an even distribution of frequencies around the class mark and also over the class interval. The assumption that all values included in any class can be regarded as equal to the mid-value of the class does not involve any serious error.

Bearing the above set of principles in mind, the following steps are followed (with reference to the example of Tables 3.4 & 3.5) :

Step 1. The range (maximum value - minimum value) is found as (19.8 - 10.3) sec = 9.5 sec (Table 3.4). Taking the number of classes to be 10, the follwoing classes (in sec.) can be formed : 10.0 – 10.9, 11.0 – 11.9,... 18.0 – 18.9 and 19.0 – 19.9, since the difference between the terminal values in each class is calculated as 9.5 sec/10 $\cong$ 1 sec.

The terminal values in each class are called *class limits*, the maximum and minimum values being called the upper and lower class limits, respectively. The classes are written in the tables in the first column. It is to be noted that the unit of class limits is taken as one-tenth of a second (as that in the case of data).

Step 2. Each datum (observation) starting from the beginning upto the end is then read and a tally mark (vertical or horizontal stroke) is put against the corresponding class in which it is supposed to lie. The strokes are then grouped in 5 for easy counting. Tally marks are entered in the second column.

Step 3. In case of data on continuous variables, if a variable value is recorded correct to 1/10 of a unit (as a second value in the first column of Table 3.4 is 12.7) then the recorded value, 12.7 can be regarded as any value between 12.65 to 12.75. Similarly, if any value is recorded correct to 1/100 of a unit as 14.95 then it can be regarded as any value between 14.945 to 14.955. Following the above mentioned rule, the class 10.0 – 10.9 can be treated as 9.95 – 10.95 and other classes can similarly be revised. The revised terminal values of the class are called *class boundaries*. Thus in presentation of data on a continuous variable, after determination of the class frequencies, the class limits are revised to define class boundaries, (upper and lower) so that no gap exists between the upper class limit of a particular class and the lower limit of the next class. So, 9.95 and 10.95 are the lower and upper class boundaries respectively of the above class. Width of the class-in terval is the difference between the upper and lower class boundaries. Class mark is half the sum of the class limits which is the same as half the sum of class boundaries. It is seen from Table 3.6 that class boundaries are expressed in one decimal place more than that of the values in the original record to avoid overlapping. Relative frequencies and cumulative frequencies (less than type and greater than type) are calculated in a similar manner (discrete case). However, in case of data on a continuous variable, it is to be noted that cumulative frequencies of less than type correspond to upper class boundaries and cumulative frequencies of greater than type correspond to the lower class boundaries. For classes of varying widths, as frequencies therein are themselves not comparable, frequency densities are compared. Frequency density is defined as the ratio of the frequency to the width of the class interval. Thus, frequency density is actually the frequency expressed per unit value of the variable. The computation of different types of frequencies *e.g.,* relative frequency, frequency density, cumulative frequencies are given in Table 3.6.

Table 3.6 — Frequency Distribution of Data obtained from Funnel Experiment

Class limits (time in Sec)	Class boundaries	Tally marks	Frequency		Cumulative frequencies	
			Absolute*	Relative	less than type	Greater than type
10.0 – 10.9	9.95 – 10.95	||||	4	0.044	4	90
11.0 – 11.9	10.95 – 11.95	~~||||~~ ||	7	0.078	11	86
12.0 – 12.9	11.95 – 12.95	~~||||~~ ~~||||~~	10	0.111	12	79
13.0 – 13.9	12.95 – 13.95	~~||||~~ ~~||||~~ ||||	14	0.156	35	69
14.0 – 14.9	13.95 – 14.95	~~||||~~ ~~||||~~ ~~||||~~ ~~||||~~	20	0.222	55	55
15.0 – 15.9	14.95 – 15.95	~~||||~~ ~~||||~~ ||	12	0.133	67	35
16.0 – 16.9	15.95 – 16.95	~~||||~~ ||||	9	0.100	76	23
17.0 – 17.9	16.95 – 17.95	~~||||~~ ||	7	0.078	83	14
18.0 – 18.9	17.95 – 18.95	||||	4	0.044	87	7
19.0 – 19.9	18.95 – 19.95	|||	3	0.033	90	3
Total			**90**	**1.000**		

* Since the class-width is 1, the absolute frequency is same as the frequency density.

3.4. Diagrammatic Representation of Frequency Type Data

Frequency distribution of a variable can be presented graphically by a suitable choice from the following type of graphs of which the first three are generally used for discrete variables and last two are used for continuous variables.

I. Column Diagram or Frequency Bar Diagram

II. Frequency Polygon

III. Step Diagram

IV. Histogram

V. Ogive

I. Column Diagram

A 'Column diagram' is used to represent a frequency distribution of a discrete variable. To represent the frequency distribution of number of letters per word (Table 3.5), a graph paper having two axes perpendicular to each other is taken and the number of letters per word are put serially (ordered) on the horizontal axis. Along the vertical axis (y-axis) different lines are erected, at the above points on the (x-axis), the length of each line being equal to the frequency of a particular number of letters (thus, 13 lines are erected at 13 points as in Fig. 3.1.

II. Frequency Polygon

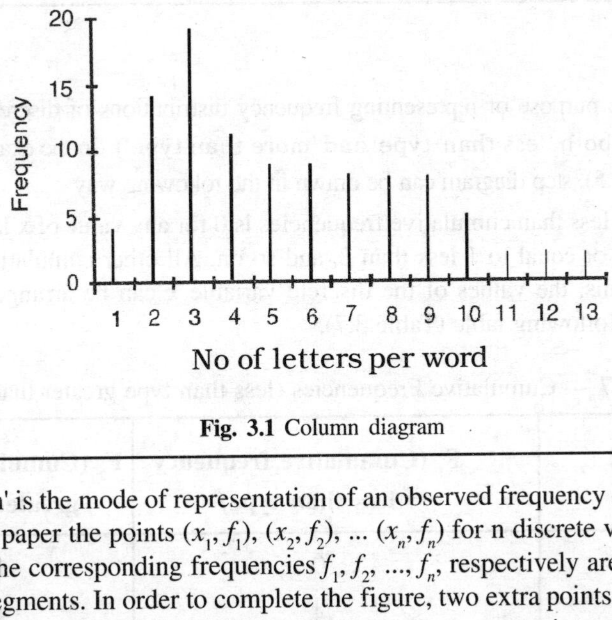

Fig. 3.1 Column diagram

A 'frequency polygon' is the mode of representation of an observed frequency distribution of a discrete variable. On a graph paper the points (x_1, f_1), (x_2, f_2), ... (x_n, f_n) for n discrete values, $x_1, x_2, x_3, ..., x_n$ of the variable having the corresponding frequencies $f_1, f_2, ..., f_n$, respectively are plotted. The points are then joined by line segments. In order to complete the figure, two extra points having co-ordinates $(x_0, 0)$ and $(x_{n+1}, 0)$ are taken on the horizontal axis, where x_0 and x_{n+1} are points on the horizontal axis just preceding x_1 and just following x_n respectively, $x_0, x_1, ..., x_n, x_{n+1}$ being equidistant points on the x-axis. The points on the graph are then joined by line segments successively and the closed area form a polygon. The graph is called a frequency polygon. In the example of the number of letters per word, the points (1,4), (2,14)..., (12,2), (13,1) along with (0,0) and (4,0) are plotted and joined subsequently as is the graph below (Fig 3.2).

If the classes are of equal widths, frequency polygon may be used to represent the frequency distribution of the continuous variable wherein the frequencies are to be plotted against mid-points of class intervals.

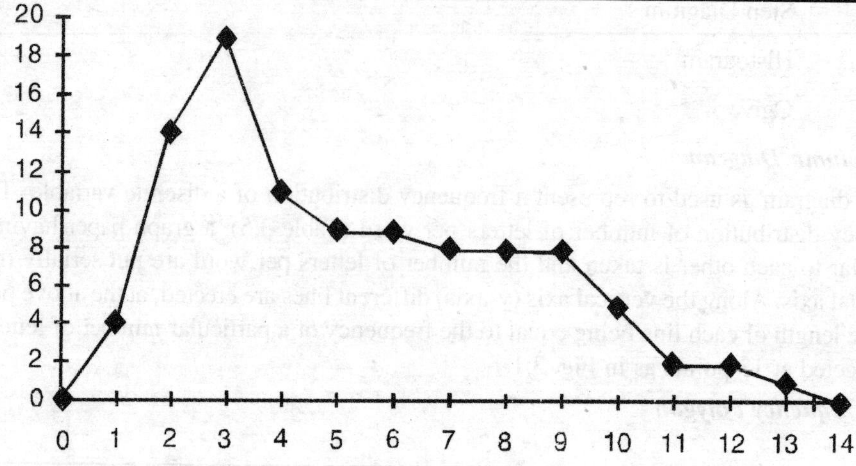

Fig. 3.2 Frequency Polygon

III. Step Diagram

This diagram serves the purpose of representing frequency distributions of discrete variable in terms of cumulative frequencies (both 'less than type' and 'more than type'). In the example of number of letters per word (Table 3.5), step diagram can be drawn in the following way.

In Table 3.5, the less than cumulative frequencies is 0 for any value of x less than 1; is 4 for any value of x greater than or equal to 1 less than 2, and so on. All other cumulative frequencies can be explained similarly. Thus, the values of the discrete variable x can be arranged under the assumed classification as in the following table (Table 3.7).

Table 3.7 — Cumulative Frequencies (less than type greater than type)

Number per word (x)	**F_1 (Cumulative frequency less than type)**	**F_2 (Cumulative frequency greater than type)**
$x < 1$	0	100
$1 \leq x < 2$	4	100
$2 \leq x < 3$	18	96
$3 \leq x < 4$	37	82
$4 \leq x < 5$	48	63
$5 \leq x < 6$	57	52
$6 \leq x < 7$	66	43
$7 \leq x < 8$	74	34

(**Table 3.7** continued)

$8 \le x < 9$	82	26
$9 \le x < 10$	90	18
$10 \le x < 11$	95	10
$11 \le x < 12$	97	5
$12 \le x < 13$	99	3
$13 \le x < 14$	100	1
$14 \le x$	100	0

When we draw a diagram based on the cumulative frequescies (both F_1 & F_2), we obtain the step diagram as in Fig, 3.3. For the step diagram showing less than type cumulative frequencies, we plot the cumulative frequencies 0, 4, 18...., 100 corresponding to the values of x as 0, 1, 2,......,13 respectively and at each point we draw a horizontal straight line as shown in Fig. 3.3. Similarly, for a step diagram showing greater than type cumulative frequency we plot the cumulative frequencies 100, 96, 82,........ 0 corresponding to the values of x as 1, 2, 3, ... 14 respectively and at each point we draw a horizontal straight line as shown in Fig. 3.3.

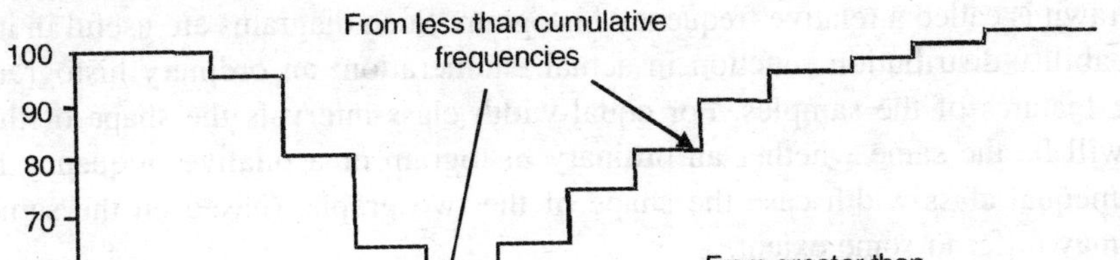

Fig. 3.3 Step Diagram

IV. Histogram

Histogram is the most commonly used form of representation of data from a continuous variable and is based on the fact that the several values of the variable are spread over the entire class interval (instead of being concentrated at a point on the horizontal axis). Thus, in a histogram, class boundaries (and not class limits) are used to demarcate class intervals on the horizontal axis, and series of rectangles are raised and placed side by side along the class intervals (i.e., classes), the lengths of the rectangles being proportional to the frequencies if class widths are equal (or to frequency densities if class-widths are unequal) using suitable scales of measurement on both axes. Histogram is a particular case of bar diagram but one can note that there exists no gap between two successive rectangles in it unlike the bar diagam. If frequency densities and plotted for the different classes, the areas of the rectangles become equal to the corresponding class frequencies. Histogram can also be used in case of grouped variate values of a discrete variable. Figure 3.4 represents the data given in Table 3.6 by means of histogram.

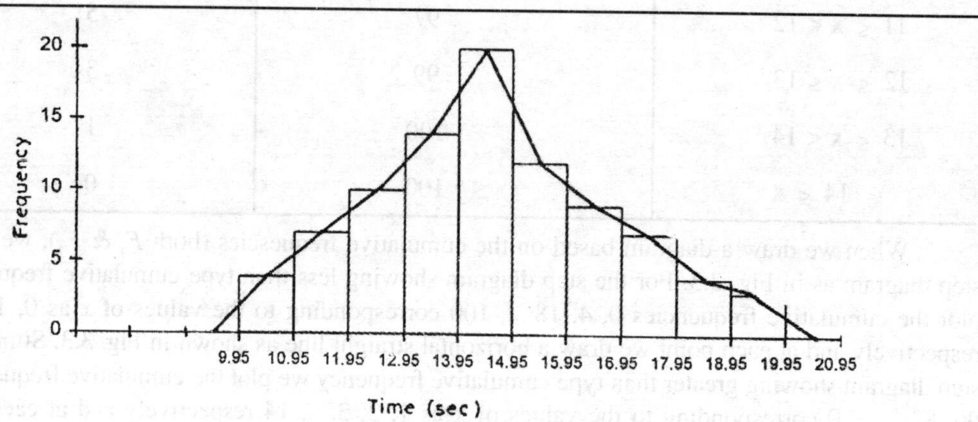

Fig. 3.4 Frequency Polygon drawn on Histogram

If instead of frequency densities, relative frequency densities are taken on the y-axis, the bar diagram so drawn is called a relative frequency histogram. Such diagrams are useful in idntifying the form of probability distribution function in actual enumeration; an ordinary histogram, however, describes the features of the samples. For equal-width class-intervals the shape of the frequency distribution will be the same whether an ordinary histogram or a relative frequency histogram is drawn. For unequal class-width case the shape of the two graphs (based on the same frequency distribution) may differ to some extent.

To identify more explicitly the points of inflexion and/or peak points of the frequency curve drawn smoothly on a histogram or a relative frequency histogram, we may take class intervals of smaller widths at the neighbourhood of those points (for which frequency densities are more) than the class interval towards the left or right tails, and hence it produces a frequency distribution of unequal class widths. In the Table 3.8, an income distribution of 2 lakh Indians is shown.

Table 3.8 —Income Distribution of Two Lakh Indians

Income (Rs, Hundred)	0–1	1–2	2–3	3–4	4–5	5–6	6–7	7–8
Relative frequency (%)	4.0	5.4	6.8	7.3	7.2	8.0	7.0	7.2

Income (Rs, Hundred)	8–9	9–10	10–12.5	12.5–15	15–20	20–25	25–35
Relative frequency (%)	7.0	5.9	12.0	7.9	6.3	4.0	3.6

The data of Table 3.8 is represented in figure 3.5 by means of relative frequency histogram wherein on every class interval on the x-axis we erect a rectangular bar whose area is equivalent to the corresponding relative frequency.

V. Ogive

Ogive is a special type of representation in case of grouped data on continuous variable. In an ogive cumulative frequencies are plotted on the y-axis and the class boundaries are plotted on the x-axis. Thus an ogive is also called a cumulative frequency diagram. In accordance with the type of cumulative frequencies plotted, less-than type ogive or more than type ogive is drawn in less than type ogive, less than cumulative frequencies are plotted against the upper boundaries of classes while in more than type ogive, more than cumulative frequencies are plotted against the lower boundaries of the classes. The points so plotted on both the diagrams are then joined by line segments. To exhibit the terminal points of the two diagrams on the x-axis, zero frequency is plotted corresponding to the lower boundary of the lowest class in the less than type ogive and zero frequency is plotted corresponding to the upper boundary of the highest class in the more than type ogive. Ogive serves an useful purpose in regard to determination of different functions of observations defined to measure certain properties of the frequency distribution (median, quantiles etc.). The ogives based on the results of 'funnel experiment' are given in Fig. 3.6.

3.5. Frequency Curves

In the histogram concerning the example (vide Table 3.4) on funnel experiments (where the variable under study is continuous) if we join the top middle points of the bars and also the mid-points of the assumed zero frequency classes (preceding the lower class boundary and following the upper class boundary of the exterme classes) by straight lines, we get a frequency polygon as shown in Fig. 3.4.

We now wish to get a smooth form of the polygon by drawing a smooth free hand curve through the vertices of the polygon and by incresing simultaneously the number of classes (thereby diminishing the class width). The curve so obtained is known as frequency curve. In fact, we draw histogram and/or frequency polygon with small/moderate number of classes. The idealised or the smooth curve as mentioned above can be drawn on a relative frequency histogram when the number of classes is increased to a large extent thereby diminishing the class width gradually, the frequency of each class however remaining finite (in such a case obviously, the total frequency becomes invariably large).

The frequency curves are classified broadly as (i) Bell shaped, (ii) J-Shaped, (or (ii) U-Shaped).

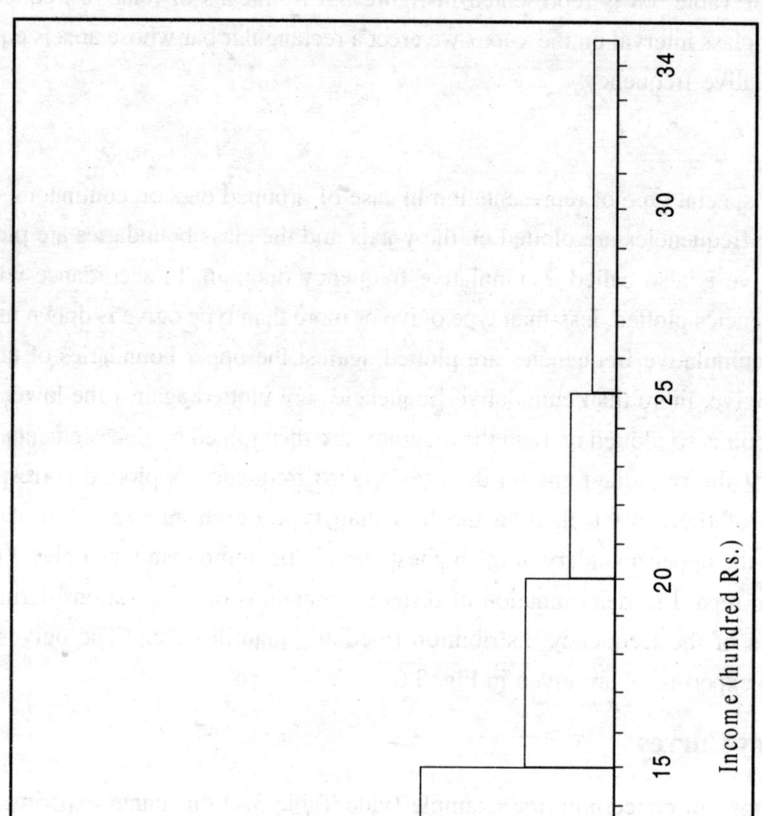

Fig. 3.5 Relative frequency Histogram

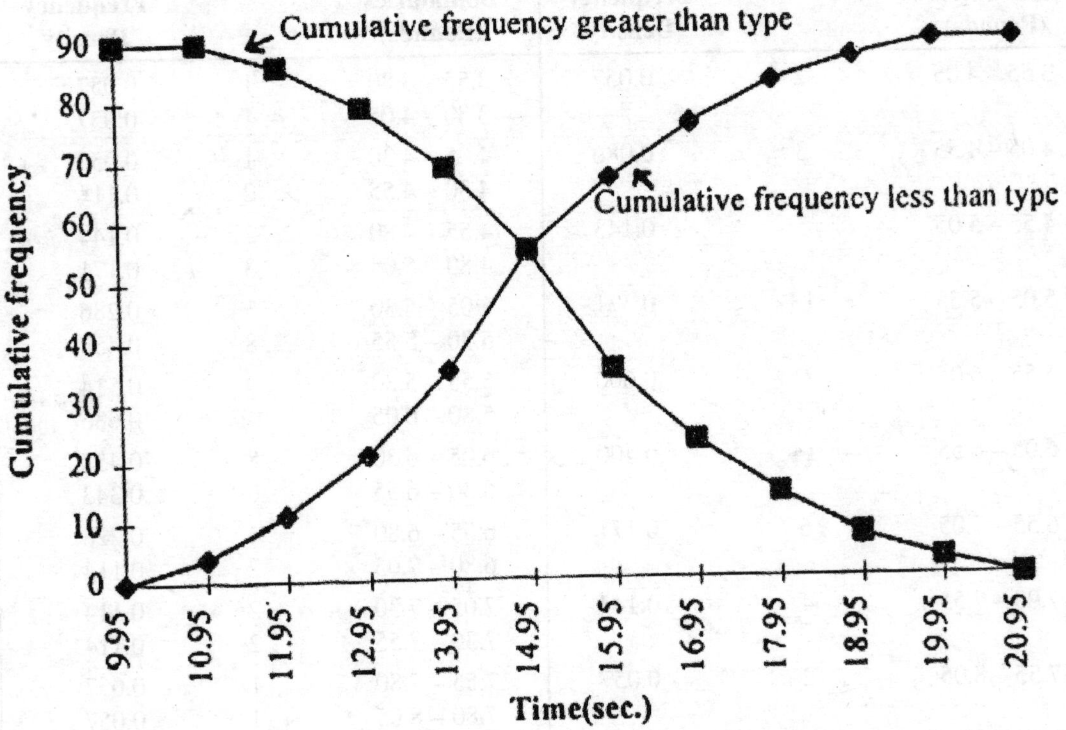

Fig. 3.6 Ogive

Bell shaped curves : Bell shaped curves are again of two types which are described below :

(a) **Symmetrical curves :** It attains a single maximum at the centre of the domain of variation of the x - variable on the horizontal axis and gradually decreases to the axis at the same rate on both sides of maximum. Empirically we may not obtain this type of curve in its exact shape but theoretically normal curve or some other curves based on the normal curve are of this type. In fact, the empirical distributions of the variables under study in controlled situations are likely to take this symmetrical shape. For example, variable under study in a production process; the errors in observations are most likely to have this shape. Table 3.9. shows the two frequency distributions of tensible strength of jute yarn taking 9 classes and 18 classes respectively on the same range of variation and also on the same number of observations (indeed from the same data). Figure 3.7 shows that the frequncy curve becomes more smooth for increased number of class intervals.

Table 3.9 – Distribution of Tensile Strength of Jute Yarn

Class Boundaries (Pounds)	Frequency	Relative Frequcncy Density	Class Boundaries (Pounds)	Frequency	Relative Frequcncy Density
3.55 – 4.05	2	0.057	3.55 – 3.80	1	0.057
			3.80 – 4.05	1	0.057
4.05 – 4.55	3	0.086	4.05 – 4.30	1	0.057
			4.30 – 4.55	2	0.114
4.55 – 5.05	5	0.143	4.55 – 4.80	2	0.144
			4.80 – 5.05	3	0.171
5.05 – 5.55	13	0.371	5.05 – 5.30	5	0.286
			5.30 – 5.55	8	0.457
5.55 – 6.05	21	0.600	5.55 – 5.80	9	0.514
			5.80 – 6.05	12	0.686
6.05 – 6.55	14	0.400	6.05 – 6.30	8	0.457
			6.30 – 6.55	6	0.343
6.55 – 7.05	6	0.171	6.55 – 6.80	4	0.228
			6.80 – 7.05	2	0.114
7.05 – 7.55	4	0.144	7.05 – 7.30	2	0.114
			7.30 – 7.55	2	0.114
7.55 – 8.05	2	0.057	7.55 – 7.80	1	0.057
			7.80 – 8.05	1	0.057
Total	70	2.000*		70	4.000*

Note : (2.0) × (class width) = (4.0) × (class width) = 1, area under a frequency curve.

*(b) **Skewed Curves :*** This type of curve is moderately asymmetrical, attains a single maximum in the domain of variation of the variable and decreases gradually to the x-axis on both sides at different rates so that it produces longer tail towards either lower values or higher values of the variables. Actually, frequency curves of most of the variables may take this shape empirically. Examples of this type of frequency distributions are given in Table 3.10 and these are shown graphically in figures 3.8 and 3.9.

Table 3.10 – Frequency Distribution of Dimension of Huckbolts* of Type *A* and Type *B*

Dimension of huckbolt (Type *A*)	Frequency	Relative Frequcncy Density	Dimension of huckbolt (Type *B*)	Frequency	Relative Frequcncy Density
43.915 – 44.145	8	0.174	11.745 – 11.785	2	0.250
44.145 – 44.375	19	0.413	11.785 – 11.825	9	1.125
44.375 – 44.605	84	1.826	11.825 – 11.865	21	2.625
44.605 – 44.835	46	1.000	11.865 – 11.905	22	2.750
44.835 – 45.065	22	0.478	11.905 – 11.945	39	4.875
45.065 – 45.295	16	0.348	11.945 – 11.985	84	10.500
45.295 – 45.525	5	0.109	11.985 – 12.025	17	2.125
Total	200	4.348		200	25.000

*Huckbolt is a special type of bolt used in manufacturing process of fan blades.

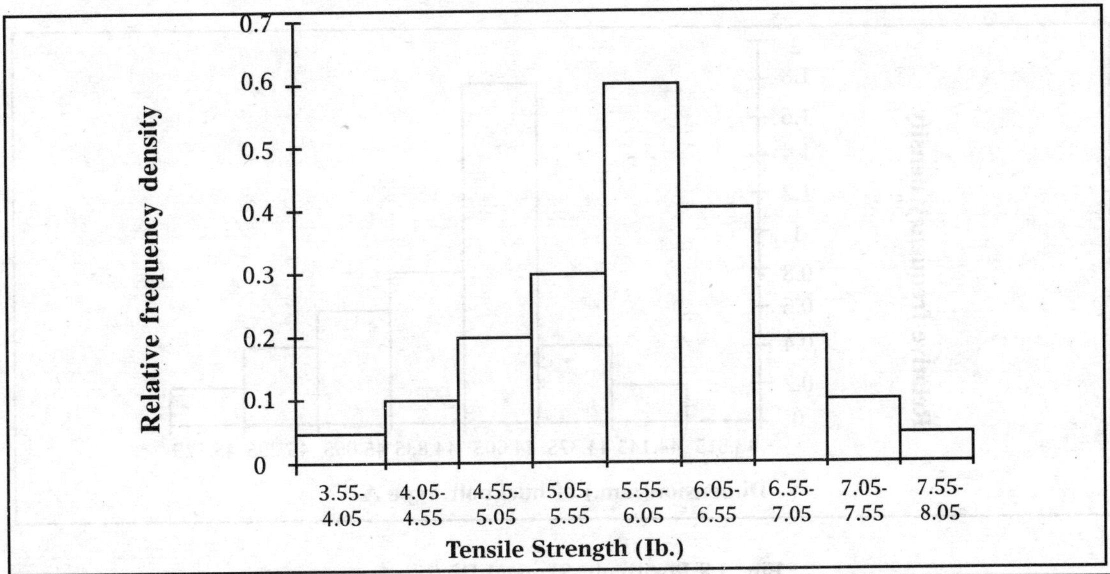

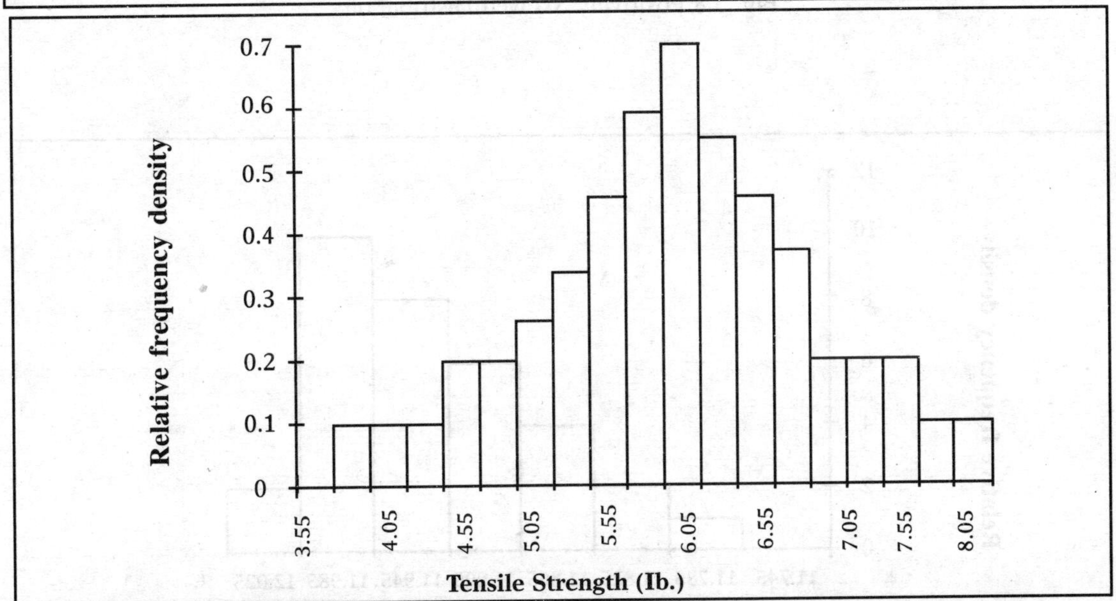

Fig. 3.7— Frequency curves showing the limiting form with smaller Class Boundaries

J - Shaped curves : This is an extremely asymmetrial type of curve. It attains maximum either to extreme left side or to extreme right side. The frequency distributions of the variables on sociological, economic and other vital phenomena may assume this type of curve. Thoretically, pareto and log-normal curves assume the form of letter *J*.

J - shaped curves are of two types, one has maximum at the extreme right *i.e.*, positive J - shaped or J-shaped curve, and the other type has maximum at the extreme left *i.e.*, negative J - shaped or Invese J - shaped curve. Tables 3.11 and 3.12 show the J - shaped frequency distributions on income and income tax respectively. These frequency distributions have been presented in figures 3.10 and 3.11.

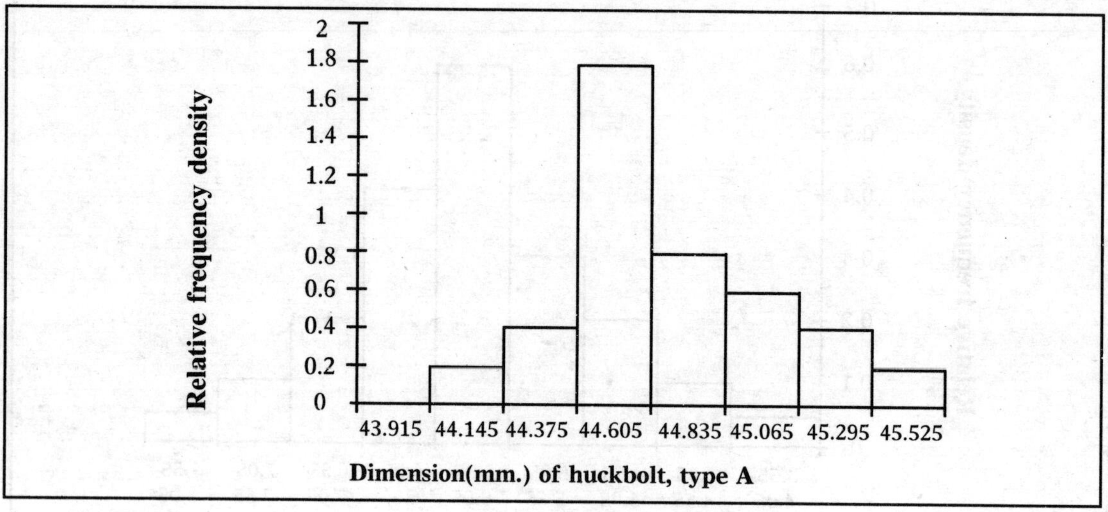

Fig. 3.8 Positively Skewed Distribution

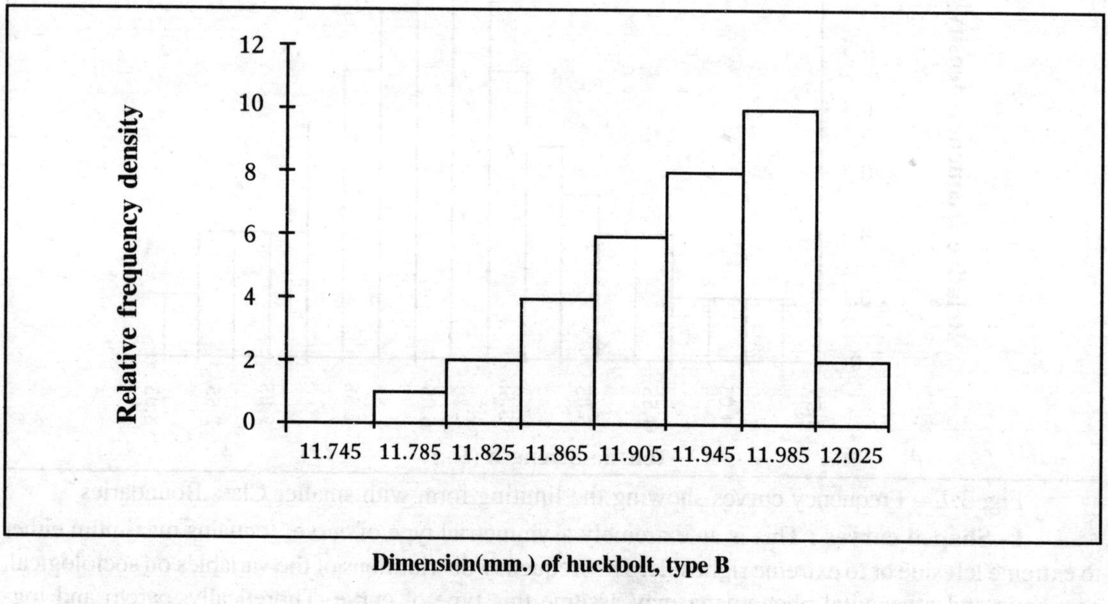

Fig. 3.9 Negatively Skewed Distribution

Tables 3.11 and 3.12 show the J - shaped frequency distributions on income and income tax respectively. These frequency distributions have been presented in figure 3.10 and 3.11.

Table 3.11 – Income Distribution of Employees of an Organisation

Monthly income (.000 Rs)	Frequency	Relative frequency density
1 – 2	45	0.45
2 – 3	30	0.30
3 – 4	12	0.12
4 – 5	6	0.06
5 – 6	4	0.04
6 – 7	2	0.02
7 – 8	1	0.01
Total	100	1.00

Table 3.12 – Distribution of Income Tax Returns from an Organisation over the Month of a financial year

Month	Taxreturn ('.000 Rs)	Relative frequency density*
Apr	3.0	
May	3.5	
Jun	6.0	
Jul	6.8	
Aug	7.0	
Sep	9.0	
Oct	12.0	
Nov	13.0	
Dec	15.0	
Jan	20.0	
Feb	23.0	
Mar	31.0	
Total	**149.3**	

*Calculation of Relative frequency density deserves no meaning, but it can be obtainable taking class width as the number of days in each month

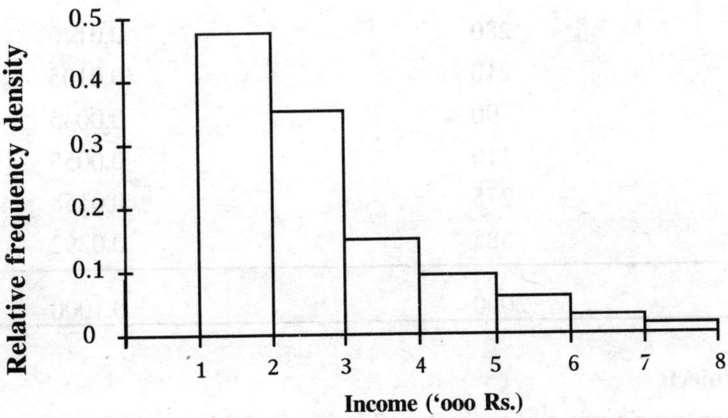

Fig. 3.10 Negatively -Shaped Distribution

U - Shaped Curves : This type of curve has only one minimum point near the centre of the domain of variation of the variable, and two maximum points lie at the extreme values of the variable. U - shaped curve looks as if one joined the two types of J - shaped curves. In the representation of distribution of some vital characteristics of human beings such type of curve may appear e.g., the age-specific distribution of death rates assumes the shape of letter U. Table 3.13 shows the distribution of

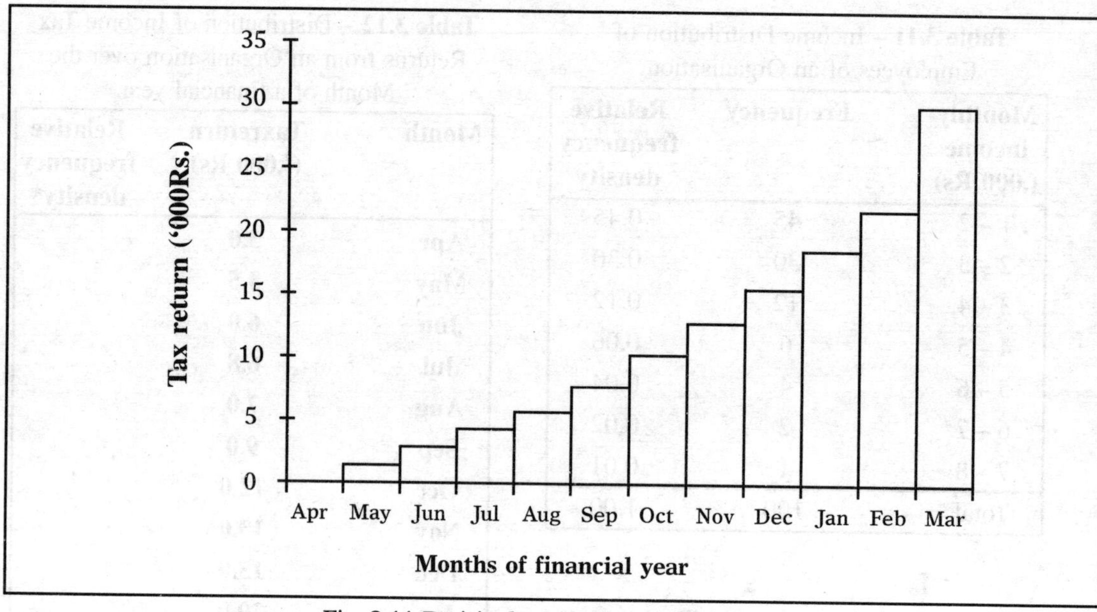

Fig. 3.11 Positively J-Shaped Distribution

number of deaths in different age-groups which occurred in a hospital over a year. The frequency curve is shown in figure 3.12.

Table 3.13 – Distribution of Number of Deaths over Age-groups

Age-group	Number of deaths	Relative frequency density
0 – 10	450	0.0225
10 – 20	280	0.0140
20 – 30	210	0.0105
30 – 40	90	0.0045
40 – 50	110	0.0055
50 – 60	275	0.0138
60 – 70	585	0.0292
Total	2000	0.1000

Exercises and Class Projects

3.1. Distinguish between attributes and variables, discrete variables and continuous variables, Histogram.

3.2 What is frequency distribution table and how is it formed for both discrete and continuous variables? Explain the importance of frequency distribution tables for statistical analysis.

3.3 Define column diagram, Histogram, Frequency Polygon, Step diagram and Ogive and describe how they are constructed, and also explain their modes of use.

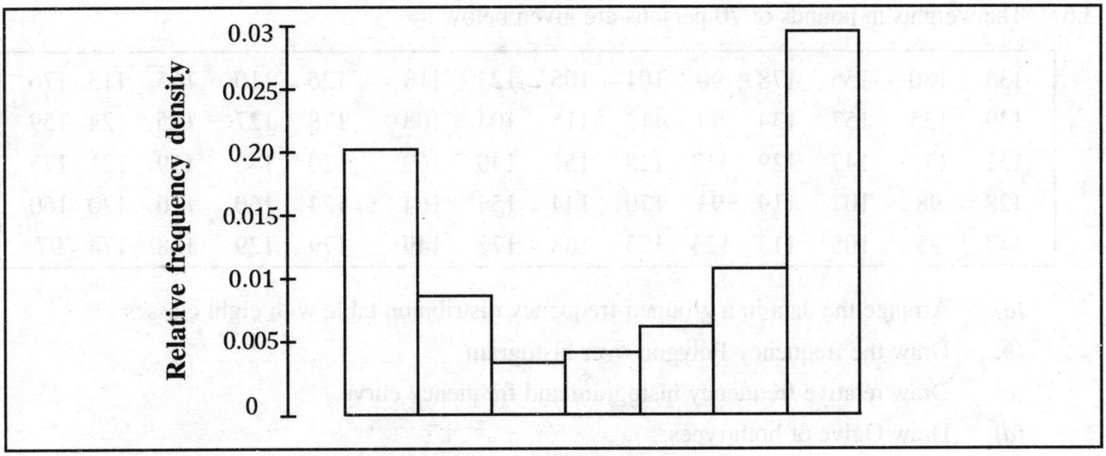

Fig. 3.12 U-Shaped Distribution

3.4. What are the principles governing the choice of

(i) Number of class intervals,

(ii) The length of the class intervals,

(iii) The mid-point of class interval

3.5 Due to machine faults at a computing centre, the daily number of computer stoppages were recorded for a period of 120 days, and these are given below :

5	3	1	5	4	5	3	3	6	5	6	8
5	5	5	3	6	4	7	4	6	5	3	5
9	7	5	4	2	5	2	6	4	4	2	3
5	9	3	2	7	9	4	2	2	3	4	8
7	1	1	7	2	4	4	4	4	4	6	9
1	1	1	4	6	4	2	9	5	3	3	3
9	5	3	3	3	5	7	2	5	5	3	5
3	5	3	3	2	3	3	6	3	2	3	4
7	1	1	4	7	5	6	8	4	5	4	7
3	1	0	2	5	3	4	5	2	5	3	0

(a) Construct a frequency distribution table of the number of stoppages per day

(b) Calculate the relative frequencies

(c) Draw a column diagram, a frequency polygon and step diagrams from the given data

(d) What is the proportion of days with more than 5 stoppages?

(e) What is the number of days with not less than 4 and not more than 6 stoppages?

3.6 The weights in pounds of 70 persons are given below :

136	180	155	178	90	101	105	124	118	126	110	165	113	176
139	135	157	134	99	112	115	104	108	178	127	145	124	159
152	175	147	129	117	128	151	140	169	120	140	119	121	175
128	98	107	119	93	170	144	154	164	174	160	176	170	160
147	95	105	113	123	133	168	172	149	179	129	168	174	97

(a) Arrange the data in a grouped frequency distributon table with eight classes.

(b) Draw the frequency Polygon over histogram

(c) Draw relative frequency histogram and frequency curve.

(d) Draw Ogive of both types.

(e) With reference to the frequency curve drawn in (c) explain its implications in the analysis of statistical data.

3.7 What is the expected shape of frequency curve for the following frequency distribution of number of lives lost in yearly epidemics?

Number of Deaths	**Number of years**
Below 24	0
25 – 49	2
50 – 74	4
75 – 99	6
100 – 149	8
150 – 199	12
200 and over	16

3.8 How a smoothened frequency curve is obtained from an empirical frequency distribution? Draw different types of asymmetrical frequency curves.

39. Determine the class intervals and construct a frequency distribution table from the following table.

Dimension (in mm) Class mark	**Cumulative frequencies ($\leq$ type)**
44.03	8
44.26	27
44.49	111
44.72	157
44.95	179
45.18	195
45.41	200

3.10 The following table shows the thickness (in cm) of 100 metal blocks with frequencies of occurrences in parenthesis (.)
3.30(1), 3.31(1), 3.32(1), 3.34(2), 3.37(1), 3.38(20, 3.40(2), 3.41(4), 3.42(1), 3.43(2), 3,44(5), 3.45(5), 3.46(15), 3.47(6), 3.48(13), 3.49(4), 3.50(9), 3.51(1), 3.52(10), 3.54(4), 3.55(1), 3.56(5), 3.59(2), 3.60(1), 3.63(10, 3.68(1). Draw two histograms from two frequency distributions taking the respective lower limits of starting classes as 3.26 and 3.27 and taking the class width as 0.05 in cash case. Comment on the two representations of the same data by drawing frequency curves.

CLASS PROJECTS

1. **Measure the palm lenghts of the students of your class. Construct a frequency distribution table with the data and represent it in a suitable diagram.**
2. **Collect the marks obtained in mathematics by the students of your school. Prepare a frequency distribution table and draw the frequency curve.**

CHAPTER 4

CHARACTERISTICS OF A FREQUENCY DISTRIBUTION AND THEIR MEASURES

4.1 Concept of Central Tendency and its Measures

It has been explained in the previous chapter that the entire mass of observations related to a variable, presented in the form of a frequency distribution table is one way of summarisation and condensation of the information contained in the set of observations. One such type of information which is also revealed through the formation of frequency distribution table is that most of the observations cluster or congregate around a central value. The above type of information from data reveals a specific characteristic of the frequency distribution and this charcteristic is known as Central Tendency of the frequency distribution. With the help of some statistical tools this characteristic can be measured. In fact, the central tendency is measured by a single value which is typical and deserves some significance since it represents the whole of the observations on one hand and on the other, it is also used for prediction purpose. This single central representative value within the domain of variation of the variable is called a Measure of Central Tendency or an average. Desirable characteristics of an ideal measure of central tendency are :

(i) It should be rigidly defind.

(ii) It should be easy to comprehend.

(iii) It should be simple to calculate.

(iv) It should be based on all observations.

(v) It should be compatible for further mathematical treatment.

(vi) It should be least affected by sampling fluctuations.

(vii) It should be least affected by the presence of extreme values (outliers).

(viii) It should be easily interpreted.

(ix) It should have potential use in statistical inference.

With a view to the above desiderata of a good measure of central tendency the following measures are used according to their suitablity :

(i) Arithmetic Mean (A.M.), *(ii)* Geometric Mean (G.M.), *(iii)* Harmonic Mean (H.M.), *(iv)* Median, *(v)* Mode and (vi) Mid-point of Range. It should be noted that the above desiderata are required to be satisfied in case of any ideal descriptive measure defined for any other property of frequency distribution also.

4.1.1 Means

(a) Arithmetic Mean (A.M.) **:** The most popular measure is the arithmetic mean. The arithmetic mean of a series of values is defined as the ratio of the sum of the given values to the total number of values in the series.

Illustration 4.1 : For the data involving 20 values on weight (in kg) as 8, 3, 3, 7, 10, 3, 5, 11, 2, 12, 6, 7, 4, 2, 8, 9, 11, 4, 3, 11, the arithmetic mean is (8 + 3 + 3+ .... + 4 + 3 + 11) kg/20 = 129 kg/ 20 = 6.45 kg.

If a variable is denoted by x and its n values , x_1, x_2, ... x_{i-1}, x_i, ..., x_n are given, then the arithmetic mean of the data on x is,

$$\bar{x} = \frac{x_1 + x_2 + ... + x_n}{n} = \frac{\sum_{i=1}^{n} x_i}{n} \quad .....(4.1)$$

This type of arithmetic mean is called simple arithmetic mean or simple average.

If in a given set of n values, x_1 occurs f_1 times, x_2 occurs f_2 times, ..., x_i occurs f_i times, ..., x_k occurs f_k times, where $\sum_{i=1}^{n} f_i = n$, then the weighted arithmetic mean of the data on x is,

$$\bar{x} = \frac{x_1 f_1 + x_2 f_2 + ... + x_i f_i + x_k f_k}{f_1 + f_2 + ... + f_i + ... + f_k} = \frac{\sum_{i=1}^{k} x_i f_i}{\sum_{i=1}^{k} f_i} = \frac{\sum_{i=1}^{k} x_i f_i}{n}, \quad f_i\text{'s are the weights} \quad ...(4.2)$$

Also, $\bar{x} = \sum_{i=1}^{k} x_i \frac{f_i}{n} = \sum x_i w_i$, where $w_i = \frac{f_i}{\sum_{i}^{k} f_i} = \frac{f_i}{n}$. Evidently, $\sum_{1}^{k} w_i = 1$

w_i is called the relative frequency corresponding to the value x_i, i =1, 2, ..., k. The arithmetic mean so obtained is called the weighted arithmetic mean or weighted mean of the variable x.

In Illustration 4.1, 2 occurs 2 times, 3 occurs 4 times, 4 occurs 2 times, 5 & 6 occur once, 7 & 8 occurs 2 times, 9 & 10 occurs once, 11 occurs 3 times, and 12 occurs once. The following table can be formed.

Table 4.1 – Frequency Table

i	1	2	3	4	5	6	7	8	9	10	11
x_i (in kg.)	2	3	4	5	6	7	8	9	10	11	12
f_i	2	4	2	1	1	2	2	1	1	3	1

The weighted arithmetic mean,

$$\bar{x} = (2 \times 2 + 3 \times 4 + 4 \times 2 + 5 \times 1 + 6 \times 1 + 7 \times 2 + 8 \times 2$$

$$+ 9 \times 1 + 10 \times 1 + 11 \times 3 + 12 \times 1) \text{ kg.}/20$$

$$= 129 \text{ kg.}/20 = 6.45 \text{ kg.}$$

Some important properties of the arithmetic mean are :

Property 1 : Sum of deviations of variate values $x_1, x_2, \ldots, x_n$ from its mean is zero, i.e.,

$$\sum_{1}^{n}(x_1 - \bar{x}) = 0, \text{ or } \sum_{i=1}^{n} d_i = 0, \text{ where } d_i = x_i - \bar{x},\ i = 1, 2, \ldots, n.$$

By definition of A.M., $\bar{x} = \dfrac{\sum x_i}{n}$, *or,* $\sum_{1}^{n} x_i = n\bar{x}$

$$\Rightarrow \sum_{1}^{n} x_i - n\bar{x} = 0, or \sum_{1}^{n} x_i - \sum_{1}^{n} \bar{x} = 0$$

$$\Rightarrow \sum_{1}^{n}(x_i - \bar{x}) = \sum_{1}^{n} d_i = 0$$

For weighted observations :

$$\text{A.M.} = \bar{x} = \frac{1}{n}\sum_{i} x_i f_i, \text{ or, } \sum_{1}^{k} x_i f_i - n\bar{x} = 0,$$

or $$\sum x_i f_i - \sum \bar{x} f_i = 0 \Rightarrow \sum_{1}^{k}(x_i - \bar{x}) f_i = \sum_{1}^{k} d_i f_i = 0$$

Property 2 : If each value of a variable x assumes a constant value c (i.e., $x_i = c$, for all i=1, 2, ... n), then $\bar{x} = c$,

By definition, $\bar{x} = \dfrac{1}{n}\sum_{1}^{n} x_i = \dfrac{1}{n}\sum_{1}^{n} c = \dfrac{1}{n} \times nc = c$

Property 3 : If each value of a variable x is deviated from a constant value *a* and divied by a constant value *c* (i.e.,$y_i = (x - a) \div c$) then the mean of the values of the transformed variable y is obtained by performing the same operation on the mean of the original values of *x* (i.e., $\bar{y} = (\bar{x} - a) \div c$)

Given $y = \dfrac{x - a}{c}$, or, $y_i = \dfrac{x_i - a}{c}$ for i = 1, 2, ..., n

$$\Rightarrow \sum y_i = \frac{1}{c}\sum (x_i - a) \Rightarrow \frac{1}{n}\sum y_i = \frac{1}{nc}\left(\sum x_i - \sum a\right)$$

i.e., $\bar{y} = \dfrac{\bar{x} - a}{c}$. Also $\bar{x} = a + c\bar{y}$. .... (4.4)

Here *a* is called the base and *c* is called the scale.

Thus, if y is a linear function of x then $\overline{y}$ will also be a linear function of $\overline{x}$ of the same type.

Property 4 : If the variable x has two sets of values, the first set having n_1 values, $x_{11}, x_{12}, \ldots\ldots x_{1n_1}$, and the second set having n_2 values, $x_{12}, x_{22}, \ldots x_{2n_2}$, then the mean of the variable x for the above mentioned set on $n=(n_1+n_2)$ values is $\overline{x}=\frac{n_1\overline{x}_1+n_2\overline{x}_2}{n_1+n_2}$, where $\overline{x}_1$ and $\overline{x}_2$ are the means of the two sets respectively.

By definition, the combined mean

$$\overline{x}=\frac{(x_{11}+x_{12}+\ldots+x_{1n_1})+(x_{21}+x_{22}+\ldots+x_{2n_1})}{n_1+n_2}=\frac{\sum_{1}^{n_1}x_{1i}+\sum_{1}^{n_2}x_{2j}}{n_1+n_2}$$

$$=\frac{n_1\overline{x}_1+n_2\overline{x}_2}{n_1+n_2}, \text{ as } \overline{x}_1=\frac{\sum_{i=1}^{n_1}x_{1i}}{n_1} \text{ and } \overline{x}_2=\frac{\sum_{j=1}^{n_2}x_{2j}}{n_2}$$

Thus $\overline{x}=\frac{n_1\overline{x}_1+n_2\overline{x}_2}{n_1+n_2}$

$$\text{or } \overline{x}=\frac{n_1\overline{x}_1+n_2\overline{x}_2}{n} \qquad \ldots (4.5)$$

Table 4.2 Data for k sets

Set	Values of x	No. of Values
1	$x_{11}, x_{12} \ldots, x_{1j}, \ldots, x_{1n_1}$	n_1
2	$x_{21}, x_{22} \ldots, x_{2j}, \ldots, x_{2n_2}$	n_2
:	:	:
:	:	:
:	:	:
i	$x_{i1}, x_{i2} \ldots, x_{ij}, \ldots, x_{in_i}$	n_i
:	:	:
:	:	:
:	:	:
k	$x_{k1}, x_{k2} \ldots, x_{kj}, \ldots, x_{kn_k}$	n_k

Here the combined mean (pooled mean) becomes,

$$\overline{x}=\frac{\sum_{j=1}^{n_1}x_{1j}+\sum_{j=1}^{n_2}x_{2j}+\ldots+\sum_{j=1}^{n_i}x_{ij}+\ldots+\sum_{j=1}^{n_k}x_{kj}}{n_1+n_2+\ldots+n_k}$$

$$= \frac{n_1\bar{x}_1 + n_2\bar{x}_2 + \ldots + n_i\bar{x}_i + \ldots + n_k\bar{x}_k}{n_1 + n_2 + \ldots + n_k}, \text{ as } \bar{x}_i = \frac{\sum_{j=1}^{n_i} x_{ij}}{n_i}, i = 1,2,\ldots,k,$$

Thus, $\bar{x} = \frac{\sum_{i=1}^{k} n_i\bar{x}_i}{\sum_{i=1}^{k} n_i}$ and $n = \sum_{i=1}^{k} n_i$

Property 5 : If for each of n individuals, values of two variables x and y are given (i.e. x_i and y_i are given for each i = 1,2, ..., n) and a new variable z is formed by adding c times the variable x with d times the variable y (*i.e.*, z is related with x and y by $z = cx + dy$, or $z_i = cx_i + dy_i$, $i = 1, 2, \ldots, n$), then the mean of the new variable (i.e. $\bar{z}$) can be obtained by applying the same relation on the means of the old variables (i.e., $\bar{z} = c\bar{x} + d\bar{y}$)

We know, $z_i = cx_i + dy_i$, $i = 1, 2, \ldots, n$

$$\Rightarrow \quad \frac{1}{n}\sum_{1}^{n} z_i = \frac{1}{n}c\sum_{1}^{n} x_i + \frac{1}{n}d\sum_{1}^{n} y_i, \text{ or } \bar{z} = c\bar{x} + d\bar{y}$$

Weighted Case :

Table 4.3 — Two sets of Variables with same frequencies

i	1	2	...	i	...	k	Total
x	x_1	x_2	...	x_i	...	x_k	
y	y_1	y_2	...	y_i	...	y_k	
f	f_1	f_2	...	f_i	...	f_k	n

Now $\quad \bar{x} = \frac{1}{n}\sum_{1}^{n} f_i x_i$ and $\bar{y} = \frac{1}{n}\sum_{1}^{k} f_i y_i$

As, $\quad z_i = cx_i + dy_i, \bar{z} = \frac{1}{n}\sum f_i z_i = \frac{1}{n}\sum f_i(cx_i + dy_i)$

$$= c \times \frac{1}{n}\sum f_i x_i + d \times \frac{1}{n}\sum f_i y_i$$

$$= c\bar{x} + d\bar{y}.$$

Property 6 : The pooled mean $\bar{x}$ of two sets of values, with means $\bar{x}_1$ and $\bar{x}_2$, lies between $\bar{x}_1$ and $\bar{x}_2$ (i.e., $\bar{x}_2 \le \bar{x} \le \bar{x}_2$, where $\bar{x}_2 \le \bar{x}_2$ is assumed with no loss of generality)

Recalling 4.5, $\bar{x} - \bar{x}_1 = \frac{n_2(\bar{x}_2 - \bar{x}_1)}{n_1 + n_2}$

$$\Rightarrow \quad \bar{x} - \bar{x}_1 \geq 0 \text{ as } \bar{x}_2 \geq \bar{x}_1. \text{ So, } \bar{x} \geq \bar{x}_1.$$

Similarly,
$$\bar{x}_2 - \bar{x} = \frac{n_2(\bar{x}_2 - \bar{x}_1)}{n_1 + n_2} \geq 0 \text{ as } \bar{x}_2 \geq \bar{x}_1$$

So
$$\bar{x}_2 \geq \bar{x}$$

Combining,
$$\bar{x}_1 \leq \bar{x} \leq \bar{x}_2$$

Equality holds if
$$\bar{x}_1 = \bar{x}_2$$

Uses of A.M. are :

The arithmetic mean possesses most of the characteristics of an ideal measure of central tendency. There are some specific situations where other types of measures of central tendency are applicable (discussed in later sections). Excepting those situations the use of arithmetic means for measuring central tendency is generally advisable.

(b) Geometric Mean (G.M) : Geometric mean of a set of n observations is the nth root of the product of the observation.

Let a variable x assume n values, $x_1, x_2, ..., ..., x_n$, then the geometric mean,

$$G.M. = G = (x_1 \times x_2 \times ... \times x_n)^{1/n} = \left(\prod_{i=1}^{n} x_i\right)^{1/n} \quad ... (4.6)$$

The computation of G is done in the following way :

Taking logarithmic on both sides of (4.6),

$$\log G = \frac{1}{n}\log(x_1 x_2 ... x_n)$$

$$= \frac{1}{n}\sum_{i=1}^{n}\log x_i,$$

or
$$G = \text{antilog}\left[\frac{1}{n}\sum_{i=1}^{n}\log x_i\right] \quad ... (4.7)$$

For grouped frequency distribution with $x_1, x_2, ..., x_k$ as the mid-values of k classes and the corresponding frequencies being $f_1, f_2, ..., f_k$, the geometric mean (G) is given by

$$G = (x_1^{f_1}\ x_2^{f_2}\ x_i^{f_i}\ x_k^{f_k})^{1/(f_1 + f_2 + ... + f_k)}$$

$$= \left\{\prod_{i=1}^{k} x_i^{f_i}\right\}^{\frac{1}{\sum f_i}} \quad ... (4.8)$$

Taking logarithm on both sides,

$$\log G = \frac{1}{\sum f_i}\log\left(\prod_{1}^{k} x_i^{f_j}\right)$$

or $$\log G = \frac{1}{\sum f_i} \sum f_i \log x_i$$

$\therefore$ $$G = \text{Antilog}\left(\frac{1}{\sum f_i} \sum f_i \log x_i\right). \quad \text{... (4.9)}$$

Geometric mean possesses the following properties.

Property 1 : If values of a variable x are all equal to c (constant), then G.M. is also equal to the constant value c.

By definition, $G = (x_1 x_2 ... x_n)^{1/n} = x = (\underbrace{c \times c \times ... c}_{n \text{ times}})^{1/n} = (c^n)^{1/n} = c.$

Property 2 : Sum of logarithms of ratios of the individual values to the geometric mean is zero, i.e., $\sum_{i=1}^{n} \log \frac{x_i}{G} = 0$

By defintion, $$G = \left\{\prod_{i=1}^{n} x_i\right\}^{1/n}$$

$\Rightarrow$ $$n \log G = \sum_{i=1}^{n} \log x_i \text{ (taking logarithm on both sides)}$$

or $$\sum_{i=1}^{n} (\log x_i - \log G) = 0 \text{ i.e., } \sum_{i=1}^{n} \log \frac{x_i}{G} = 0 \quad \text{...(4.10)}$$

For weighted case,

$$\sum_{i=1}^{k} f_i \log \frac{x_i}{G} = 0 \quad \text{... (4.11)}$$

Property 3 : In case of two sets of observations, if the G.M. of the first set of n_1 observations is G_1 and the G.M of the second set of n_2 observations is G_2, then the G.M of the combined set of $(n_1 + n_2)$ observations is

$$G = \text{antilog}\left[\frac{n_1 \log G_1 + n_2 \log G_2}{n_1 + n_2}\right] \quad \text{... (4.12)}$$

By definition, $n_1 \log G_1 = \sum_{i=1}^{n} \log x_i, n_2 \log G_2 = \sum_{i=1}^{n_2} \log y_i$

and $$G = (x_1 \times x_2 \times ... \times x_{n_1} \times y_1 \times y_2 \times ... \times y_{n_2})^{\frac{1}{n_1 + n_2}}$$

So, $$\log G = \frac{1}{n_1+n_2}\left[\sum_{i=1}^{n_1}\log x_i + \sum_{i=1}^{n_2}\log yi\right] = \frac{n_1 \log G_1 + n_2 \log G_2}{n_1+n_2}$$

Taking antilog,

$$G = \text{antilog}\left[\frac{n_1 \log G_1 + n_2 \log G_2}{n_1+n_2}\right]$$

Generalising (for data in Table 4.2),

$$\text{G} = \text{antilog}\left[\frac{\sum_{i=1}^{n} n_i \log G_i}{\sum_{i=1}^{k} n_i}\right],$$

where G_i is the G.M. of the observations of the i-th set and G is the G.M of the pooled data.

Property 4 : The combined G.M. (G) of two sets of values with respective G.M's, G_1 and G_2, lies between G_1 and G_2 (*i.e.*, $G_1 \le G \le G_2$, where $G_1 \le G_2$ is assumed).

$$G_1 \le G_2 \Rightarrow \log\ G_1 \le \log\ G_2$$

Now, $$\log G - \log G_1 = \frac{n_1 \log G_1 + n_2 \log G_2}{n_1+n_2} - \log G_1$$

$$= \frac{n_2(\log G_2 - \log G_1)}{n_1+n_2} \ge 0 \qquad (\because \log G_2 \ge \log G_1)$$

$$\Rightarrow \log G \ge \log G_1$$

Again, $$\log G_2 - \log G = \frac{n_1(\log G_2 - \log G_1)}{n_1+n_2} \ge 0 \text{ as before}$$

$\Rightarrow \log G \le \log G_2$

Combining, $\log G_1 \le \log G \le \log G_2$

$\Rightarrow G_1 \le G \le G_2$

Equality holds when $G_1 = G_2$.

Property 5 : If for each of n individuals, values of two variables x and y are given (*i.e.*, x_i and y_i are given for each i = 1, 2, ..., n) and a new variable z is formed by multiplying the variable x by the variable y (*i.e.*, $z = xy$), then the geometric mean of the new variable z can be obtained by multiplying the geometric mean of x by the geometric mean of y (*i.e.*, $G_z = G_x\ G_y$).

From the definition we know,

$$G_x^n = \prod_{i=1}^{n} x_i, \quad G_y^n = \prod_{i=1}^{n} y_i,$$

So the n-th power of the G.M. of z.

$$G_z^n = z_1 \times z_2 \times ... \times z_n$$
$$= (x_1 y_1)(x_2 y_2) .... (x_n y_n)$$
$$= (x_1 x_2 ... x_n)(y_1 y_2 ... y_n)$$
$$= \prod_{i=1}^{n} x_i \times \prod_{i=1}^{n} y_i = G_x^n G_y^n = (G_x G_y)^n$$

So $\quad G_z = G_x G_y$.

The above property also holds when the data are arranged as in Table 4.3.

There are certain **demerits of Geometric Mean**, which are :

1. It is difficult to understand and to interpret.
2. Its value cannot be calculated (as real) if any one of the values in the set is negative.
3. If two variables are in additive relationship, then unlike arithmetic means, corresponding geometric means do not follow additive relationship.
   (*i.e.*, if $y = a + bx$, then $G_y \neq a + bG_x$, where G_x and G_y are respective geometric means)
4. G.M. becomes zero if any of the values in the set is zero.

Uses of Geometric Mean (G.M) are :

1. Geometric mean is fairly used for averaging rates and ratios.
2. G.M. is used for construction of different index number formulae.

(*c*) *Harmonic Mean (H.M.)* : Harmonic mean of a set of observations is the reciprocal of the arithmetic mean of the reciprocals of the observations, *i.e.*, the H.M. of $x_1, x_2, ... x_n$ is,

$$\text{H.M.} = H = \frac{1}{\frac{1}{n}\left(\frac{1}{x_1} + \frac{1}{x_2} + ... + \frac{1}{x_n}\right)} = \frac{n}{\frac{1}{x_1} + \frac{1}{x_2} + ... + \frac{1}{x_n}} \qquad ...(4.13)$$

For the case of grouped frequency distribution with x_i' s as the mid-values of the classes and f_i's as the corresponding frequencies (i = 1, 2, ..., k), the harmonic mean is,

$$\text{H} = \frac{1}{\frac{1}{\sum f_i} \sum_{i=1}^{k} \frac{f_i}{x_i}} = \frac{\sum^{k} f_i}{\sum_{i=1} \frac{f_i}{x_i}} \qquad ...(4.14)$$

The properties of **Harmonic mean** are :

Property 1 : If a set contains values each of which is a constant say c, then the harmonic mean of the set of values is equal to the constant value c.

Property 2 : The sum of deviations of the reciprocals of a given set of values from the reciprocal of harmonic means is equal to zero $\left(i.e., \sum_{i=1}^{n}\left(\frac{1}{x_i} - \frac{1}{H}\right) = 0\right)$

Form the definition,

$$H = \frac{n}{\sum \frac{1}{x_i}}, \quad \text{or,} \quad \sum \frac{1}{x_i} - \frac{n}{H} = 0$$

$$\Rightarrow \sum_{i=1}^{n}\left(\frac{1}{x_i} - \frac{1}{H}\right) = 0 \quad \text{... (4.15)}$$

For grouped data, the expression (4.15) can be written as, $\sum_{i=1}^{n} f_i\left(\frac{1}{x_i} - \frac{1}{H}\right) = 0$ ... (4.16)

Property 3 : If two variables are in additive relationship, then, unlike the arithmetic means, corresponding harmonic means do not follow additive relationship (*i.e.*, if $y = a + bx$, then $H_y \neq a + bH_x$, where H_x and H_y are the respective harmonic means).

But if $a = 0$ then $H_y = bH_x$.

Here $\quad y_i = a + bx_i, i = 1, 2, ..., n$

By definition of H.M.,

$$H_y = \frac{n}{\sum \frac{1}{a + bx_i}} \neq a + bH_x.$$

But if $\quad a = 0,$

then,

$$H_y = \frac{n}{\sum_{i=1}^{n} \frac{1}{bx_i}} = \frac{bn}{\sum \frac{1}{x_i}} = bH_x. \quad \text{...(4.17)}$$

Property 4 : If in case of two sets of observations, the H.M. Of the first set of n_1 observations is H_1 and the H.M. of the second set of n_2 observations is H_2, then the H.M. of the combined set of $(n_1 + n_2)$ observations is

$$H = \frac{n_1 + n_2}{\frac{n_1}{H_1} + \frac{n_2}{H_2}}. \quad \text{...(4.18)}$$

For k sets of observations, i-th set containing n_i observations, the harmonic mean of the combined data is

$$H = \frac{n_1 + n_2 + ... + n_k}{\frac{n_1}{H_1} + \frac{n_2}{H_2} + ... + \frac{n_k}{H_k}} = \frac{\sum_{1}^{k} n_i}{\sum_{i=1}^{k} \frac{n_i}{H_i}},$$

where H_i is the harmonic mean of the i-th set, $i = 1, 2, ..., k$,

Property 5 : The combined H.M. of two gropus of observations with respective H.M's, H_1 and H_2 lies between H_1 and H_2 (*ie.,* $H_1 \le H \le H_2$, where with no loss of generality $H_1 \le H_2$).

By definition of combined H.M.,

$$\frac{1}{H} = \frac{1}{(n_1+n_2)}\left(\frac{n_1}{H_1}+\frac{n_2}{H_2}\right) = \frac{n_1 H_2 + n_2 H_1}{(n_1+n_2)H_1 H_2}$$

Now, $$\frac{1}{H} - \frac{1}{H_1} = \frac{n_1 H_2 + n_2 H_1}{(n_1+n_2)H_1 H_2} - \frac{1}{H_1} = \frac{n_2(H_1 - H_2)}{(n_1+n_2)H_1 H_2} \le 0 \; (\because H_1 \le H_2)$$

$$\therefore \quad \frac{1}{H} \le \frac{1}{H_1} \Rightarrow H \ge H_1$$

Again $$\frac{1}{H} - \frac{1}{H_2} = \frac{n_1 H_2 + n_2 H_1 - n_1 H_1 - n_2 H_1}{(n_1+n_2)H_1 H_2}$$

$$= \frac{n_2(H_2 - H_1)}{(n_1+n_2)H_1 H_2} \ge 0$$

$$\therefore \quad \frac{1}{H} \ge \frac{1}{H_2} \Rightarrow H \le H_2$$

Combining, $H_1 \le H \le H_2$ ... (4.19)

Demerits of Harmonic Mean are :

1.It is difficult to understand and to interpret.

2.If two variables are in additive relationship, then, unlike arithmetic means corresponding harmonic means do not follow additive relationship (i.e., if $y = a + bx$, then H.M. $(y) \neq a + b \times$ H.M. (x).

3. H. M. becomes undefined if any of the values in the set is zero.

Uses of Harmonic Mean (H.M.) are :

The following examples will bear out the usefulness of harmonic mean as well as the arithmetic mean.

Example 4.1 : A sprinter runs at a speed of 10 miles/hr., 12 miles/hr, and 14 miles/hr. to cover the same distance say 50 miles, on three occasions. Find out the average speed of the sprinter.

Following the definition of average speed, we can write —

$$\text{Average Speed} = \frac{\text{Total distance run by the sprinter}}{\text{Total time taken}}$$

$$= \frac{(3 \times 50)\text{ miles}}{(50/10 + 50/12 + 50/14)\,\textit{hrs.}} = \frac{3}{1/10 + 1/12 + 1/14}\text{ miles/hr}$$

= The H.M. of the three given speeds *i.e.,* H.M. of 10 miles/hr, 12 miles/hr and 14 miles/hr.

The above example exhibits the use of simple H.M.

Example 4.2 : If in Exmaple 4.1 the different distances to be covered by the sprinter on three occasions be 50 miles, 60 miles and 70 miles respectively at the above mentioned speeds then the the average speed of the sprinter can be obtained as follows :

$$\text{Average Speed} = \frac{\text{Total distance covered by the sprinter}}{\text{Total time}}$$

$$= \frac{(50+60+70)}{50/10+60/12+70/14}\ \text{miles/hr}$$

$= $ The weighted H.M. of the three speeds

10 miles/rh, 12 miles/hr and 14 miles/hr with weights 50 miles, 60 miles and 70 miles, respectively.

Example 4.3 : If in Example 4.1, the sprinter runs for equal time periods (instead of equal distances), say 5 hr. on three occasions, the average speed will be as follows :

$$\text{Average Speed} = \frac{\text{Total distance run}}{\text{Total time}}$$

$$= \frac{(10\times 5+12\times 5+14\times 5)}{3\times 5\ \text{hrs.}}\ \text{miles}$$

$$= \frac{10+12+14}{3}\ \text{miles/hr.}$$

$= $ The simple A.M. of the three given speeds.

Example 4.4. : If in Example 4.3 the sprinter runs for 5 hr., 6 hr. and 8 hr. on the three occasions the speeds 10 miles/hr., 12 miles/hr and 14 miles / hr respectively, the average speed will be as follows :

$$\text{Average Speed} = \frac{\text{Total distance run}}{\text{Total time}}$$

$$= \frac{10\times 5+12\times 6+14\times 8}{5+6+8}\ \text{miles/hr.}$$

$= $ The weighted A.M. of the three speeds

10 miles/hr., 12 miles/hr. and 14 miles/hr., respective weights being 5 hr., 6 hr. and 8 hr.

The above examples illustrate that speed of a man, price rate of a commodity, density of a liquied, etc. represent rates in the form of 'P units of x variable per unit of y variable', where x represents distance, price, mass etc. and y represents time, quantity, volume etc.

Exmaples 4.1 and 4.2 reveal that, if the given rates (P units of x variable per unit of y variable) are to be averaged for a fixed or different values of x variable, we have to use the simple or weighted harmonic mean of the rates (i.e., of the P values). Examples 4.3 and 4.4 reveal that if the rates are to be averaged for a fixed or different values of y variables, we have to use simple or weighted arithmetic mean of the rates.

4.1.2 Relationship between A.M., G.M. and H.M.

Proposition 4.1 : For the variate values x_1, x_2, ..., x_n, all being greater than zero, the A.M., G.M. and H.M. and related by the following inequality.

A.M. ≥ G.M. ≥ H.M. ...(4.20)

Proof : By definition, for n variate values.

$$\text{A.M.} = \frac{x_1 + x_2 + ... + x_n}{n}, \text{ G.M.} = (x_1 \times x_2 + ... \times x_n)^{1/n}$$

$$\text{and H.M.} = \frac{n}{\frac{1}{x_1} + \frac{1}{x_2} + ... + \frac{1}{x_n}}$$

Let us prove the proposition for $n = 2$.

It is obvious that $(\sqrt{x_1} - \sqrt{x_2})^2 \geq 0$

$$\Rightarrow \quad x_1 + x_2 \geq 2(x_1x_2)^{1/2}, \text{ or } \frac{x_1 + x_2}{2} \geq (x_1x_2)^{1/2}$$

So A.M. ≥ G.M. is true for n = 2.

Now replacing x_1 by $\frac{x_1 + x_2}{2}$ and x_2 by $\frac{x_3 + x_4}{2}$ in the inequality A.M. ≥ G.M. for $n = 2$ we can write

$$\frac{\frac{x_1 + x_2}{2} + \frac{x_3 + x_4}{2}}{2} \geq \left\{\left(\frac{x_1 + x_2}{2}\right)\left(\frac{x_3 + x_4}{2}\right)\right\}^{1/n}$$

$$\text{or,} \quad \frac{x_1 + x_2 + x_3 + x_4}{4} \geq \left\{(x_1x_2)^{1/2}(x_3x_4)^{1/2}\right\}^{1/2} \quad (\because \text{A.M.} \geq \text{G.M. for } n = 2)$$

$$\text{or,} \quad \frac{x_1 + x_2 + x_3 + x_4}{4} \geq (x_1x_2x_3x_4)^{1/4}$$

Thus A.M. ≥ G.M. is also true for $n = 4$. proceeding in this way, we can prove that the inequality, A.M. ≥ G.M., holds for $n = 2^k$, where k is any positive integer.

It remains to prove that the inequality A.M. ≥ G.M. holds for $n \neq 2^k$. For $n \neq 2^k$, let us take the minimum k such that $n < 2^k$, and along with the given n values, $x_1, x_2, ..., x_n$, let us also consider $(2^k - n)$ other positive quantities, $x_n + 1, x_n + 2, ..., x_{2k}$, all equal to the A.M. of the given n values, *i.e.,*

$$x_i = A = \sum_{i=1}^{n} x_i \div n, \text{ for } i = n+1, n+2, ..., 2^k$$

Let us also denote, G.M. = $G = \left\{\prod_{1}^{n} x_i\right\}^{1/n}$, and

$$\text{H.M.} = H = \frac{n}{\sum_{i=1}^{n} \frac{1}{x_i}}$$

From above it is seen that A.M. ≥ G.M. is valid for $x_1, x_2, ..., x_{2^k}$.

Therefore, $$\frac{x_1+x_2+...+x_n+x_{n+1}+...+x_{2^k}}{2^k} \geq \{x_1x_2...x_nx_{n+1}...x_{2^k}\}^{1/2^k}$$

or, $$\frac{nA+(2^k-n)A}{2^k} \geq \left\{G^n A^{2^k-n}\right\}^{1/2^k}$$

or, $$A \geq \left\{\frac{G^n}{A^n}.A^{2^k}\right\}^{1/2^k}$$

or, $$1 \geq \left\{\frac{G}{A}\right\}^{n/2^k}$$

$$\Rightarrow \qquad A \geq G. \qquad ...(4.21)$$

Thus A.M. ≥ G.M. holds for any n positive quantities

Now $\frac{1}{x_1}, \frac{1}{x_2}, ..., \frac{1}{x_n}$, all being greater than zero, are the reciprocals and applying the inequality (4.21), we get

$$\frac{\frac{1}{x_1}+\frac{1}{x_2}+...+\frac{1}{x_n}}{n} \geq \left\{\frac{1}{x_1}\times\frac{1}{x_2}\times...\times\frac{1}{x_n}\right\}^{1/n}$$

or, $$\frac{1}{\dfrac{n}{\frac{1}{x_1}+\frac{1}{x_2}+...+\frac{1}{x_n}}} \geq \left\{\frac{1}{x_1}\times\frac{1}{x_2}\times...\times\frac{1}{x_n}\right\}^{\frac{1}{n}}$$

or, $$\frac{1}{H} \geq \frac{1}{G}$$

$$\Rightarrow \qquad H \leq G$$

So, $$\text{H.M.} \leq \text{G.M.} \qquad ...(4.22)$$

Combining the inequalities (4.21) and (4.22) we can write,

$$A \geq G \geq H$$

Hence A.M. ≥ G.M. ≥ H.M.

Note 4.1 A.M. = G.M. = H.M. = c,

if $x_i = c$ (constant) for all $i = 1, 2, ..., n$

4.1.3 Median, Mode and Midpoint of Range

Median : The median of a set of n observations, $x_1, x_2, ..., x_n$ is the middle most value when the observations are arranged either in ascending (non-decreasing) or in descending (non-increasing) order of magnitude and is denoted by M_e. If n is an odd number, there is a unique-middle value which divides the data into two equal halves of numbers of observations, and that unique value is the median. In fact, median is the $(n + 1)/2$-th observations in an ordered series of observations. If n is an even number, there are two middle values below and above which we have equal number of observations, and then the average of the middle values is the median. In fact, in case of an even number of observations, two middle values are the $\frac{n}{2}th$ and $\left(\frac{n}{2}+1\right)$- th observations in the ordered series and the average of these two middle values is the median.

Example 4.5 : Find the median of the following data

x (number of letters per word) :	8	3	3	7	10	3	5	11	12	2	6

The arranged data (ascending order) :

2, 3, 3, 3, 5, 6, 7, 8, 10, 11, 12

5 observations (2, 3, 3, 3, 5) — 6 — 5 observations (7, 8, 10, 11, 12)

Thus the middle value is $\frac{n+1}{2}th = \frac{11+1}{2}th$ = 6th observation. The 6*th* observation has value 6, so the median number of letters per word is 6.

If the last observation is deleted then the two middle most observations will be the $\frac{n}{2}th$ and $\left(\frac{n}{2}+1\right)th$ observations. The data after deletion are :

2, 3, 3, 3, 5, 7, 8, 10, 11, 12

4 observations (2, 3, 3, 3) — 5, 7 — 4 observations (8, 10, 11, 12)

Now the middle most values are 5 and 7, and hence the median is (5 + 7)/2 = 6. In case of grouped continuous frequency distribution, the median is that value on the real line ($-\infty < x$ (real) $< \infty$) the perpendicular on which divides the frequency curve into two equal halves.

Calculation of median from frequency distribution

I. Discrete case :

Steps : From cumulative frequency (less than type)	*Steps :* From cumulative frequency (greater than type)
1. Prepare cumulative frequency table (less than type)	1. Prepare cumulative frequency table (greater than type)
2. Calculate *n*/2, *n* = total frequency	2. Calculate *n*/2, *n* = total frequency
3. If $CF(i-1) < \frac{n}{2} < CF(i)$, then $M_e = x_i$, *CF* = cumulative frequency (l.t.t.) corresponding to x_i and if $n/2 = CF(i)$ *i.e.*, an integer, then $M_e = \frac{x_i+x_{i+1}}{2}$	3. If $CF^*(i) < \frac{n}{2} < CF^*(i-1)$, then $M_e = x_i - 1$, $CF^*(i)$ is more than type cumulative frequency corresponding to x_i and if $n/2 = CF^*(i)$ *i.e.*, an integer, then $M_e = \frac{x_i+x_{i+1}}{2}$

II. Continuous case :

The following steps are required to determine the median.

1.Prepare a column for cumulative frequencies (less than type).

2. Calculate $n/2$ and let $CF(i)$, denote the cumulative frequency (less than type) of the i-th class.

3. If $CF(i-1) \leq n/2 \leq CF(i)$, then we conclude that the median (M_e) lies within the class boundaries of the i-th class, *i.e.*, $x_l < M_e < x_u$, x_l and x_u being lower and upper class boundaries of the i-th class respectively, n being the total frequency, f_{med} being the frequency of the median class.

4. Assume that all the variate values in the median class are evenly distributed over the whole class (*i.e.*, cumulative frequency is a linear function of the variate value between x_l and x_u so that the principle of linear interpolation (rule of proportional parts) can be applied).

Thus we can write,

$$\frac{M_e - x_l}{x_u - x_l} = \frac{\frac{n}{2} - CF(i-1)}{f_{med}} \quad \text{... (4.23)}$$

$$\Rightarrow \quad M_e = x_l + \frac{\frac{n}{2} - CF(i-1)}{f_{med}} \times w$$

where w = width of median class = $x_u - x_l$.

Following similar line,

$$\frac{x_u - M_e}{x_u - x_l} = \frac{CF(i) - \frac{n}{2}}{f_{med}}$$

$$\Rightarrow \quad M_e = x_u - \frac{CF(i) - \frac{n}{2}}{f_{med}} \times w \quad \text{... (4.24)}$$

Median can also be determined using cumulative frequency of greater than type. If CF^* (i) denotes the cumulative frequency (greater than type) of the i-th class and if CF^* (i) $\leq \frac{n}{2} \leq CF^*(i-1)$ then we conclude that i-th class is the median class (x_l and x_u being the lower and upper class boundaries of the class).

Again assuming the rule of proportional parts to hold we get,

$$\frac{M_e - x_l}{x_u - x_l} = \frac{CF^*(i-1) - \frac{n}{2}}{f_{med}}$$

$$\Rightarrow \quad M_e = x_l + \frac{CF^*(i-1) - \frac{n}{2}}{f_{med}} \times w \quad \text{... (4.25)}$$

Also, $$\frac{x_l - M_e}{x_u - x_e} = \frac{\frac{n}{2} - CF^*(i)}{f_{med}}$$

$$M_e = x_u - \frac{\frac{n}{2} - CF^*(i)}{f_{med}} \times w \qquad \text{... (4.26)}$$

Median possesses the following properties.

Property 1 : If each value of a variable x assumes a constant value c (*i.e.*, $x_i = c$, $i = 1, 2, ..., n$), then the median of x is c.

Property 2 : If two variables x and y are related linearly (*i.e.*, $y = ax + b$, a and b, being constants, *i.e.*, for n variate values $y_i = ax_i + b$, $i = 1, 2, ..., n$), then $M_e(y) = aM_e(x) + b$, where $M_e(x)$ and $M_e(y)$ are the medians of x and y variables respectively.

Proof : With no loss of generality, let the values of x be arranged in asscending order of magnitude. Now corresponding to each value of x, the transformation, $y = ax + b$ will give rise to a unique value of y. As the values of x are arranged in ascending or descending order of magnitude, the corresponding values of y will be arranged either in ascending order or in descending order of magnitude (because of the monotonic property of linear function). In the x series, $[n/2]$ observations lie below and above $M_e(x)$, median of x-values and corresponding to $M_e(x)$, the value in the y-series is $aM_e(x) + b = y^*$, say. Now in the y-series, $[n/2]$ observations lie below and above y^*.

So y^* = Median of $y = M_e(y)$

Hence, $M_e(y) = aM_e(x) + b$.

(Here n is total number of observations and $[n/2]$ denotes the greatest integer contained in it).

Property 3 : Combined median (M) of two sets of values always lies between the medians of the respective sets, M_1 and M_2

(i.e., $M_1 \leq M_2$, where WNLG $M_1 \leq M_2$ is assumed).

Proof : Suppose we have two sets of observations having medians M_1 and M_2. If M is the median of the combined set then we have to prove $M_1 \leq M \leq M_2$.

Let n_1 and n_2 be the numbers of observation of the two sets respectively. Since M_1 is the median of the first set, $[n_1/2]$ observations will lie below M_1 when the observations are arranged in non-decreasing order of magnitude. Also $[n_1/2]$ observations will lie above M_1. Here $[x]$ denotes the greatest integer contained in x. Similarly, since M_2 is the median of the second set, $[n_2/2]$ observation will lie below and above M_2. Finally, since M is the median of the conbined set, $\left[\frac{n_1+n_2}{2}\right]$ observations will lie below and above M when the observations of the combined set are arranged in non-decreasing order of magnitude. The following two cases are possible.

Case I : $M_1 = M_2$

Suppose $M_1 = M_2 = M'$ and $\left[\frac{n_2}{2}\right]$ observations from the second set lie below M'. So in all $\left[\frac{n_1}{2}\right]+\left[\frac{n_2}{2}\right]$ observations lie below M'. Now since $\left[\frac{n_1}{2}\right]+\left[\frac{n_2}{2}\right] \leq \left[\frac{n_1+n_2}{2}\right]$, we get $M' \leq M$. ... (1)

Similarly, $\left[\frac{n_1}{2}\right]$ observations from the first set lie above M' and $\left[\frac{n_2}{2}\right]$ observations from the second set lie above M' and hence in all $\left[\frac{n_1}{2}\right]+\left[\frac{n_2}{2}\right]$ observations lie above M' and also since $\left[\frac{n_1}{2}\right]+\left[\frac{n_2}{2}\right] \leq \left[\frac{n_1+n_2}{2}\right]$, we get,

$$M' \geq M \qquad \text{... (2)}$$

From (1) and (2) we can write M' = M

and hence the property follows when $M_1 = M_2$.

Case II : $M_1 < M_2$

To prove $M_1 \leq M \leq M_2$ (excluding the case $M_1 = M_2$), let us proceed as follows.

In the combined set the number of observations below M_1 coming from the second set $= \left[\frac{n_2}{2}\right] - k$, say, k (≥ 0) is the number of observations from the second set between M_1 and M_2. So in the combined set below M_1 there are $\left[\frac{n_1}{2}\right]+\left[\frac{n_2}{2}\right]-k$ observations $\leq \left[\frac{n_1+n_2}{2}\right]-k$ observations $\leq \left[\frac{n_1+n_2}{2}\right]$ observations.

Thus the number of observations below M_1 is less than or equal to the number of observations below M.

$$\text{Hence, } M_1 \leq M \qquad \text{... (3)}$$

Again, in the combined set the number of observations above M_2 from the first set $\left[\frac{n_1}{2}\right]-k'$, say, $k'(\geq 0)$ is the number of obserbaions between M_1 and M_2 from the first set. So in the combined set above M_2 there are $\left[\frac{n_1}{2}\right]-k'+\left[\frac{n_2}{2}\right]$ observations $\leq \left[\frac{n_1+n_2}{2}\right]-k'$ observations $\leq \left[\frac{n_1+n_2}{2}\right]$ observations.

Thus the number of observations above M_2 is less than or equal to the number of observations abvoe M.

$$\text{Hence } M \leq M_2 \qquad \text{... (4)}$$

Now combining (3), (4), we get

$M_1 \leq M \leq M_2$ (excluding the case $M_1 = M_2$)

The case $M_1 \geq M_2$ can be dealt with in a similar line as outlined above.

Property 4 : Sum of absolute deviations of values of a variable about median is least. (Proof is given in Section 4.2.2.)

Property 5 : If $z = cx + dy$, c and d being of same sign, then

$M_e(z) = cM_e(x) + dM_e(y)$.

Proof : It can be Proved in the line of proof of property 2.

Demerits of Median are :

1. It is not based on all observations.
2. It is amenable to mathematical treatments of specific types only.
3. It is mostly affected by sampling fluctuations.

$(\because v(M_e) = \pi\sigma^2/2n > \frac{\sigma^2}{n} = (\bar{x}))$, n and σ being sample size and population standard deviation respectively.

The median is used in preference to other measures of central tendency in the following cases :

1. When the data set contains extreme observations either at both ends or at any one of the ends (or for frequency distribution having open ends or undefined classes).

2. For non-existence of arithmetic mean in some theoretical distributions (like Cauchy distribution) use of median is advocated in estimation problem from such distribution.

Mode : Mode is that value of the variable which is observed maximum number of times, i.e., mode is defined as a variate value with, (a) highest frequency in case of ungrouped data on discrete or continousus variable, or (b) highest frequency density for grouped continuous frequency distribution.

For example, the mode of the series of observations, 8, 3, 3, 7, 10, 3, 5, 11, 2 is 3 since it occurs with maximum frequency 3.

The location of mode on the x-axis can be obtained from frequency curve as follows :

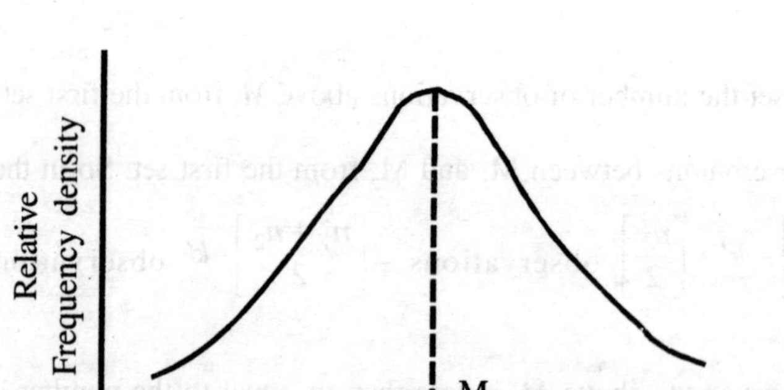

Fig. 4.1 A Frequency curve with indication of mode

Graphically, the mode M_0 lies on the x-axis corresponding to the highest frequency density/relative frequency density.

In case of data laid out in the form of a grouped frequency distribution table, the calculation of mode is done through the following steps :

Step I : Identify the class which corresponds to the highest frequency. The identified class is called the modal class.

Step II : Identify the frequencies of the modal class just preceding and the class just following as f_0, f_{-1} and f_1 respectively.

Step III : Denote by x_l, x_u and w the lower class boundary, upper class boundary and class width $(x_u - x_l)$ of the modal class respectively.

Step IV : Calculate M_0 using the formula,

$$M_0 = x_l + \frac{f_0 - f_{-1}}{2f_0 - f_{-1} - f_1} \times w \qquad \text{... (4.27)}$$

The following geometrical representations (Fig. 4.2 and Fig. 4.3) provide a proof of the formula.

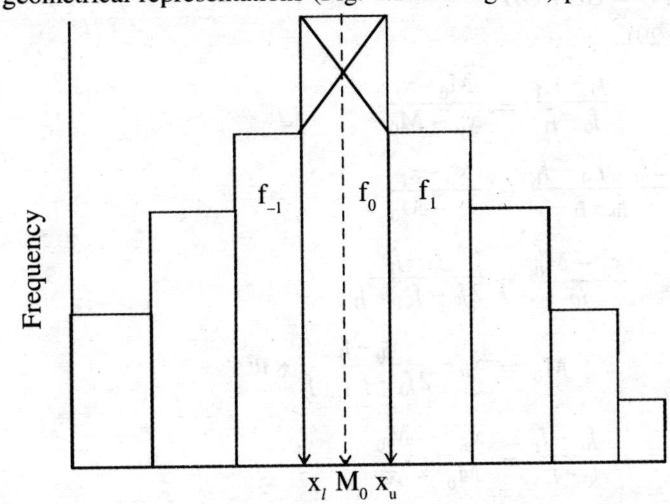

Fig. 4.2 Geometrical determination of mode from histogram

In the above figure, the mid-value, $\frac{x_l + x_u}{2}$, of the modal class is the mode, since $f_{-1} = f_1$.

But if $f_{-1} \neq f_1$, it is supposed that mode would be close to x_l if $f_{-1} > f_1$ and to x_u if $f_1 > f_{-1}$.

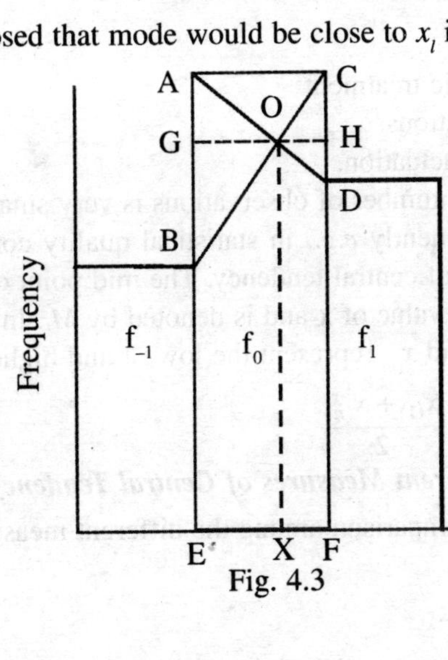

Fig. 4.3

Form Figure 4.3

$$AB = f_0 - f_{-1}$$
$$AE = CF = f_0 \qquad \text{... (4.28)}$$
$$CD = f_0 - f_1$$
$$OG = EX = M_0 - x_l$$
$$HO = XF = x_u - M_0$$

From the similar triangles *AOB* and *COD, AB / CD = AO / OD.*

Also, from the similar triangles *AOG* and *HOD, AO / OD = GO / OH.*

Combining, *AB / CD = GO / OH* ... (4.29)

Using (4.28) in (4.29)

$$\frac{f_0 - f_{-1}}{f_0 - f_1} = \frac{M_0 - x_l}{x_u - M_0} \qquad \text{... (4.30)}$$

$$\Rightarrow \quad \frac{2f_0 - f_{-1} - f_1}{f_0 - f_1} = \frac{x_u - x_l}{x_u - M_0}$$

or
$$\frac{x_u - M_0}{w} = \frac{f_0 - f_1}{2f_0 - f_{-1} - f_1} \qquad \text{... (4.31)}$$

or
$$M_0 = x_u - \frac{f_0 - f_1}{2f_0 - f_{-1} - f_1} \times w$$

Alternatively,
$$\frac{f_0 - f_1}{f_0 - f_{-1}} = \frac{x_u - M_0}{M_0 - x_l}$$

$$M_0 = x_l + \frac{f_0 - f_{-1}}{2f_0 - f_{-1} - f_1} \times w \qquad \text{... (4.32)}$$

Demerits of Mode are :

1. It may not be unique
2. It is not amenable to algebraic treatment
3. It is not based on all observations
4. It is affected by sampling fluctuation.

Mid Point of Range : When the number of observations is very small and an average value is to be found very quickly or more frequently *e.g.,* in statistical quality control in production, the mid point of range is used as a measure of central tendency. The mid point of range is defined as half of the sum of lowest value and highest value of x and is denoted by M_r. In the set of *n* observations on the variable **x**, $x_1, x_2, ..., x_n$, let $x_{(1)}$ and $x_{(n)}$ represent the lowest and highest values respectively, then

$$M_r = \frac{x_{(1)} + x_{(n)}}{2}$$

4.1.4 Comparisons Among Different Measures of Central Tendency

The following table presents a comparison among the different measures of location with respect to different desiderate.

Table 4.4. — Comparison Table

Desiderrata \ Measures	Arith– metic Mean	Geo– metric Mean	Har– monic Mean	Median	Mode	Mid point of Range
1. Rigidly Defined	*y*	*y*	*y*	*y*	*y**	*y*
2. Easy to understand *y*	*N*	*N*	*y*	*y*	*y*	
3. Easy to calculate	*y*	*N*	*N*	*y*	*y*	*y*
4. Based on all observations	*y*	*y*	*y*	*N*	*N*	*N*
5. Further Mathematical Treatment possible	*y*	*y*	*y*	*y*	*N*	*y*
6. Least Affected by Samping Fluctuation	*y*	*y*	*y*	*N*	*N*	*N*
7. Least Affected by Extreme Values	*N*	*N*	*N*	*y*	*y*	*y*
8. Used in Statistical Inference	*y*	*N*	*N*	*y*	*N*	*y*
9. Used for the Data Sets containing Extreme Values	*N*	*N*	*N*	*y*	*y*	*y*
10. Used for the Frequency Distribution with undefined class Intervals	*N*	*N*	*N*	*y*	*y*	*N*
11. Graphical Location Possible	*N*	*N*	*N*	*y*	*y*	*y*
12. Used for the Data Sets with zero or negative values	*y*	*N*	*N*	*y*	*y*	*y*

y : YES N : NO

* Mode is not rigidly defined for unequal class intervals.

4.1.5. Other Measures of Location : Quartiles, Deciles & Percentiles

When the observations in a series, occurring with frequencies, are arranged either in ascending or in descending order of magnitude, median represents a value below and above which there are $\frac{n}{2}$ observations, n being the total number of observations. A slight extension of the idea helps to identify two more points, Q_1 and Q_3, below which there exist $\frac{n}{4}$ and $\frac{3n}{4}$ observations, respectively. Q_1 and Q_3 are called the first and Quartiles respectively. Thus Q_1, Q_2 (*i.e.*, M_e) and Q_3 together divide the full set of ordered observations into four equal parts. In fact, one fourth, half and three fourth of the total number of observations lie below Q_1, Q_2 and Q_3, respectively. Similarly, 9 (nine) decile points, namely, $Q_1, Q_2, ..., D_9$ (i.e., nine deciles) divide the ordered set of observations into 10 equal parts, the i-th decile D_i being a point below which there are $\frac{i}{10}$th of the total number of observations (above which there are $\left(1-\frac{i}{10}\right)$th of the total number of observations).

Finally, ninety nine (99) percentile points $(P_1, P_2, ..., p_{99})$ divide the ordered series of observations into hundred equal parts, the i-th percentile P_i being a point below which there are $\left(\frac{i}{100}\right)$th of the total number of observations i = 1, 2, ..., 99.

In the following, general relations among quartiles, deciles and percentiles are given :

$$Q_1 = P_{25}, Q_2 = \text{Median} = D_5 = P_{50}$$

$$Q_3 = P_{75}, D_k = P_{10k}, k = 1, 2, ..., 9.$$

More generally, in an ordered series series of n observations corresponding to p-th quantile, F_p, there exist np observations below F_p and $(1-p)n$ observations above F_p, p being any proportion. Thus $F_{0.175}$ is a point below which threre are $0.175n$ observations and thus, the first quartile, $Q_1 = F_{0.25}$; 2nd quartile (median), $Q_2 = F_{0.50}$; third quartile $Q_3 = F_{0.75}$; i-th decile, $D_i = F_{0.1i}$, i = 1, 2, ..., 9 ; i-th percentile, $P_i = F_{0.01i}$, i = 1, 2, ..., 99.

Illustration 4.2 : The scores of 31 students who appeared at a Mathematical aptitude test are :

90, 128, 106, 93, 116, 152, 136, 156, 105, 125, 132, 143, 133, 158, 134, 146, 137, 155, 92, 97, 104, 121, 113, 119, 127, 140, 161, 122, 101, 111.

Identify $Q_1, Q_2, Q_3, D_5, D_9, P_{12}, P_{85}$ and $F_{0.80}$ scores.

The ordered values from smallest to largest are ;

90, 92, 93, <u>97</u>, 101, 104, 105, <u>106</u>, 111, 113, 116, 119, 121, 122, 125, <u>127</u>, 128, 132, 133, 134, 136, 137, 140, <u>143</u>, <u>146</u>, 150, <u>152</u>, <u>155</u>, 156, 158, 161.

The positional values of the **underlined scores** are identified serially as

$$P_{12}, Q_1, Q_2, \equiv D_5, Q_3, F_{0.80}, P_{85} \text{ and } D_9.$$

For grouped frequency distribution the formulae for quartiles, deciles and percentiles are given below :-

$$Q_1 = x_l\,(Q_1) + \frac{\frac{n}{4} - CF(x_l(Q_1))}{f_{Q_1}} \times w$$

$$Q_2 = \text{Median} = x_l\,(Q_2) + \frac{\frac{n}{2} - CF(x_l(Q_2))}{f_{Q_2}} \times w$$

$$Q_3 = x_l\,(Q_3) + \frac{\frac{3n}{4} - CF(x_l(Q_3))}{f_{Q_3}} \times w$$

$$D_k = x_l\,(D_k) + \frac{\frac{kn}{10} - CF(x_l(D_k))}{f_{D_k}} \times w, k = 1, 2, ..., 9$$

$$P_i = x_l(P_i) + \frac{\frac{in}{100} - CF(x_l(P_i))}{f_{P_i}} \times w, \; i = 1, 2, ..., 99$$

where $x_l(Q_1)$, $x_l(Q_2)$, $x_l(Q_3)$, $x_l(D_k)$, $x_l(P_i)$, are the lower class boundaries of 1st, 2nd, 3rd quartiles, k-th decile and i-th percentile classes respectively; f_{Q_1}, f_{Q_2}, f_{Q_3}, f_{D_k}, and f_{P_i} are the corresponding class frequencies, $CF(x_l(Q_1))$, $CF(x_l(Q_2))$, $CF(x_l(Q_3))$, $CF(x_l(D_k))$, and $CF(x_l(P_i))$ are the cumulative frequencies upto the corresponding lower class boundaries; w and n are class width and total frequency respectively.

Note : The above formulae can be recasted using the upper class boundaries and the cumulative frequencies (less than type) as in case of median.

4.2 Concept of Dispersion and its measures

In the previous section it has been pointed out that an average presents a value of significance representing the whole set of observations. It can be seen as well that two sets of observations may possess different variabilities (or extent of variation) though they may have same average value. From the following dot diagrams of two sets of observations we may examine how the observation are strewn around the same central (average) value = 5.9 ≈ 6.

It can be seen that the observations in the first set Figure 4.4(a) are less dispersed than those in the second set Figure 4.4(b) though the average values are equal. This necessiates to the proposition of another characteristic of grouped frequency distribution, namely, Dispersion (or Scatter) owing to which observations in a data set tend to vary (be dissimilar) among themselves.

The following measures of dispersion are commonly used :

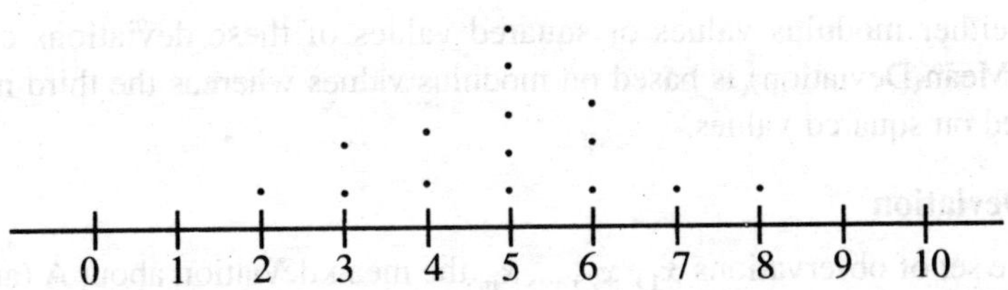

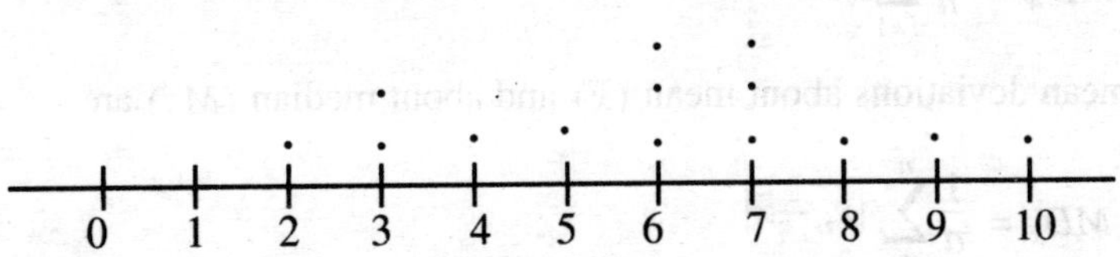

(i) Range, (ii) Mean Deviation, (iii) Standard Deviation and (iv) Quartile Deviation or Semi Inter-quartile Range.

4.2.1. Range

The simplest measure of dispersion is range. The range is defined as the difference between largest and smallest observations in a given set of values. In a set of n observations, $x_1, x_2, ..., x_n$, let $x_{(n)}$ and $x_{(1)}$ denote the largest and smallest observations respectively, then the Range = $x_{(n)} - x_{(1)}$.

Illustration 4.3 : From Figure 4.4 (a), we can identify the largest and smallest values as 9 and 3 respectively, whereas from Figure 4.4 (b) those are 10 and 2, respectively.

Thus

Range (1st data set) $= 9 - 3 = 6$

and, Range (2nd data set) $= 10 - 2 = 8$

Uses : (i) Range can be an appropriate measure of dispersion if the number of observations is small (3 to 5) or the observations are homogeneous. In statistical quality control, range is extremely useful in preparation of control charts to check the quality of a process by taking the observations in subgroups (subgroup size is usually varied from 3 to 5 observations).

(ii) In weather reporting, i.e., in reporting temperature, humidity, barometric pressure etc. range is used.

(iii) To know the variation in stock market prices, range is also useful.

In short, range is meaningfully used by business/industrial people and also by laymen in studying variation.

For a given set of n observations, $x_1, x_2, ..., x_n$ with mean $\bar{x}$, the dispersion can be measured from the deviations of the type $(x_1 - \bar{x}), (x_2 - \bar{x}), ..., (x_n - \bar{x})$.

The simple mean of these deviations, i.e., $\frac{1}{n}\sum(x_i - \bar{x})$ does not serve the purpose, since $\sum(x_i - \bar{x}) = 0$ (large positive and negative deviations may add up to zero). Thus to make the deviations free from signs, either modulus values or squared values of these deviations can be taken. The second measure (Mean Deviation) is based on modulus values whereas the third measure (Standard Deviation) is based on squared values.

4.2.2 Mean Deviation

From the above set of observations $x_1, x_2, ..., x_n$ the mean deviation about A (an arbitrary value), denoted by MD_A is defined as follows :

$$MD_A = \frac{1}{n}\sum_{i=1}^{n} |x_i - A|$$

Similarly, the mean deviations about mean $(\bar{x})$ and about median (M_e) are

$$MD_{\bar{x}} = \frac{1}{n}\sum_{1}^{n} |x_i - \bar{x}|$$

and $$MD_{Me} = \frac{1}{n}\sum_{1}^{n} |x_i - M_e|, \text{ respectively.}$$

For grouped frequency distribution with k class-marks, $x_1, x_2, ..., x_k$ with corresponding frequencies $f_1, f_2, ..., f_k$, the mean deviation about A is

$$MD_A = \frac{1}{n} \sum_1^k f_i |x_i - A|, \; n = \sum_1^n f_i$$

$MD_{\bar{x}}$ and MD_{M_e} can be obtained by replacing A by $\bar{x}$ and M_e, respectively.

Illustration 4.3 : The scores of a test on 10 boys are 128, 106, 93, 116, 152, 136, 156, 105, 125 and 132. To calculate MD_{M_e} and $MD_{\bar{x}}$ we proceed as follows :

Table 4.5 — The calcullation of MD_{M_e} and $MD_{\bar{x}}$

S. No.	Ordered Score	$\lvert x_i - 124.9\rvert$	$\lvert x_i - 126.5\rvert$
1	93	31.9	33.5
2	105	19.9	21.5
3	106	18.9	20.5
4	116	8.9	10.5
5	125	0.1	1.5
6	128	3.1	1.5
7	132	7.1	5.5
8	136	11.1	9.5
9	152	27.1	25.5
10	156	32.1	29.5
Total	**1249**	$\sum\lvert x_i - \bar{x}\rvert = 159.2$	$\sum\lvert x_i - M_e\rvert = 159$

$$\bar{x} = 1249/10 = 124.9, \; M_e = (125 + 128)/2 = 126.5,$$

$$MD_{M_e} = 159/10 = 15.9 \text{ and } MD_{\bar{x}} = 159.2/_{10} = 15.92$$

Proposition 4.1 : Mean deviation about median is least i.e., $MD_{M_e} \leq MD_A$ where A is any arbitrary variate value.

Proof : With no loss of generality, we can arrange the given n observations in ascending order of magnitude and denote them as $x_{(1)} \leq x_{(2)}, ..., x_{(n)}$; $x_{(1)}$ and $x_{(n)}$ being the smallest and largest observations, respectively. It can be seen that

$$\frac{1}{n} \sum_{i=1}^{n} |x_i - A| = \frac{1}{n} \sum_{i=1}^{n} |x_{(i)} - A|$$

Case I : $n = 2k$ (even)

The numerator of MD_A becomes

$$|x_{(1)} - A| + |x_{(2)} - A| + \ldots + |x_{(k)} - A| + |x_{(k+1)} - A| + \ldots + |x_{(2k-1)} - A| + |x_{(2k)} - A|$$

Which consists of k pairs of terms as

$$|x_{(1)} - A| + |x_{(2k)} - A|,\ |x_{(2)} - A| + |x_{(2k-1)} - A|, \ldots,\ |x_{(k)} - A| + |x_{(k-1)} - A|.$$

The sum $|x_{(1)} - A| + |x_{(2k)} - A|$ is least if A is any value between $x_{(1)}$ and $x_{(2k)}$. Similarly, $|x_{(2)} - A| + |x_{(2k-1)} - A|$ is least when A lies between $x_{(2)}$ and $x_{(2k-1)}$, and so on.

Finally, $|x_{(k)} - A| + |x_{(k+1)} - A|$ is minimum when A lies between $x_{(k)}$ and $x_{(k+1)}$. Thus the numerator of MD_A is least $\left(\text{i.e., } MD_A = \frac{1}{n}\sum_{1}^{n} |x_i - A| \text{ is least} \right)$, whre A lies between $x_{(k)}$ and $x_{(k+1)}$ as then each pair of term least. The median value, $M_e = \left(\frac{x_{(k)} + x_{(k+1)}}{2} \right)$ lies between $x_{(k)}$ and $x_{(k+1)}$ definitely, and hence $\frac{1}{n}\sum |x_i - M_e| \le \frac{1}{n}\sum |x_i - A|$.

Case II : $n = 2k + 1$ (odd)

The numerator of MD_A becomes

$$|x_{(1)} - A| + |x_{(2)} - A|,\ |x_{(2)} - A| + \ldots + |x_{(k)} - A| + |x_{(k+1)} - A| + |x_{(k+2)} - A| + \ldots + |x_{(2k-1)} - A| + |x_{(2k)} - A|$$

As before,

$$|x_{(1)} - A| + |x_{(2k)} - A| \text{ is minimum if } x_{(1)} \le A \le x_{(2k)}$$

$$|x_{(2)} - A| + |x_{(2k-1)} - A| \text{ is minimum if } x_{(2)} \le A \le x_{(2k-1)}$$

$$|x_{(k)} - A| + |x_{(k+2)} - A| \text{ is minimum if } x_{(k)} \le A \le x_{(k+2)}$$

and, $|x_{(k+1)} - A|$ is minimum if $A = x_{(k+1)}$.

Now, if $A = x_{(k+1)}$ each of the above sums is minimum and hence

$MD_A = \frac{1}{n}\sum_1^n |x_i - A|$ will be minimum if $A = x_{(k+1)}$, the median value. Thus in either case we see that MD_A is least if A = median. Hence the proposition.

Uses : Mean deviation has very limited use. It is easily understandable owing to its computational simplicity. In business reporting it is generally used. In textile field a well known relative measure of dispersion is named as *U%,* where (mean deviation is used)

$$U\,\% = \frac{\text{Mean deviation about mean}}{\text{Mean}} \times 100$$

4.2.3. Standard Deviation

As mentioned earlier that squared deviations provide a means of measuring the dispersion and as such the positive square root of the average of the squared diviations about *A* (an assumed mean), called the root-mean-square deviation about A, is considered as a measure of variability. Symbolically,

$$RMSD_A = \sqrt{\frac{1}{n}\sum (x_i - A)^2} \text{ for ungrouped data}$$

$$= \sqrt{\frac{1}{n}\sum f_i (x_i - A)^2}\text{ , for grouped data}$$

If $A = \bar{x}$, i.e., if the deviations are taken about the mean, then the corresponding root-mean-squrae deviation (RMSD) is called the standard Deviation. In fact, the standard deviation (s) is the root-mean square-deviation about mean or $RMSD_{\bar{x}}$. Thus s or Standard Deviation (S.D.)

$$= \sqrt{\frac{1}{n}\sum (x_i - \bar{x})^2}\text{ , for ungrouped data}$$

$$= \sqrt{\frac{1}{n}\sum f_i (x_i - \bar{x})^2}\text{ , for grouped frequency distribution.}$$

For computational purposes, the above expressions can be written in a simpler form as derived below :

In fact, $\quad \sum (x_i - \bar{x})^2 = \sum^n (x_i^2 - 2\bar{x}x_i + \bar{x}^2)$

$$= \sum x_i^2 - 2\bar{x}\sum x_i + \sum \bar{x}^2$$

$$= \sum x_i^2 - n\bar{x}^2, \quad \left[\because \sum x_i = n\bar{x}\right]$$

and for grouped frequency distribution we have,

$$\sum_{i=1}^{n}(x_i-\bar{x})^2 f_i = \sum_{1}^{k} x_i^2 f_i - n\bar{x}^2, \; n\bar{x} = \sum_{1}^{k} x_i f_i \text{ and } n = \sum_{1}^{k} f_i$$

Hence,
$$s = \sqrt{\frac{1}{n}\sum x_i^2 - \bar{x}^2}, \text{ for ungrouped data}$$

$$= \sqrt{\frac{1}{n}\sum_{1}^{k} x_i^2 f_i - \bar{x}^2}, \text{ for grouped frequency distribution.}$$

Variance : The square of the standard deviation (s^2) is called the variance of a given set of observations and it is denonted by v, υ, s^2 etc. Alternatively, the standard deviation is the positive square root of the variance.

Some properties of variance and hence that of standard deviation are :

Property 1 : If a variable x assumes a constant value only, then the variance and hence standard deviation of x is equal to zero.

Let $x_i = c$, for $i = 1, 2, \ldots, n$ then $\bar{x} = c$, so for each $x_i, x_i - \bar{x} = 0$. Hence $s^2 = \frac{1}{n}\sum(x_i - \bar{x})^2 = 0$.

Property 2 : If $y = a + bx$, then $\upsilon(y) = b^2\,\upsilon(x)$ and hence $s_y = |b| s_x$ where s_x and s_y represent standard deviation of x and y respectively.

Here for i-th variate value, $i = 1, 2, \ldots, n$

and hence $\bar{y} = a + b\bar{x}$.

$\therefore\; y_i - \bar{y} = b(x_i - \bar{x}), \text{ or } (y_i - \bar{y})^2 = b^2 (x_i - \bar{x})^2$

Summing over all i, $\sum(y_i - \bar{y})^2 = b^2 \sum(x_i - \bar{x})^2$

$$\Rightarrow \sqrt{\frac{1}{n}\sum(y_i - \bar{y})^2} = \sqrt{\frac{b^2}{n}\sum(x_i - \bar{x})^2}$$

or, $s_y = |b| s_x$

Property : 3 For two sets of observations with standard deviations s_1 and s_2, the standard deviation (s) of the combined set becomes,

$$s = \sqrt{\frac{n_1 s_1^2 + n_2 s_2^2 + n_1(\bar{x}_1 - \bar{x})^2 + n_2(\bar{x}_2 - \bar{x})^2}{n_1 + n_2}}, \text{ where } \bar{x}_1$$

and $\bar{x}_2$ are the means of the two sets based on n_1 and n_2 observations respectively and $\bar{x}$ is the mean of the combined set.

The combined mean $(\bar{x})$ for n_1+n_2 values is given by,

$$\bar{x} = \frac{n_1\bar{x}_1+n_2\bar{x}_2}{n_1+n_2}, \; x_{11}, x_{12}, ..., x_{1n_1} \text{ and } x_{21}, x_{22}, ..., x_{2n_2}$$

being the two sets of observatrions.

The combined variance s^2 can be written as

$$s^2 = \frac{1}{n_1+n_2}\left\{\sum_{i=1}^{n_1}(x_{1i}-\bar{x})^2+\sum_{j=1}^{n_2}(x_{2j}-\bar{x})^2\right\}$$

$$\text{Now} \quad \sum(x_{1i}-\bar{x})^2 = \sum\{(x_{1i}-\bar{x}_1)+(\bar{x}_1-\bar{x})\}^2$$

$$= \sum\left\{(x_{1i}-\bar{x}_1)^2+2(\bar{x}_1-\bar{x})(x_{1i}-\bar{x}_1)+(\bar{x}_1-\bar{x})^2\right\}$$

$$= \sum(x_{1i}-\bar{x}_1)^2+n_1(\bar{x}_1-\bar{x})^2. \left[\because \sum(x_{1i}-\bar{x}_1)=0\right]$$

$$= n_1 s_1^2+n_1(\bar{x}_1-\bar{x})^2.$$

Similarly,

$$\sum_{j=1}^{n_2}(x_{2j}-\bar{x})^2 = n_2 s_2^2+n_2(\bar{x}_2-\bar{x})^2$$

Thus, the combined variance of the two groups can be written as,

$$s^2 = \frac{n_2 s_1^2+n_2 s_2^2+n_1(\bar{x}_1-\bar{x})^2+n_2(\bar{x}_2-\bar{x})^2}{n_1+n_2}$$

and hence the combined standard deviation becomes,

$$s = \sqrt{\frac{n_1 s_1^2+n_2 s_2^2+n_1(\bar{x}_1-\bar{x})^2+n_2(\bar{x}_2-\bar{x})^2}{n_1+n_2}}$$

Note : The above result can be obtained under a more general setting containing P sets of observation, $\bar{x}_i$ and s_i being the mean and standard deviation of the i-th set based on n_i observations, i = 1, 2, ..., P, then the standard deviation of x in all the sets taken together is given by

$$s = \sqrt{\left(\frac{\sum n_i s_i^2+\sum n_i(\bar{x}_i-\bar{x})^2}{\sum n_i}\right)}$$

$$\text{where} \quad \bar{x} = \frac{\sum^{p} n_i x_i}{\sum_{i=1} n_i}$$

Proposition 4.2 : Root-mean-square deviation about $\overline{x}$ is least.

i.e., $$\sqrt{\frac{1}{n}\sum(x_i-\overline{x})^2} \le \sqrt{\frac{1}{n}\sum(x_i-A)^2}$$

$$\sum_{1}^{n}(x_i-A)^2 = \sum\{(x_i-\overline{x})+(\overline{x}-A)\}^2$$

$$= \sum_{i=1}^{n}(x_i-\overline{x})^2 + n(\overline{x}-A)^2,$$

$$\ge \sum_{1}^{n}(x_i-\overline{x})^2, \quad [\because n(\overline{x}-A)^2 \ge 0],$$

the equality sign holds only when A = $\overline{x}$.

Thus we have

$$\sqrt{\frac{1}{n}\sum(x_i-\overline{x})^2} \le \sqrt{\frac{1}{n}\sum(x_i-A)^2}$$

and hence the proposition.

Uses : Standard deviation is the most widely used measure of dispersion as its satisfies almost all requisites of a good measure.

Merits of Standard Deviation :

In general, standard deviation satisfies the following desiderata of a good measure :

(i) It is rigidly defined and based on all the observation.

(ii) It is amenable to further mathematical treatment.

(iii) It is least affected by sampling fluctuation.

(iv) It has potential use in statistical inference.

(v) It is easily interpreted.

Demerits of Standard Deviation are :

(i) Standard deviation is affected by presence of extreme values in the data, and

(ii) It cannot be calculated when the frequency distribution has open end or underfined classes.

4.2.4. Quartile Deviation

Like range (the difference between maximum and minimum observation), the difference between 3rd quartile and 1st quartile gives an idea of the magnitude of dispersion existing in a series of observation. The range confers maximum weights on the extreme values and no weight on the intermediate values. To eliminate these difficulties partically, the measures of dispersion or spread based on quartiles are defined as follows :

(a) Inter-quartile range = $Q_3 - Q_1$

(b) Semi inter-quartile range, or

$$\text{Quartile deviation} = \frac{Q_3 - Q_1}{2}$$

Uses : (i) To get some degree of protection from wild observation, as in reports on income, inter-quartile range or semi inter-quartile range is usually preferred as a measure of spread, and (ii) When the class intervals are unequal or not defined properly (e.g., in open-end class intervals) these measures are quite helpful.

4.2.5. Measures Based on Mutual Differences of Observations

A statistician, C. Gini, suggested some measures of dispersion in which mutual differences among the observations had been utilised rather than utilising the deviations of the observations from a central value. Some other statisticians believe that the choice of a central value to measure the dispersion is arbitrary and also any measure of dispersion should depend on the mutual differences, $(x_i - x_j), i, j = 1, 2, \ldots, n$ among the observartions relating to a variable x, therefore, it is necessary to develop measures of dispersion based on mutual differences.

The first of such measures is the mean of the absolute values of all mutual differences. This measure is called Gini's Mean Difference and it is given by,

$$\Delta_1 = \frac{1}{n^2}\sum_{i=1}^{n}\sum_{j=1}^{n}|x_i - x_j|, \text{ for discrete data}$$

$$= \frac{1}{n^2}\sum_{i=1}^{k}\sum_{j=1}^{k}f_i f_j|x_i - x_j|, \text{ for grouped data and } \sum f_i = \sum f_j = n.$$

The second such measure is given by,

$$\Delta_2 = \frac{1}{n^2}\sum_{i=1}^{n}\sum_{j=1}^{n}(x_i - x_j)^2, \text{ for discrete data}$$

$$= \frac{1}{n^2}\sum_{i=1}^{k}\sum_{j=1}^{k}f_i f_j(x_i - x_j)^2, \text{ for grouped data and } \sum f_i = \sum f_j = n$$

and it eliminates the difficulties involved in the first measure in regard to mathematical treatment.

Uses : Economists use this measure generally to know the extent of concentration of wealth, income, wage etc, through the expression $G = \frac{\Delta_1}{2\bar{x}}$ which is a relative measure of dispersion and is known as Gini's Coefficient of Concentration.

4.2.6. Relationship Among Measures of Dispersion

(i) Standard deviation is not less than mean deviation about $\bar{x}$ *i.e.*,

$S.D. \not< MD_{\bar{x}}$.

Proof : Let $x_1, x_2 \ldots, x_n$ be a set of n observations, we have to prove

$$S.D. \not< MD_{\bar{x}}.$$

or, $$(S.D.)^2 \not< (MD_{\bar{x}})^2$$

or, $$(S.D.)^2 \geq (MD_{\bar{x}})^2, \text{ i.e., } \frac{1}{n}\sum(x_i - \bar{x})^2 \geq \left(\frac{1}{n}\sum|x_i - \bar{x}|\right)^2$$

Let $|x_i - \bar{x}| = y_i$ and then we have to prove

$$\frac{1}{n}\sum y_i^2 \geq \left(\frac{1}{n}\sum y_i\right)^2$$

$$\text{or,}\quad \frac{1}{n}\sum y_i^2 - \left(\frac{1}{n}\sum y_i\right)^2 \geq 0$$

$$\text{or,}\quad \frac{1}{n}\sum y_i^2 - \bar{y}^2 \geq 0 \; i.e., s_y^2 \geq 0,$$

which is always true and hence it is proved that $S.D. \nless MD_{\bar{x}}$.

(ii) For the n observations relating to the variable $x, x_1, x_2, ..., x_n$, prove

$$\frac{R^2}{2n} \leq s^2 \leq \frac{R^2}{4}.$$

Proof : Let $x_{(1)}$ be the smallest observation and $x_{(n)}$ be the largest observation, then $R = x_{(n)} - x_{(1)}$. We know from the definition of variance that

$$ns^2 = \sum_{i=1}^{n}(x_i - \bar{x})^2$$

$$\leq \sum\left(x_i - \frac{x_{(1)} + x_{(n)}}{2}\right)^2, \quad \left(\because \sum(x_i - \bar{x})^2 \leq \sum(x_i - A)^2 \text{ here } A = \frac{x_{(1)} + x_{(n)}}{2}\right)$$

$$= \sum_{x_i \leq \frac{x_{(1)} + x_{(n)}}{2}}{}_1 \left(x_i - \frac{x_{(1)} + x_{(n)}}{2}\right)^2 + \sum_{x_i > \frac{x_{(1)} + x_{(n)}}{2}}{}_2 \left(x_i - \frac{x_{(1)} + x_{(n)}}{2}\right)^2$$

$$\leq \sum{}_1\left(x_{(1)} - \frac{x_{(1)} + x_{(n)}}{2}\right)^2 + \sum{}_2\left(x_{(n)} - \frac{x_{(1)} + x_{(n)}}{2}\right)^2$$

$$= \sum{}_1 \frac{R^2}{4} + \sum{}_2 \frac{R^2}{4} = \left(\sum{}_1 + \sum{}_2\right)\frac{R^2}{4} = \sum_{i=1}^{n}\frac{R^2}{4} = \frac{nR^2}{4}$$

Thus we get $\quad s^2 \leq \dfrac{R^2}{4}$

Again, $\quad ns^2 = \sum(x_i - \bar{x})^2$

$$\geq (x_{(1)} - \bar{x})^2 + (x_{(n)} - \bar{x})^2$$

$$\geq \left(x_{(1)} - \frac{x_{(1)} + x_{(n)}}{2}\right)^2 + \left(x_{(n)} - \frac{x_{(1)} + x_{(n)}}{2}\right)^2,$$

$$\left(\because \sum (x_i - A)^2 \geq \sum (x_i - \text{Average})^2, \text{ here } A = \bar{x}, \text{ Average } = \frac{x_{(1)} + x_{(n)}}{2}\right)$$

$$= \frac{R^2}{4} + \frac{R^2}{4} = \frac{R^2}{2}$$

Hence, $s^2 \geq \dfrac{R^2}{2n}$

Thus we have $\dfrac{R^2}{2n} \leq s^2 \leq \dfrac{R^2}{4}$

(iii) variance is half of the Gini's mean difference based on squares of the mutual differences. i.e.,

$$s^2 = \frac{\Delta_2}{2}$$

Proof : $\displaystyle\sum_i \sum_j (x_i - x_j)^2 = \sum_i \sum_j \{(x_i - \bar{x}) + (x_j - \bar{x})\}^2$

$$= \sum_i \sum_j (x_i - \bar{x})^2 + \sum_i \sum_j (x_j - \bar{x})^2, \ \left(\because \sum (x_i - \bar{x}) = 0\right)$$

$$= n \sum_i (x_i - \bar{x})^2 + n \sum_j (x_j - x)^2$$

$$= n \times ns^2 + n \times ns^2 = 2n^2 s^2$$

or $\displaystyle s^2 = \frac{1}{2n^2} \sum_i \sum_j (x_i - x_j)^2 = \frac{\Delta_2}{2}$

Thus the property holds.

4.2.7. Comparison Among Measures of Dispersion

The following table presents a comparative status of the different measures of dispersion with respect of the satisfaction of different desiderata.

Table 4.6 — Comparison Table

Desiderrata \ Measures	Range	Mean Deviation	Standard Deviation	Quartile Deviation	Gini's Mean Differeence Δ_1	Gini's Mean Differeence Δ_2
1. Rigidly defined	y	y	y	y	y	y
2. Easy to comput	y	N	N	N	N	N
3. Easy to comprehend	y	y	N	y	y	N
4. Based on all observations	N	y	y	N	y	y
5. Further mathematical Treatment possible	N	N	y	N	N	y
6. Least affected by sampliing fluctuation	N	y	y	N	y	y

(**Table 4.6** continued)

Desiderrata \ Measures	Range	Mean Deviation	Standard Deviation	Quartile Deviation	Gini's Mean Differeence Δ_1	Gini's Mean Differeence Δ_2
7. Least affected by extreme values	*y*	*N*	*N*	*y*	*N*	*N*
8. Used in statistical inference	*N*	*N*	*y*	*N*	*N*	*y*
9. Used for the data set containing extreme values	*N*	*y**	*N*	*y*	*N*	*N*
10. Used for the frequency distribution with undefined class Intervalls	*N*	*y*	*N*	*y*	*N*	*N*

y : YES N : NO

* It is true for the mean deviation about median

Note : The students are advised to express the ideas contained in the tables (Table 4.4 and Table 4.6) in descirptive form for examination purpose.

4.3 Measures of Relative Dispersion

All measures of dispersion possess the same unit as that of the variable. With the unit related measure of dispersion we cannot compare the variability features of two or more sets of observations expressed in different units. So unit free measures of dispersion are necessary for purpose of comparison. The above consideration leads to the relative or absolute measures of dispersion which are usually expressed as percentages of central tendency measures. These are :

Coefficient of Range, Coefficient of Mean Deviation, Coefficient of Variation, Coefficient of Quartile Deviation and Gini's Coefficient of Concentration.

The most extensively used measure of relative dispersion is coefficient of variation,

$$C.V. = \frac{s}{\bar{x}} \times 100; \ \bar{x} \neq 0,$$

s and $\bar{x}$ being standard deviation and mean, respectively.

The above measure (C.V.) of relative dispersion gives an idea of relative dispersion in case of two sets of data (having different units), and moreover it enables us to obtain a true picture of relative accuracy when two sets of data expressed in the same unit are compared. The above can be visualised through the following example. Repeated measurements are taken to measure two screws having diameters 1 cm & 5 cm, respectively. Ten measurements are taken in each case. The standard deviation comes out to be '0.2 cm' in each set of data. Obviously, the data in the second set was recorded more precisely (provided unbiased measurements under two situations give the values of means as 1 cm and 5 cm respectively) as in this set the same value of variability is observed even when its mean value is five times. The value of the coefficient of variation is also found to be smaller in the second data set (1/5th of the value of C.V. in the first set). Thus this measure of relative dispersion gives a true picture of relative precision.

Other measures of relative dispersion are :

(i) Coefficient of range = $\dfrac{x_{(n)} - x_{(1)}}{x_{(n)} + x_{(1)}} \times 100$,

$x_{(n)}$ and $x_{(1)}$ being the highest and lowest observations in a set of observations relating to a variable x.

(ii) Coefficient of mean deviation = $\dfrac{\text{Mean deivation about median}}{\text{Median}} \times 100$

$$= \frac{\text{Mean deivation about mean}}{\text{Median}} \times 100$$

(iii) Coefficient of Quartile deviation

$$= \frac{Q_3 - Q_1}{Q_3 + Q_1} \times 100$$

This measure is used in the case of open end classes where s and $\bar{x}$ cannot be computed.

(iv) **Gini's coefficient of concentration.**

A relative measure of dispersion based on Gini's mean difference is named as Gini's coefficient of concentration and symbolically,

$G = \dfrac{\Delta_1}{2\bar{x}}$ and coefficient of variation (C.V.) is equal to $\dfrac{\sqrt{\Delta_2}}{\sqrt{2\bar{x}}}$, where Δ_1 and Δ_2 are two representations of Gini's mean difference.

Note : Measures of relative dispersion (*e.g.*, C.V.) are also called for when two series, widely different in values, are compared. In such cases though the standard deviations may not be the same but their C.V's may be the same. Thus, deflation by mean makes the two series amenable to comparison with respect to variability.

4.4. Lorenz Curve or Curve of Concentration

Sometimes economists/ statisticians are interested in the following query : "What percentage of wealth/ income/ wage is concentrated among what precentage of individuals in a particular section of people". From a frequency distribution table we can get an answer to the above query by means of some algebraic calculation. To measure the concentration graphically, Max O. Lorenz proposed a curve which is known as Lorenz curve or curve of concentration. From the curve we can determine the extent of maldistribution or the nature of concentration of wealth/ income/ wage among the particular community under survey.

The following procedure is adopted to draw the curve from a frequency distribution table :

(i) Prepare a column multiplying mid-value with the corresponding frequency, *i.e.*, $x \times f$, for each of the k classes.

(ii) Prepare a cumulative frequency (less than type) column and a column for cumulative of total variable value (*i.e.*, cumulative of $x \times f$ values) of each class,

(iii) Express all the cumulative values both for frequencies and variable totals into the corresponding percentage figures.

(iv) Denote percentages corresponding to the cumulative frequencies as $F(x)$, i.e., for the p-th class, $F(x_p) = 100 \sum_{1}^{p} f_i \div \sum_{1}^{k} f_i$ and that for the cumulative of total variable values as $G(x)$, *i.e.*, for the p-th class,

$$G(x_p) = 100 \sum_{1}^{p} x_i f_i \div \sum_{1}^{k} x_i f_i, \; p = 1, 2, ..., k.$$

(v) Next plot $(F(x), G(x))$ values and join these plotted points so as to make a curve which is known as Lorenz curve. This curve shows the line of actual concentration of the variable.

The line $F(x) = G(x), \forall x$, which is diagonally drawn from south-west origin to north-east corner, represents the line of equal distribution. The area bounded by the curve and the line is called the area of concentration. As this area becomes larger, the extent of maldistribution or degree of concentration becomes more and more. Thus the area of concentration serves as an indicator of the degree of concentration. Twice of this area is Gini's coefficient of concentration.

In a study of wealth distribution the line of equal distribution reveals that any particular proportion of individuals receives the same proportion of wealth, *i.e.*, 20 percent of the population possess 20 percent of the total wealth, 60 percent of the population posses 60 percent of the total wealth etc. The curve of concentration may reveal that 5 percent of the population possess 40 percent of total wealth, 50 percent of the population receive only 10 percent of the total wealth etc. Thus, the curve of concentration indicates the nature of maldistribution of wealth/income/wage etc. The Table 4.7 and the Figure 4.5 explain the construction procedure of the Lorenz curve.

Table 4.7 — Calculations of F (x) and G (x) for drawing Lorenz curve from a given frequency distribution

Annual Income ('000)	**Number of staff in a Gove. office (*f*)**	**Mid Value (*x*)**	**Total Income (*f.x*)**	**Cumulative Frequency**	**Cumulative total Income**	**% Cumulative Frequency (*F* (*x*))**	**% Cumulative Total Income (*G* (*x*))**
25 – 50	7	37.5	262.5	7	262.5	7.7	3.9
50 – 70	50	62.5	3125.0	57	3387.5	62.6	40.3
75 – 100	24	87.5	2100.0	81	2487.5	89.0	81.4
100 – 125	6	112.5	675.0	87	6162.5	95.6	91.5
125 – 150	3	137.5	412.5	90	6575.0	98.9	97.6
150 – 175	1	162.5	162.5	91	6737.5	100.0	100.0
Total	**91**		**6737.5**				

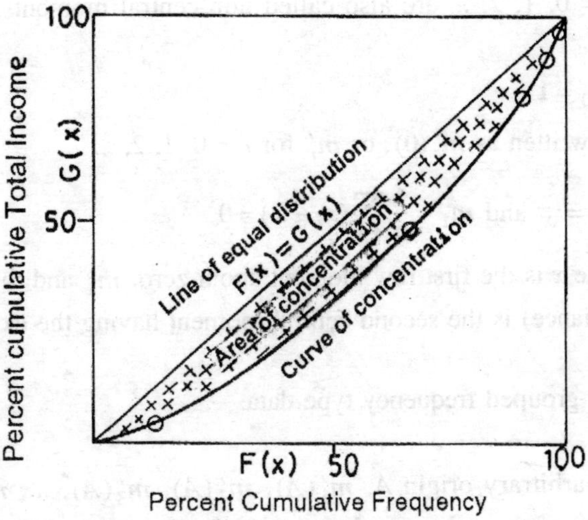

Fig. 4.5 — Lorenz Curve

4.5. Non-central and Central Moments

The two important measures namely mean and variance have been discussed in the preceding section. They are referred to as measures of central tendency and dispersion respectively. For the ungrouped data $(x_i, i = 1, 2, ..., n)$ and for data arranged under grouped frequency distribution (x_i occurring with frequency f_i, $i = 1, 2, ..., k$, $\sum_{i=1}^{k} f_i = n$, x_i being class-mark for i-th class), the expressions of means and variances are :

$$\bar{x} = \frac{1}{n}\sum_{1}^{n} x_i, \text{ or } \bar{x} = \frac{1}{n}\sum_{i=1}^{k} x_i f_i = \text{mean}$$

$$s^2 = \frac{1}{n}\sum_{1}^{n}(x_i - \bar{x})^2, \text{ or } s^2 = \frac{1}{n}\sum_{1}^{k}(x_i - \bar{x})^2 f_i = \text{variance}$$

The most general way to measure the descriptive characteristics of any dimension of a series of data of the above mentioned type is to compute quantities of the following types :

$$m_r'(A) = \frac{1}{n}\sum_{i=1}^{n}(x_i - A)^r, \text{or } m_r'(A) = \frac{1}{n}\sum_{i=1}^{k}(x_i - A)^r f_i, r = 0, 1, 2, ...$$

In fact, $m_r'(A)$ is called the r-th Raw Moment of x about the arbitrary origin A.

When the origin of a moment is taken at the arithmetic mean of a variable (*i.e.*, $A = \bar{x}$), it is called a central moment. Thus the expression of the r-th Central Moment, m_r is

$$m_r = \frac{1}{n}\sum_{i=1}^{n}(x_i - \bar{x})^r, \text{or } m_r = \frac{1}{n}\sum_{i=1}^{k}(x_i - \bar{x})^r f_i, r = 0, 1, 2, ...$$

The moments, m'_r, $r = 0, 1, 2, \ldots$ are also called non-central moments. For any real value A (including zero)

$$m'_0(A) = m_0 = 1$$

For A = 0, $m'_r(A)$ are written as $m'_r(0)$, or m'_r for $r = 0, 1, 2, \ldots$

Obviously, $m'_1 = \text{mean} = \bar{x}$ and $m_1 = \frac{1}{n}\sum(x_i - \bar{x}) = 0$.

Thus mean of a variable x is the first raw moment about zero, m'_1 and the first central moment, $m_1 = 0$ whereas m_2 (= variance) is the second central moment having the expression, $m_2 = s^2 = \frac{1}{n}\sum_{i=1}^{k}(x_i - \bar{x})^2 f_i$ for grouped frequency type data.

When moments about arbitrary origin A, $m'_1(A)$, $m'_2(A)$, $m'_3(A)$, ..., $m'_r(A)$ are given, it is possible to obtain the value of the r-th central moment of x, m_r in terms of the above noncentral moments.

Now, $(x_i - \bar{x})^r f_i = \{(x_i - A) - (\bar{x} - A)\}^r f_i$

$$= \left\{(x_i - A)^r - \binom{r}{1}(x_i - A)^{r-1}(\bar{x} - A) + \binom{r}{2}(x_i - A)^{r-2}(\bar{x} - A)^2\right.$$

$$\left. + \binom{r}{3}(x_i - A)^{r-3}(\bar{x} - A)^3 + \ldots + (-1)^{r-1}\binom{r}{r-1}(x_i - A)(\bar{x} - A)^{r-1} + (-1)^r(\bar{x} - A)^r\right\} f_i$$

Summing the expressions over $i = 1, 2, \ldots, k$ and dividing by n, where $n = \sum_{i=1}^{k} f_i$ we get,

$$m_r = m'_r(A) - \binom{r}{1} m'_{r-1}(A) m'_1(A) + \binom{r}{2} m'_{r-2}(A) m'^2_1(A) - \binom{r}{3} m'_{r-3}(A) m'^3_2(A) + \ldots$$

$$+ (-1)^{r-1} r m'_1(A)\, m'_1(A)^{r-1} + (-1)^r (m'_1(A))^r,$$

$$\left(\text{since } \bar{x} - A = \frac{1}{n}\sum_{i=1}^{k}(x_i - A) f_i - m'_1(A)\right)$$

Taking A = 0, m_r can alternatively be written as

$$m_r = m'_r - \binom{r}{1} m'_{r-1} m'_1 + \binom{r}{2} m'_{r-2}(m'_1)^2 + \ldots + (-1)^{r-1} r (m'_1)^r + (-1)^r (m'_1)^r.$$

In particular, putting $r = 1, 2, \ldots$ in the above relation, we get,

$$r = 1;\quad m_1 = m'_1 - 1 \times m'_0 \times m'_1 = m'_1 - m'_1 = 0$$

$$r = 2;\quad m_2 = m'_2 - 2 m'_1 m'_1 + 1 \times m'_0 (m'_1)^2$$

$$= m_2' - (m_1')^2 = s^2 = \text{variance}$$

or, $$m_2 = m_2'(A) - (m_1'(A))^2 = s^2,$$

$$r = 3; \quad m_3 = m_3' - 3m_2' m_1' + 3m_1'(m_1')^2 - (m_1')^3$$

$$= m_3' - 3m_2' m_1' + 2(m_1')^3$$

$$r = 4; \quad m_4 = m_4' - 4m_3' m_1' + 6m_2'(m_1')^2 - 4m_1'(m_1')^3 + (m_1')^4$$

$$= m_4' - 4m_3' m_1' + 6m_2'(m_1')^2 - 3(m_1')^4.$$

The above formulae enable us to find out the first four central moment of x if the first four noncentral moments are known. To find out the reverse relation, i.e., to obtain noncentral (raw) moments from central moments, the following The r-th noncentral moment,

$$m_r'(A) = \frac{1}{n}\sum_{1}^{k}(x_i - A)^r f_i = \frac{1}{n}\sum_{1}^{k}\{(x_i - \bar{x}) + (\bar{x} - A)\}^r f_i$$

$$= \frac{1}{n}\sum_{i=1}^{k}\left\{(x_i-\bar{x})^r + \binom{r}{1}(x_i-\bar{x})^{r-1}(\bar{x}-A) + \binom{r}{2}(x_i-\bar{x})^{x-2}(\bar{x}-A)^2 + \ldots + \binom{r}{r-1}(x_i-\bar{x})(\bar{x}-A)^{r-1} + \binom{r}{r}(\bar{x}-A)^r\right\} f_i$$

$$= m_r + \binom{r}{1}m_{r-1}\, m_1'(A) + \binom{r}{2}m_{r-2}(m_1'(A))^2 + \ldots + r(m_1'(A))^{r-1} m_1(A) + (m_1'(A))^r$$

In particular, for $A = 0$

$$m_r' = m_r + \binom{r}{1}m_{r-1}\, m_1' + \binom{r}{2}m_{r-2}(m_1')^2 + \ldots + r(m_1')^{r-1} m_1 + (m_1')^r$$

Putting r = 1, 2, 3, 4 we get,

r = 1; $m_1' = m_1'$, since $m_1 = 0$

r = 2; $m_2' = m_2 + 2m_1 m_1' + (m_1')^2 = m_2 + (m_1')^2, (\because m_1 = 0)$

r = 3; $m_3' = m_3 + 3m_2 m_1' + 3m_1(m_1')^2 + (m_1')^3$

$= m_3 + 3m_2 m_1' + (m_1')^3, (\because m_1 = 0)$

r = 4; $m_4' = m_4 + 4m_3 m_1' + 6m_2(m_1')^2 + 4m_1(m_1')^3 + (m_1')^4$

$= m_4 + 4m_3 m_1' + 6m_2(m_1')^2 + (m_1')^4, (\because m_1 = 0)$

To guard against any computational error, Charlier proposed the following checks :

$$\sum_{i=1}^{k}(x_i + 1)\, f_i = \sum x_i f_i + \sum f_i; \quad \sum (x_i + 1)^2 f_i = \sum x_i^2 f_i + 2\sum x_i f_i + \sum f_i$$

$$\sum_{1}^{k}(x_i+1)^3 f_i = \sum x_i^3 f_i + 3\sum x_i^2 f_i + 3\sum x_i f_i + \sum f_i$$

$$\sum_{1}^{k}(x_i+1)^4 f_i = \sum x_i^4 f_i + 4\sum x_i^3 f_i + 6\sum x_i^2 f_i + 4\sum x_i f_i + \sum f_i .$$

To calculate moments upto 4th order, only the 4th Charlier check may be made.

Result : If from the variable x, a new variable y is formed, where $y = (x - c)/d$, then it can be shown that $m_r(y) = \frac{1}{d^4} m_r(x)$.

The moments of the new variable are independent of change of origin (base).

4.6. Sheppard's Corrections for Moments

The computation of moments from data arranged in grouped frequency distribution is made under the assumption that frequencies are evenly distributed over the class interval or the observations falling in a class (*i.e.*, the f_i values falling in the *i*-th class) are all equal to the class-mark (for *i*-th class it is x_i) while, in fact, the observations may be really unequal. The above mentioned assumption (approximation) introduces some errors called errors due to grouping, which can be corrected by adjusting suitably the calculated moments of different orders under certain conditions to be satisfied by data distribution. These corrections were obtained for equi-sized classes by Sheppard and are known as Sheppard's Correction for Moments. For the first four moments (central and non-central), these corrections are as follows :

$$m_1'(\text{corrected}) = m_1'$$

$$m_2'(\text{corrected}) = m_2' - \frac{w^2}{12}$$

$$m_3'(\text{corrected}) = m_3' - \frac{w^2}{4} m'$$

$$m_4'(\text{corrected}) = m_4' - \frac{w^2}{4} m_2' + \frac{7}{240} w^4$$

and $$m_2(\text{corrected}) = m_2 - w^2/12$$

$$m_3(\text{corrected}) = m_3$$

$$m_4(\text{corrected}) = m_4 - \frac{w^2}{2} m_2 + \frac{7}{240} w^4$$

where w is the width of the class interval. It is to be noted that corrections are not needed for symmetrical or moderately asymmetrical type of frequency distributions (having equisized classes) where the width of the class interval is not greater than 1/20th of the range, as in such cases the assumption of even distribution of frequencies over the classes does not induce grouping error. The following conditions are to be fulfilled for ensuring validity of Sheppard's correction :

1. The underlying variable should be continuous, the associated frequency curve of the distribution should be continuous and should have high-order contact at each end of the range of the variable (i.e., the frequencies decrease gradually and smoothly to zero at both ends). The verification of the conditions requires a sufficiently large number of observations, in which case the sampling errors of moments also remain small. Thus the total frequency should be sufficiently large for valid application of Sheppard's correction.
2. The width of the class interval should not be too small and the number of classes should not be too large. In case of class width having a small value (number of classes becomes large accordingly) the magnitude of the correction needed to each moment become insignificant.

As a working rule, Sheppard's correction is applied when the total frequency is more than 1000 and the number of classes is less than 20.

4.7. Definition of Skewness and Kurtosis and their Measures

The discussion in the previous sections reveals that a series of data arranged in grouped frequency distribution exhibit properties, namely, central tendency and dispersion. With the help of their measures of location and dispersion, it is possible to differentiate between two frequency distributions. However, situations may arise where the measures of location and dispersion fail to discriminate between the two frequency distributions as is evident from the following example. Though the means and variances are same for both the frequency distributions, their histograms suggest evidence of disparity in the shapes of the two distribution :

Example 4.6 : In the following two frequency distribtions of 20 individuals (over a certain range of values of a variable) are given.

Class boundaries	:	5 – 10	10 – 15	15 – 20	20 – 25	25 – 30	30 – 35
Frequency distribution I	:	1	3	6	6	3	1
Frequency distribution II	:	1	4	3	9	2	1

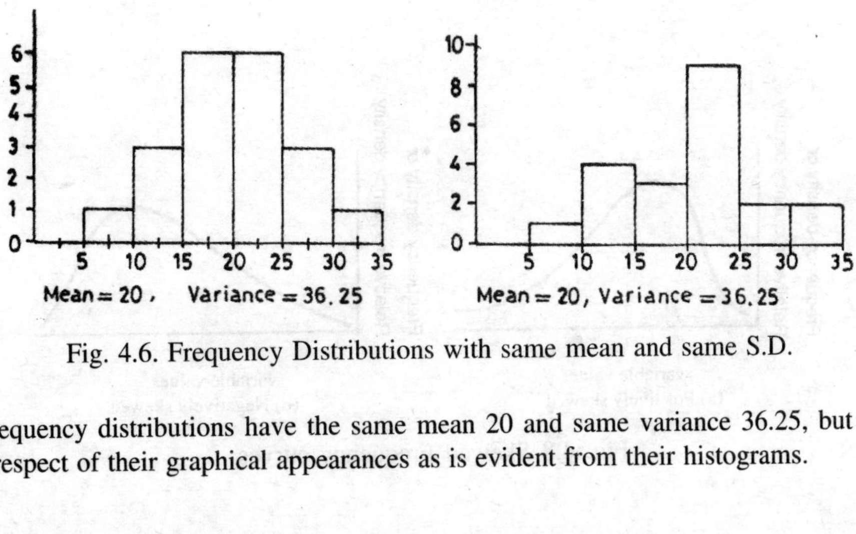

Fig. 4.6. Frequency Distributions with same mean and same S.D.

Both frequency distributions have the same mean 20 and same variance 36.25, but they differ widely in respect of their graphical appearances as is evident from their histograms.

Finally, two other properties, namely, skewness and kurtosis are envisaged to explain this specific feature of the frequency distribution. A detailed account of these properties is given below.

Skewness : Literally skewness means "lack of symmetry" or "departure from symmetry" but the skewness of frequency distribution means the extent of its departure from symmetry.

The frequency distribution of a discrete variable x is said to be symmetrical about the value x_0 if the frequency at the value $x_0 - h$ is the same as the frequency at the value $x_0 + h$, whatever h may be. The frequency curve of a continuous variable is said to be symmetrical about x_0 if the

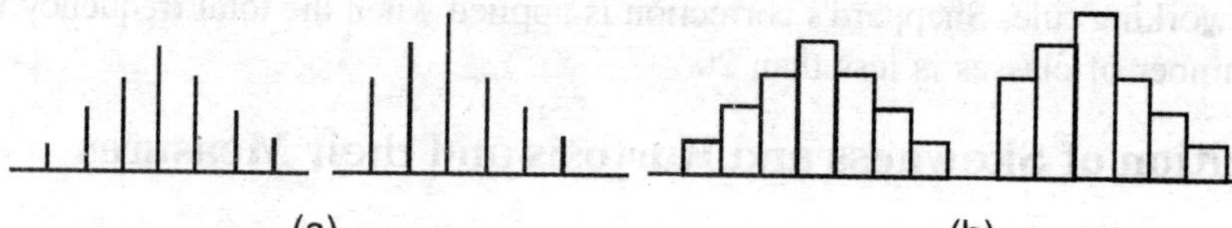

(a) Discrete variable (b) Continuous variable

Fig. 4.7 Symmetrical and asymmetrical frequency distributions

frequency density at $x_0 - h$ is equal to the frequency density at the value $x_0 + h$, whatever h may be.

A distribution which is not symmetrical is called an asymmetrical or skewed distribution. For skewed distributions, frequencies are spread over a longer range of values at either of the ends. The property, skewness, gives an idea of the degree or the extent of departure from symmetry in case of a frequency distribution or curve. The skewness is positive or negative according as the longer tail of the frequency curve lies towards the right or left, i.e., towards higher or lower values of the variable (if frequencies are spread over a longer range of values at right end, the right tail of the frequency curve will be longer).

Positively skewed and negatively skewed frequency curves are shown in Fig. 4.8.

The definition of symmetric frequency distribution implies the following relations :

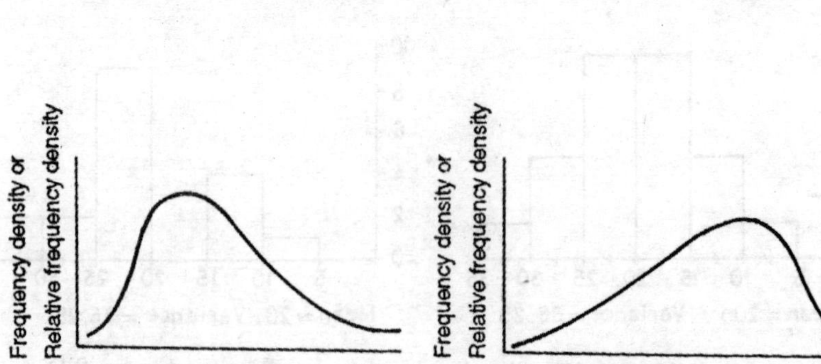

(a) Positively skewed (b) Negatively skewed

Fig. 4.8 Skewed Frequency curves

(i) Mean = Median = Mode

(ii) $Q_2 - Q_1 = Q_3 - Q_2$, the symbols having usual significance.

(iii) All odd order central moments are zero, *i.e.*, $m_{2r+1} = 0, \forall\ r = 0, 1, 2, ...,$

$(i.e.,\ m_1 = m_3 = m_5 = ... = 0)$.

If for a frequency distribution any of the above three conditions is violated, it becomes skewed.

Measures of Skewness : For a positively skewed distribution,

Mean > Median > Mode and $Q_3 - Q_2 > Q_2 - Q_1$.

For a negatively skewed distribution,

Mean < Median < and $Q_3 - Q_2 < Q_2 - Q_1$.

So Absolute Measures of skewness are,

(1) $\bar{x} - M_e$, (2) $\bar{x} - M_0$, (3) $(Q_3 - Q_2) - (Q_2 - Q_1)$ and,

(4) $(P_{90} - P_{50}) - (P_{50} - P_{10})$ or $(D_9 - D_5) - (D_5 - D_1)$

For positively / negatively skewed distributions, the above expressions attain positive/ negative values. By a measure of skewness a relative measure (i.e., unit free measure) is considered and such measures are known as coefficients of skewness, named after their proposers, is given in the following :

I. Karl Pearson's coefficient of skewness

This measure is based on central tendency. The formula for the measure is

$$S_{k_1} = (\bar{x} - M_0) \div s$$

For moderately skewed distribution the empirical relation $\bar{x} - M_0 = 3(x - M_e)$ holds, and as such $S_{k_1}' = 3(\bar{x} - M_e) \div s$ can also be taken as a measure for such distribution.

As $|\bar{x} - M_0|$ and $|\bar{x} - M_e|$ are less than the standard deviation s, so $-1 < S_{k_1} < 1$ and $-3 < S_{k_1}' < 3$.

$$\left(\text{Actually, } |\bar{x} - M_e| = \frac{1}{n}\left|\sum(x_i - M_e)\right| \le \frac{1}{n}\sum|x_i - M_e| \le \frac{1}{n}\sum|x_i - \bar{x}| \le s\right).$$

II. Bowley's measure of skewness

This measure is based on the quartiles, Q_1, Q_2 and Q_3. The formula for the coefficient is,

$$S_{k_1} = \frac{(Q_3 - Q_2) - (Q_2 - Q_1)}{(Q_3 - Q_2) + (Q_2 - Q_1)} = \frac{Q_3 - 2Q_2 + Q_1}{Q_3 - Q_1}$$

$$= \frac{\text{Absolute Measure}}{\text{Quartile Range}}$$

Bowley's coefficient of skewness is useful,

(i) When the frequency distribution has open-end classes or unequal class intervals.

(ii) When mode or mean does not provide a representative value (figure) for the frequency distribution

Bowley's coefficient lies between – 1 and + 1, since each of $(Q_2 - Q_1)$ and $(Q_3 - Q_2)$ is greater than zero, and $|a-b| < |a+b|$

$$\Rightarrow \left|\frac{a-b}{a+b}\right| < 1 \Rightarrow |S_{k_1}| < 1, \text{ taking } a = Q_2 - Q_1 \text{ and } b = Q_3 - Q_2.$$

The serious limitation of Bawley's coefficient of skewness lies in the fact that it is based on the central 50% values of the frequency distribution and ignores the rest 50% values. To overcome the limitation Kelly suggested a measure based on 1st & 9th deciles or 10th & 90th percentiles.

III. Kelly's Coefficient of skewness

This measure is based on diciles or percentiles and the formulae are :

$$Sk_3 = \frac{D_1 + D_9 - 2D_5}{D_9 - D_1}, \text{ or } Sk_3 = \frac{P_{10} + P_{90} - 2P_{50}}{P_{90} - P_{10}}$$

where $D_5 = P_{50}$ = median and $D_9 - D_1 = P_{90} - P_{10}$ = percentile range. It can be shown that $-1 < Sk_3 < 1$.

The values of all the three coefficients are zero, positive and negative for symmetrical, positively skewed and negatively skewed distributions, respectively.

IV. Moment measures of skewness

Relative measures of skewness can also be obtained with the help of moments. In fact, odd order central moments are used for the purpose, the measures of skewness can be made unit free by deflating these odd order central moments by appropriate powers of standard deviation. The simplest of these measures is

$$Sk_4 = g_1 = \frac{m_3}{s^3} = \frac{m_3}{m_2^{2/3}}$$, (x values are not constant) and m_3 being third order central moment.

Also $b_1 = g_1^2 = m_3^2 / m_2^3$ is used, when the sign of skewness is known. $Sk_4 > = < 0$ according as the distribution is positively skewed, symmetrical or negatively skewed, respectively.

Kurtosis : The properties of frequency distributions as described in the earlier sections are not sufficient for comparison of two frequency distributions. Two frequency distributions may have the same mean and same variance and may be equally skewed (i.e., the said distributions may have the same degree of departure from symmetry in the same direction) but still one of the two distributions may be more peaked than the other. In such cases, the peaks are compared against the bell-shaped peak of a normal distribution with same mean and same standard deviation. The degree of peakedness or steepness (deviation from bell shaped peak) is referred to as the amount of Kurtosis existing in the distribution. The Greek word Kurtosis means steepness and thus the distributions are classified under three categories, namely, platykurtic (flat peaked), mesokurtic (bell shaped peak) and leptokurtic (sharp peaked). The three types of distributions are shown below :

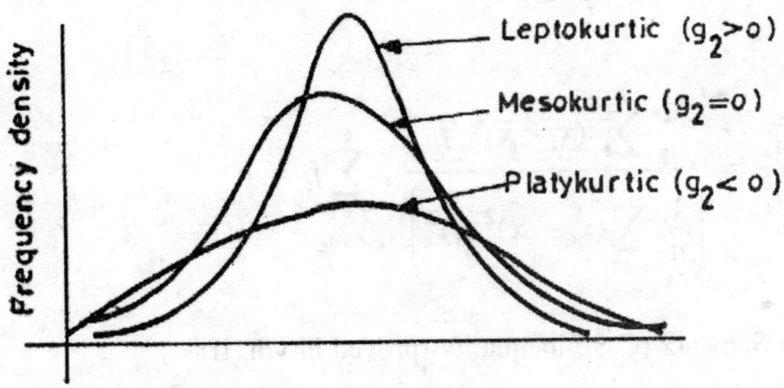

Fig. 4.9 Three symmetrical distributions with same means and variances but with varying degree of peakedness

Thus, kurtosis is a discriminating indicator among symmetric distributions with same means and Standard deviations. The most extensively used measures of kurtosis is the moment measure which is unit free and is defined as $g_2 = \frac{m_4}{m_2^2} - 3$, the notations having usual significance.

For a normal or mesokurtic distribution (discussed in second volume) the value of g_2 is zero. While a positive value of g_2 indicates a high concentration of the values around its central value (with existence of high tails), a negative value of g_2 indicates low concentration of values around the central value (with existence of low tails), the basis of comparison being normal distribution with same mean and same standard deviation in both cases. The situations are depicted clearly in Fig. 4.9 under the three types, leptokurtic ($g_2 > 0$), mesokurtic ($g_2 = 0$) and platykurtic ($g_2 < 0$) distributions, respectively.

The coefficient, $b_2 = \frac{m_4}{m_2^2} = g_2 + 3$ is also used as a measure of kurtosis and is called Pearson's measure of Kurtosis.

A measure of skewness based on b_1 and b_2 coefficients is also available in literature and is defined as

$$Sk_5 = \frac{\sqrt{b_1}\,(b_2+3)}{2(5b_2-6b_1-9)}$$

4.7.1. Some inequalities concerning b_1 and b_2 coefficients :

(a) $b_1 \geq 0$

Proof : $b_1 = \frac{m_3^2}{m_2^3} \geq 0$, as both numerator and denominator are positive.

(*b*) $b_2 \geq 1$

Proof : $$b_2 = \frac{m_4}{m_2^2} = \frac{\frac{1}{n}\sum^{k}(x_i-\bar{x})^4 f_i}{\left(\frac{1}{n}\sum_{i=1}(x_i-\bar{x})^2 f_i)^2\right)}, \quad \sum_{i=1}^{k} f_i = n.$$

Recalling Cauchy Schwarz (C.S.) inequality (proved in vol. II)

$\sum u_i^2 \sum v_i^2 \geq \left(\sum u_i\ v_i\right)^2$, and substituting $u_i = (x_i-\bar{x})^2\sqrt{f_i}$, $v_i = \sqrt{f_i}$, $i = 1, 2, ..., k.$ in the inequality, we get

$$\sum_{i=1}^{k}((x_i-\bar{x})^2\sqrt{f_i})^2\sum_{i=1}^{k}(\sqrt{f_i})^2 \geq \left(\sum(x_i-\bar{x})^2\sqrt{f_i}\sqrt{f_i}\right)^2$$

or $$\frac{1}{n}\sum_{1}^{k}(x_i-\bar{x})^4 f_i \geq \left(\frac{1}{n}\sum(x_i-\bar{x})^2 f_i\right)^2$$

or $$m_4 \geq m_2^2$$

$$\Rightarrow \quad b_2 \geq 1$$

(c) $b_2 \geq b_1$

Proof : To show $\frac{b_2}{b_1} \geq 1$, we have to establish $\frac{m_4}{m_2^2} \geq \frac{m_3^2}{m_2^3}$, or $m_4 m_2 \geq m_3^2$. Recalling C.S. inequality and putting $u_i = (x_i-\bar{x})^2\sqrt{f_i}$ and $v_i = (x_i-\bar{x})\sqrt{f_i}$, for all *i*, we get,

$$\sum_{1}^{k}(x_i-\bar{x})^4 f_i \sum_{1}^{k}(x_i-\bar{x})^2 f_i \geq \left(\sum_{1}^{k}(x_i-\bar{x})^2\sqrt{f_i}\,(x_i-x)\sqrt{f_i}\right)^2$$

or $$\frac{1}{n}\sum_{1}^{k}(x_i-\bar{x})^4 f_i \times \frac{1}{n}\sum(x_i-\bar{x})^2 f_i \geq \left(\frac{1}{n}\sum(x_i-\bar{x})^3 f_i\right)^2$$

$$\Rightarrow m_4 \times m_2 \geq m_3^2 \Rightarrow b_2 \geq b_1$$

(d) $b_2 \geq 1 + b_1$

Proof : Let the variable *x* be standardised as,

$$t_i = \frac{(x_i - x)}{s}, \quad i = 1, 2, ..., k \text{ and } \sum_{i=1}^{k} f_i = n$$

Obviously, $\sum t_i f_i = 0, \sum t_i^2 f_i = n, \sum t_1^3 f_i = n\sqrt{b_1}$

and, $\sum t_i^4 f_i = nb_2$

Now, $$b_2 - b_1 - 1 = \frac{1}{n}\sum t_i^4 f_i - \left(\frac{1}{n}\sum t_i^3 f_i\right)^2 - 1$$

$$= \frac{1}{n^2}\left(n\left(\sum t_i^4 f_i - n\right) - \left(\sum t_i^3 f_i\right)^2\right)$$

$$= \frac{1}{n^2}\left(n\sum (t_i^4 - 1) f_i - \left(\sum t_i^3 f_i\right)^2\right)$$

Now, $$\sum (t_i^4 - 1) f_i = \sum (t_i^2 - 1)^2 f_i$$

as, $$\sum (t_i^2 - 1)^2 f_i = \sum t_i^4 f_i - 2\sum t_i^2 f_i + \sum f_i$$

$$= \sum t_i^4 f_i - 2n + n = \sum t_i^4 f_i - n$$

$$= \sum (t_i^4 - 1) f_i .$$

and, $$\sum t_i^3 f_i = \sum t_i (t_i^2 - 1) f_i ,$$

as, $$\sum t_i (t_i^2 - 1) f_i = \sum t_i^3 f_i - \sum t_i f_i = \sum t_i^3 f_i , \left(\because \sum t_i f_i = 0\right)$$

So, $$b_2 - b_1 - 1 = \frac{1}{n^2}\left(n\sum (t_i^4 - 1) f_i - \left(\sum t_i^3 f_i\right)^2\right)$$

$$= \frac{1}{n^2}\left(\sum t_i^2 f_i \sum (t_i^2 - 1)^2 f_i - \left(\sum t_i (t_i^2 - 1) f_i\right)^2\right) \geq 0$$

(In C.S. inequality substituting $u_i = t_i\sqrt{f_i}$ and

$$\upsilon_i = (t_i^2 - 1)\sqrt{f_i} , \; i = 1, 2, ..., k, \text{ we get}$$

$$\sum t_i^2 f_i \sum (t_i^2 - 1)^2 f_i - \left(\sum t_i (t_i^2 - 1) f_i\right)^2 \geq 0)$$

Thus $b_2 \geq b_1 + 1$ has been proved.

Examples

Example 4.1 compute Mean, Median and Mode for a discrete variable (vide Table 3.5)

Solution

Table — Calculation of mean of number of letters per word

Number of letters per words (x_i)	Frequency (f_i)	($f_i x_i$)	Cumulative Frequencies	
			Less than CF (i)	More than CF^* (i)
1	4	4	4	100
2	14	28	18	96
3	19	57	37	82
4	11	44	48	63
5	9	45	57	52
6	9	54	66	34
7	8	56	74	34
8	8	64	82	26
9	8	72	90	18
10	5	50	95	10
11	2	22	97	5
12	2	24	99	3
13	1	13	100	
Total	$n = 100$	533		

Arithmetic Mean = 533/100 = 5.33 $\cong$ 5 letters per word

Median is the average of $\frac{n}{2}$(= 50)-th and $\left(\frac{n}{2}+1\right)$th (=51$st$) observations.

Since, 48 ($\equiv CF$ (4)) < 50 < 57 ($\equiv CF$ (5)) in less than type cumulative frequencies and 43 ($\equiv CF^*$ (6)) < 50 < 52 ($\equiv CF^*$ (5)) in more than type cumulative frequencies, the median will be 5, *i.e.,* 5 letters per word.

Mode = 3, since 3 apperas with highest frequency in the given series of observation.

Example 4.2 : Compute the Mean, Median and Mode for a continuous variable (vide Table 3.6)

Solution :

Table : Calculation of mean time taken in a funnel experiment.

Class Boundaries (sec)	*Mid— Value (x_i) (sec)	$d_i = x_i$ – 14.45	f_i	$f_i d_i$		Cumulative Frequency Less Than	Cumulative Frequency More Than
9.95 – 10.95	10.45	– 4	4	– 16		4	90
10.95 – 10.95	11.45	– 3	7	– 21		11	86
11.95 – 10.95	12.45	– 2	10	– 20		21	79
12.95 – 10.95	13.45	– 1	14	– 14		35	69
13.95 – 10.95	14.45	0	20	0	– 71	55	55
14.95 – 10.95	15.45	1	12	12		67	35
15.95 – 10.95	16.45	2	9	18		76	23
16.95 – 10.95	17.45	3	7	21		83	14
17.95 – 10.95	18.45	4	4	16		87	7
18.95 – 10.95	19.45	5	3	15		90	3
					82		
Total			$n = 90$		11		

* Mid value of a class = (Sum of the upper and lower class boundaries of a class ÷ 2). It is also called class-mark.

Class width (w) = 1 sec.

Arithmetic mean, $$\bar{x} = \bar{d} + A = \frac{\sum f_i d_i}{n} + 14.45$$

$$= \frac{11}{90} + 14.45 = 14.57 \text{ sec.}$$

Median :

(*a*) Using less than type cumulative frequency and lower class boundary of the median class —

n/2 = 90/2 = 45, So median class is (13.95 sec – 14.95 sec)

$$M_e = l + \frac{\frac{n}{2} - C.F.(l)}{f_{med}} \times w = \left(13.95 + \frac{45-35}{20} \times l\right) \text{ sec}$$

$$= 14.45 \text{ sec}$$

(*b*) Using less than type cumulative frequency and upper class boundary —

$$M_e = u - \frac{C.F.(u) - \frac{n}{2}}{f_{med}} \times w = \left(14.95 + \frac{55-45}{20} \times l\right) \text{ sec}$$

$$= 14.45 \text{ sec}$$

(*c*) Using more than type cumulative frequency and upper class boundary —

$$M_e = L + \frac{CF^*(L) - \frac{n}{2}}{f_{med}} \times w = \left(13.95 + \frac{55-45}{20} \times l\right) \text{sec}$$

$= 14.45$ sec

(*d*) Using more than type cumulative frequency and upper class boundary —

$$M_e = U - \frac{n/2 - CF^*(U)}{f_{med}} \times w = \left(14.95 + \frac{45-35}{20} \times l\right) \text{sec}$$

$= 14.45$ sec

Note that all the four computing formulae for median give the same value, 14.45 sec.

Mode :

Modal class (13.95 sec – 14.95 sec) corresponds to the highest frequency, $f_0 = 20$; f_{-1} = frequency of the perceding class = 14 and f_1 frequency of the following class = 12. Now,

$$M_0 = l + \frac{f_0 - f_{-1}}{2f_0 - f_{-1} - f_1} \times w = \left(13.95 + \frac{20-14}{2 \times 20 - 14 - 12} \times l\right) \text{sec}$$

$= 14.38$ sec

Example 4.3 : (Open end or undefined class intervals).

In the following frequency distribution of population over the different age groups

Age group (years)	below 15	15 – 25	25 – 40	40 – 60	above 60	Total
Population ('000)	20	12	50	30	10	122

Calculate average age, median age and modal age.

Solution :

In the above problem lower limits of the first class and upper limit of the last class are not known. This is a frequency distribution with open end or undefined class intervals. In such cases, we have to assume the corresponding class limits for calculation of the arithmetic mean. But in absence of any knowledge on the characteristics of data, such assumption in regard to the lower and upper class limits does not have much relevance. Owing to this fact use of arithmetic having open end classes or undefined class intervals. However, assuming the class limits the mean value has been calculated below :

Median : Here n/2 = 61, the median class is (25 – 40), since cumulative frequency (l.t.t.) upto this class is 82 and that of the preceding class is 32. So,

$$M_e = \left(25 + \frac{61-32}{50} \times 15\right) = 33.7 \text{ year.}$$

Mode : (25 – 40) year age group is the modal class having highest frequency density.

So, $$M_0 = 25 + \frac{50-12}{100-32-12} \times 15 = 34.83 \text{ year.}$$

Mean : Let us take the terminal class limits as (0 – 15) year and (60 – 75) year respectively. The mid-values are (7.5, 20, 32.5, 50 and 67.5) year with the corresponding frequencies – 20, 12, 50, 30 and 10.

$$A.M. = \frac{\sum x_i f_i}{n} = 4190/122 = 34.3 \text{ year.}$$

The class interval of the terminal classes has been taken as 15 year, since 15 year is the average class interval as determined form the rest of the classes, (10 + 15 + 20)/3 = 15)

Example 4.4 : (Mistake correction)

The marks obtained by 100 students are distributed as follows with an average mark 33.7.

Marks	0–10	10–20	20–30	30–40	40–50	50–60	60–70	70–80	80–90	90–100	Total
No. of Student	10	15	25	18	13	8	5	3	2	1	100

On scrutiny the following mistakes in tabulation are detected for 8 students.

Entry Marks :	16	63	45	68	26	20	18	18
Corrected Marks :	61	36	54	86	36	21	48	8

Calculate the average mark directly and also by correcting the frequency distribution table.

Solution :

I : Direct calculation of average mark

Corrected total marks = Incorrect total marks – total of entry marks + total corrected marks of incorrect entry

= 3370 – 278 + 350 = 3442

Corrected average marks = 3442/100 = 34.42

II : By correcting the frequency distribution table.

Marks	0–10	10–20	20–30	30–40	40–50	50–60	60–70	70–80	80–90	90–100	Total
Mid-value (x_i)	5	15	25	35	45	55	56	57	58	95	
Frequency (f_i)	10	15	25	18	13	8	5	3	2	1	100
Entry Mark (Tally) (E_i)					(3)			(2)			(1)				(2)				
Corrected Mark (Tally) (c_i)		(1)			(1)	|| (2)		(1)		(1)		(1)			(1)		
Corrected Frequency ($F_i = f_i - E_i + C_i$)	11	12	24	20	13	9	4	3	3	1	100
$d_i = (x_i - 45)/10$	–4	–3	–2	–1	0	1	2	3	4	5	
$d_i F_i$	–44	–36	–48	–20	00	9	8	9	12	5	–105

Now $A.M.\ (d) = \bar{d} = \sum d_i F_i \div 100 = -105 \div 100 = -1.05$

$\therefore \quad A.M.\ (x) = \bar{x} = 45 + 10 \times \bar{d} = 45 - 10.5 = 34.50$

The corrected average mark = 34.50

Example 4.5 : (missing frequencies)

The following table shows the distribution of families according to their expenditure. Number of families corresponding to the expenditure groups (10 – 20) and (30 – 40) are missing. Determine the missing frequencies with the help of the following sets of information.

(a) $n = 100$, the total number of families and the average expenditure, $\bar{x} = 25.6$

(b) Median of the distribution = 25 and Mode = 24

(c) $n = 100$ and Mode = 25 and

(d) $n = 101$ and Median = 25.

Expenditure (x, '00 Rs.)	0–10	10–20	20–30	30–40	40–50
Number of Families	14	—	27	—	15

Solution :

Let the number of families in the income groups (10–20) and (30–40) be f_1 and f_2 respectively.

(a) From the given two condition we can write

$$14 + f_1 + 27 + f_2 + 15 = 100$$

$$\Rightarrow f_1 + f_2 = 44 \qquad \text{... (1)}$$

and,

$$5 \times 14 + 15 \times f_1 + 25 \times 27 + 35 \times f_2 + 45 \times 15 = 25.6 \times 100$$

$$\Rightarrow \quad 15f_1 + 35f_2 = 1140 \text{ or, } 3f_1 + 7f_2 = 228 \qquad \text{... (2)}$$

Solving (1) and (2), we get, $f_1 = 20$ and $f_2 = 24$.

(b)

Expenditure	Frequency	Cumulative Frequency (Less than type)
0 – 10	14	14
10 – 20	f_1	$14 + f_1$
20 – 30	27	$41 + f_1$
30 – 40	f_2	$41 + f_1 + f_2$
40 – 50	15	$56 + f_1 + f_2$

Since both median and mode lie in the class (20 – 30), so it is the median as well as the modal class. Thus using the formulae for median and mode respectively we get,

$$25 = 20 + \frac{(56 + f_1 + f_2) \div 2 - (14 + f_1)}{27} \times 10$$

$$= 20 + \frac{10\,(28 + f_2 - f_1)}{27 \times 2} \Rightarrow f_1 - f_2 = 1 \qquad \text{... (3)}$$

Again, $24 = 20 + \frac{27 - f_1}{54 - f_1 - f_2} \times 10 \Rightarrow 3f_1 - 2f_2 = 27$ ... (4)

Solving (3) and (4) we get, $f_1 = 24$ and $f_2 = 25$.

(c) Since mode = 25 and the mid-value for the modal class, (20 – 30) is also 25, the frequencies, f_1 and f_2 of the classes contiguous to the modal class should have a common value.

Thus we have, $f_1 = f_2$ ... (5)

and, $f_1 + f_2 = 100 - 14 - 27 - 15 = 44$ ... (6)

Combining (5) and (6), we ger $f_1 = f_2 = 22$

(d) Median = 25 $\Rightarrow f_1 - f_2 = 1$ (using the formulae for median) ... (7)

Also, $n = 101\ f_1 + f_2 = 45$ ... (8)

Solving (7) and (8), we get, $f_1 = 23$ and $f_2 = 22$

Example 4.6 : (Geometric Mean)

The weighted geometric mean of three numbers 229, 275 and 125 is 203. The weights of 1_{st} and 2_{nd} numbers are 2 and 4 respectively. Find the weight of the third.

Solution :

Let the weight of the third number be w

By definition of geometric mean, we know

$$\text{G.M.} = 203 = (229^2 \times 275^4 \times 125^w)^{1/6+w}$$

$$\therefore \quad \log 203 = \frac{1}{6+w}(2 \log 229 + 4 \log 275 + w \log 125)$$

or $\quad w(\log 203 - \log 125) = 2 \log 229 + 4 \log 275 - 6 \log 203$

or $\quad w(2.3075 - 2.0969) = 2 \times 2.3598 + 4 \times 2.4393 - 6 \times 2.3075$

or $\quad w = 0.6318 \div 0.2106 = 3$

So the weight of the third number is 3.

Example 4.7 : (Mistake correction)

The geometric mean and harmonic mean of 10 observations of a variable were calculated as 16.2 and 16, respectively. It was later found that one of the observations was wrongly recorded as 12.9, in fact, it was 21.9. Calculate the correct G.M. and H.M.

Solution :

By definition of G.M., the product of the 10 original observations is $(16.2)^{10}$. The product of the all correct observation will be,

$$\frac{(16.2)^{10}}{12.9} \times 21.9$$

Thus the corrected geometric mean $= \left(\frac{(46.2)^{10} \times 21.9}{12.9}\right)^{1/10} = G$, say,

Now, $\quad \log G = (10 \log 16.2 + \log 21.9 - \log 12.9) \div 10$

$$= (10 \times 1.2095 + 1.3404 - 1.1106) \div 10 = 1.2325$$

or G = antilog 1.2325 = 17.06

Thus the corrected G.M. = 17.06

Again, the sum of the reciprocal of original observations,

n/H.M. = 10/16 (by definition)

Now sum of the reciprocal of observations after correction

$$= 10/16 - 1/12.9 + 1/21.9$$

So the corrected H.M. $= \dfrac{10}{10/16 - 1/12.9 + 1/21.9} = 16.86$

Example 4.8 : (Miscellaneous averaging)

(a) A motor car covered a distance of 50 miles in four different times - in the first time at 50 mph, in the second time at 20 mph, in the third time at 40 mph and in the fourth time at 25 mph. Calculate the average speed and explain the choice of that average.

(b) A machine is assumed to depriciate 40% in value in the 1st year, 25% in the 2nd year and 10% per annum for the next two years, each percentage being calculated on the diminishing value. What is the average percentage depreciation per year?

(c) A man travelled from one city to another. The distances between the two cities was 4 miles. He drove his car at 40 mph. After travelling one mile by car he travelled 1.5 miles by a bicycle at 10 mph and lastly he covered the remaining distrance on foot at 4 mph. Find the average speed (mph) of that person to cover the whole distance.

(d) The mean age of a combined group of men and women is 30 years. If the mean age of the group of men is 32 and that of the group of the women is 27, find the percentage of men and women in the group.

(e) The average monthly sales for the 1st eleven months of the year by a salesmen was Rs. 12,000, but due to his illness during the last month, the average monthly sale for the whole year came down to Rs. 11,375. What was the amount of last month's sale by that salesman?

(f) In a moderately skewed distribution the arithmetic mean = 24.6 and the mode = 26.1. Find the value of the median.

(g) A cyclist spends 3 hours at an average speed of 8 mph, next 2 hours at 9 mph and the last 4 hours at 2 mph. Find the average speed of the cyclist in the entire journey.

Solution :

(a) Here average speed is to be determined by simple harmonic mean. Since the given different speeds (mph) are accounted for constant distance (50 miles). So the average speed

$$= \frac{1}{\frac{1}{4}(1/50 + 1/20 + 1/40 + 1/25)} = 29.63 \text{ mph}$$

(b) Let the machine value be M and the vareage of successive depreciation rates be 100*x*. So we can wirte

$$M\,(1 - 0.40)\,(1 - 0.25)\,(1 - 0.10)2 = M\,(1 - x)^4$$

$$\Rightarrow \quad x = 1 - (0.60 \times 0.75 \times (0.90)^2)^{1/4}$$

$= 1 - 0.78 = 0.22$

Hence the average depreciation rage

$= 100x = 100 - \text{G.M.}\ ((100 - 40), (100 - 25), (100 - 10), (100 - 10))$

$= 22\%$

(c) Here the average speed is to be determined by weighted harmonic mean, since the given different speeds 40 mph, 10 mph and 4 mph are accounted for (different) distances as 1 mile, 1.5 miles and 1.5 milse respectively. So the average speed

$$= \frac{1}{\frac{1}{4}(1/40 + 1.5/10 + 1.5/4)} = 7.27 \text{ m.p.h.}$$

Here weights are $w1 = 1$, $w_2 = w_3 = 1.5$, $w_1 + w_2 + w_3 = 4$.

(d) Here grand mean $\bar{x} = 30$ year, mean age $= \bar{x}_1 = 32$ year, women's mean age = 27 year. Let number of men be = n_1 and number of women = n_2. We know

$$\bar{x} = \frac{n_1\bar{x}_1 + n_2\bar{x}_2}{n_1 + n_2} \Rightarrow (n_1 + n_2)\,30 = n_1 \times 32 + n_2 \times 27$$

or $\quad 2n_1 = 3n_1$

or $\quad \frac{n_1}{n_2} = \frac{3}{2} = \frac{60}{40}$

So the group has 60% men and 40% women

(e) Let the monthly sale for the last month be x. For the grand mean, we write

$$11{,}375 = (11 \times 12{,}000 + x)/12$$

$\therefore \quad x = 4500$

(f) For a moderately skewed distribution it can be empirically verified that

$$\text{Mean} - \text{Mode} = 3\,(\text{Mean} - \text{Median})$$

$$24.6 - 26.1 = 3\,(24.6 - \text{Median})$$

$\Rightarrow \quad \text{Median} = 25.1.$

(g) Here the average speed is to be determined by weighted arithmetic mean since the given different speeds (mph) are accounted for different times as 3 hrs, 2 hrs and 4 hrs. So the average speed = $(3 \times 8 + 2 \times 9 + 4 \times 4) / (3 + 2 + 4) = 5.56$ mph.

Example 4.9 : (Quartiles, Deciles, Percentiles or Quantiles)

The following table gives the frequency distribution of marks of 65 students.

Mark	0–4	4–8	8–12	12–15	15–18	18–21	21–25	25–	Total
Number of Students	10	12	18	7	5	3	4	6	65
Cumulative Frequency (< type)	10	22	40	47	52	55	59	65	

Calculate,

(i) First and third quartilies (i.e., Q_1 and Q_3)

(ii) The mark below which there are 26 students, and identify the mark as a particular measure of location.

(iii) The mark above which there 15 students, and identify the mark as a particular measure of location.

(iv) Number of students who secured more than 16 marks.

(v) Number of students who secured marks between (10 – 15).

Solution :

(i) Here $n/4 = 65/4 = 16.25$. So the 1st quartile class is (4 – 8), since $10 < 16.25 < 22$ (verified from cumulative frequency row)

$$\therefore Q_1 = 4 + (16.25 - 10)\ 4/12 = 6.08.$$

Again, $3n/4 = 48.75$. So the 3rd quartile class is (15 – 18), since $47 < 48.75 < 52$.

$$\therefore Q_3 = 15 + 3\ (48.75 - 47)/5 = 16.05.$$

(ii) Since $22 < 26 < 40$, the required mark will lie in the class (8 – 12) and by interpolation the mark will be determined as,

$$8 + (12 - 8) \times (26 - 22)/18 = 8.89 \cong 9.$$

So below 9 there are 26 students and since $26 \div 65 = 0.4$, the mark 9 can be identified as the 4th decile, D_4.

(iii) Here below the required mark there are 50 (= 65 – 15) students and since $47 < 50 < 52$, the mark will lie in the class (15 – 18). By interpolation, the required mark is $15 + (18 - 15) \times (50 - 47)/5 = 16.8$. So above 16.8 = 17, there are 15 students and since $50 \div 65 = 0.77$. The mark 17 can be regarded as the 77th percentile, P_{77}.

(iv) The mark 16 lies in the class (15 –18).

So, the number of students who have obtained more than 16 marks = number of students who have obtained more than 18 marks + number of students whose marks lie between (16 – 18) $= n_1 + n_2$. say, Now $n_1 = 3 + 4 + 6 = 13$ (from the given frequency distribution table). and,

$$n_2 = \frac{\text{frequency of the class}(15-18)}{\text{Class width}} \times (18-16)$$

$$= 5/3 \times 2 = 3.3$$

So the number of students = 13+3.3=16.3 = 16 (approx.)

(v) The mark 10 lies in the class (8 – 12) and the mark 15 is the upper class boundary of the class (12 – 15). So the number of students who have obtained marks lying between 10 and 15.

$= N_1 + N_2$, say, where

$$N_1 = \text{frequency of the class } (12 - 15) = 7$$

N_2 = number of students who have obtained marks lying between 10 and 12 = (8/4) (12–10) = 9

Hence the number of students = 7 + 9 = 16.

Example 4.10 Prove that arithmetic mean and median of first n natural numbers are equal.

Solution :

$$A.M = \frac{1+2+...+n}{n} = \frac{n+1}{2}$$

$$Median = \left(\frac{n+1}{2}\right) - th\ observation = \frac{n+1}{2}, if\ n = odd$$

$$= A.M. of\ \frac{n}{2}\ th\ and\ \left(\frac{n}{2}+1\right) - st\ observation$$

$$= \frac{\frac{n}{2}+\left(\frac{n}{2}+1\right)}{2}, if\ \ n\ is\ even$$

$$= \frac{n+1}{2}$$

So, for first n natural numbers, mean = median $= \frac{n+1}{2}$, whether n is even or odd.

Example 4.11 : Calculate the weighted mean of the first n natural numbers, the weights of the integer i being *(a)* i, *(b)* i – 1 and *(c)* i + 1.

Solution :

(a) Here $x_i = i$ and $f_i = i$, for all $i = 1, 2, ..., n$

So, $$\sum x_i f_i = \sum_{i=1}^{n} i^2 = \frac{n(n+1)(2n+1)}{6}$$

$$\therefore \quad \bar{x} = \frac{n(n+1)(2n+1)}{6 \times n} = \frac{(n+1)(2n+1)}{6}$$

(b) Here $x_i = i$ and $f_i = i - 1$, for all $i = 1, 2 ...., n$

So, $$\sum f_i x_i = \sum i(i-1) = \sum i^2 - \sum i = \frac{n(n+1)(2n+1)}{6} - \frac{n(n+1)}{2}$$

$$\therefore \bar{x} = \frac{(n+1)(2n+1)}{6} - \frac{n+1}{2} = \frac{n^2-1}{3}$$

(c) Here $x_i = i$ and $f_i = i + 1$, for all $i = 1, 2, ... n$

So, $$\sum x_i f_i = \sum i(i+1) = \frac{n(n+1)}{2}\left(\frac{2n+1}{3}+1\right)$$

$$\therefore \quad \bar{x} = \frac{(n+1)(n+2)}{3}$$

Example 4.12 : In a group of wage earners, 20%, 40%, 70% and 80% of the wage earners receive less than Rs. 50, Rs. 120, Rs. 300 and Rs. 350, respectively and 5% above Rs. 400. Calculate the A.M. assuming the range of salary as Rupee 0 to 500.

Solution : Let the total number of wage earners be n. Based on the given information the following table can be prepared.

Wage Class (Rs.)	No. of persons	Class boundary	Frequency (fi)	Midvalue (xi)
Below 50	0.2 n	0 – 50	0.2 n	25
Below 120	0.4 n	50 – 120	0.2 n	75
Below 300	0.7 n	120 – 300	0.3 n	210
Below 350	0.8 n	300 – 350	0.1 n	325
Above 400	0.05 n	350 – 400	0.15 n	375
		400 – 500	0.5 n	450
		Total	n	

Now $\sum x_i f_i = 194.25n,$ $\quad\therefore$ A.M. = 194.25 n ÷ n = Rs. 194.25

Example 4.13 Let x be a variable assuming the values 1, 2, ... n and let $F_1, F_2, ..., F_n$ be the corresponding more than cumulative frequencies. Show that

$$\bar{x} = \sum_{i=1}^{n} F_i \div F_1$$

Solution :

Variable (x_i)	Cumulative Frequency (more than type)	Frequency (*fi*)	*xi fi*
1	F_1	F_1-F_2	F_1-F_2
2	F_2	F_2-F_3	2 (F_2-F_3)
3	F_3	F_3-F_4	3(F_3-F_4)
4	F_4	F_4-F_5	4 (F_4-F_5)
.			
.			
.			
$n-2$	F_{n-2}	$F_{n-2} - F_{n-1}$	$(n-2)(F_{n-2} - F_{n-1})$
$n-1$	F_{n-1}	$F_{n-1} - F_n$	$(n-1)(F_{n-1} - Fn)$
n	F_n	F_n	nF_n
Total		F_1	$F_1 + F_2 + ... + Fn$

$\therefore \quad \bar{x} = \sum x_i f_i \div n = \sum F_i \div F_1$ (Proved)

Example 4.14 : The age distribution of 4,488 Bengalee males is given below :

Age (yr)	Frequency	Age (yr)	Frequency	Age (yr)	Frequency
0	156	5 – 9	472	40 – 49	522
1	121	10 – 14	434	50 – 59	370
2	111	15 – 14	407	60 – 69	213
3	106	20 – 24	383	70 – 79	80
4	103	25 – 29	357	80 – 89	11
		30 – 34	335	90 – 99	1
		35 – 39	306	Total	4488

Compute mean age from the grouped means and also directly.

Solution : The given age distribution be classified into 3 groups : Gr 1 - ungrouped age distribution, Gr II grouped age distribution with class width 5 yr and Gr. III grouped age distribution with class width 10 yr.

For Gr - I $\quad \bar{x}_1 = \sum_1 x_i f_i / \sum_1 f_i = 1073/597\, yr = 1.7973\, yr$

For Gr - II $\quad \bar{x}_2 = \sum_2 x_i f_i / \sum_2 f_i = 55538/2694\, yr = 20.6154\, yr$

For Gr - III $\quad \bar{x}_3 = \sum_3 x_i f_i / \sum_3 f_i = 64116.5/1197\, yr = 53.5643\, yr$

The grand mean, $\bar{x} = \dfrac{597 \times 1.7973 + 2694 \times 20.6154 + 1197 \times 53.5643}{597 + 2694 + 1197}\, yr$

$= 26.90$ yr

By direct calculation,

we have, $\sum x_i f_i = 120727.5$

$\sum f_i = 4488$

$\therefore \quad \bar{x} = 120727.5/4488 = 26.90 yr.$

Example 4.15 : Calculate all measures of dispersion, all measures of skewness and all measures of kurtosis on the discrete data given in the Table 3.5.

Solution :

Table – Calculation of Measures of Dispersion and other Measures on Data on Number of Letters per Word

Number of Letters per word (xi)	*Frequency (f_i)*	xf_i	$x_i^2 f_i$	$x_i^3 f_i$	$x_i^4 f_i$	$(x_i+1)^4 f_i$*	$\lvert x_i - \bar{x}\rvert$	*Cumulative Frequency (≤)*	$\lvert x_i - \bar{x}\rvert f_i$
1	4	4	4	4	4	64	4.33	4	17.32
2	14	28	56	112	224	1134	3.33	18	46.62
3	19	57	171	513	1539	4864	2.33	37	44.27
4	11	44	176	704	2816	6875	1.37	48	14.63
5	9	45	225	1125	5625	11664	0.33	57	2.97
6	9	54	324	1944	11664	21609	0.67	66	6.03
7	8	56	392	2744	19208	32768	1.67	74	13.36
8	8	64	512	4096	32768	52488	2.67	82	21.36
9	8	72	648	5832	52488	80000	3.67	90	29.36
10	5	50	500	5000	50000	73205	4.67	95	23.35
11	2	22	242	2662	29282	41472	5.67	97	11.34
12	2	24	288	3456	41472	57122	6.67	99	13.34
13	1	13	169	2197	28561	38416	7.67	100	7.67
Total	n =100	533	3707	30389	275651	421681			251.62

* This column is required for earlier check.

Here $\bar{x} = \sum x_i f_i / n = 533/100 = 5.33 \, and \, Mode = M_0 = 3$

Now, $\sum (x_i+1)^4 f_i = \sum x_i^4 f_i + 4\sum x_i^3 f_i + 6\sum x_i^2 f_i + 4\sum x_i f_i + \sum f_i$

= 275651 + 4 × 30389 + 6 × 3707 + 4 × 533 + 100

= 421681 (checked)

Range = 13 – 1 = 12

$MD_{\bar{x}} = 251.62/100 = 2.52 = Mean\, deviation\, about\, \bar{x}$

Variance $= \dfrac{\sum x_i^2 f_i}{n} - \left(\dfrac{\sum x_i f_i}{n}\right)^2$ = 3707/100 – (533/100)^2 = 8.66 = m_2

Standard deviation (S.D.) = $\sqrt{8.66} = 2.94 = s$.

Third Moment, $m_3 = \frac{1}{n}\sum (x_i - \bar{x})^3 f_i$

$$= \left(\sum x_i^3 f_i - 3\bar{x}\sum x_i^2 f_i + 2\times n\times \bar{x}^3\right) \div 100$$

$$= \left(30389 - 3\times 5.33\times 3707 + 2\times 100\times 5.33^3\right)\div 100$$

$$= 13.98$$

Fourth Moment, $m_4 = \frac{1}{n}\sum (x_i - \bar{x})^4 f_i$

$$= \frac{1}{100}\left(\sum x_i^4 f_i - 4\bar{x}\sum x_i^3 f_i + 6\bar{x}^2\sum x^2 f_i - 3n\bar{x}^4\right)$$

= (275651 – 4 × 5.33 × 30389 + 6 x 5.33^2 × 3707 – 3 × 100 × 5.33^4) ÷ 100 =175.09.

1st Quartile, Q_1 = 25th observation

= 3 (from cumulative frequency column)

2nd Quartile, $Q_2 = \dfrac{\left[\frac{n}{2}th + \left(\frac{n}{2}+1\right)th\right] observation}{2}$

$$= \frac{[50th + 51st]\,observation}{2} = (5+5)/2 = 5 = Median$$

3rd Quartile $Q_3 = \frac{3n}{4}th(=75th)\,observation$

= 8

Measure of skewness based on central values :

$$Sk_1 = \frac{\bar{x} - M_0}{s}, \; or \; Sk_1 = \frac{3(\bar{x} - M_e)}{s} = 0.79 \text{ or } = 3 \times 0.33 / 2.94 = 0.34$$

Measured of skewness based on quartiles :

$$Sk_2 = \frac{Q_3 - 2Q_2 + Q_1}{Q_3 - Q_1} = 0.20$$

Kelly's coefficient of skewness (based on percentiles or deciles) :

$$Sk_3 = \frac{D_1 + D_9 - 2D_5}{D_9 - D_1} = \frac{P_{10} + P_{90} - 2P_{50}}{P_{90} - P_{10}} = (2+9-10)/(9-2) = 1/7 = 0.14$$

Measure based on moments :

$$Sk_4 = \frac{m_3}{m_2^{3/2}} = g_1 = \frac{13.98}{\sqrt{8.66^3}} = \frac{13.98}{25.48} = 0.55$$

Coefficient of kurtosis, $g_2 = \frac{m_4}{m_2^2} - 3 = -0.66$

Since the frequency distribution is severely positively skewed, calculation of coefficient of kurtosis is meaningless.

Now the coefficient of skewness based on g_1 and g_2 is

$$Sk_5 = \frac{\sqrt{b_1}(b_2+3)}{2(5b_2 - 6b_1 - 9)}, \text{ where } b_1 = g_1^2 \text{ and } b_2 = g_2 + 3$$

$$= 2.93/1.70 = 1.72$$

Example 4.16 : Calculate the measures of dispersion, the measures of skewness and the measure of kurtosis on the continuous data given in the Table 3.6.

Solution :

Table – Calculation of Measures of Dispersion and other Measures on Data from Funnel experiment

Mid value l x_i	Frequency f_i	$d_i = x_i - 14.45$	$d_i f_i$	$d_i^2 f_i$	$d_i^3 f_i$	$d_i^4 f_i$	$(d_i+1)^4 f_i$	$\lvert d_i - \bar{d}\rvert f_i$	cumulative Frequency (≤ type)
10.45	4	–4	– 16	64	– 256	1024	924	16.48	4
11.45	7	– 3	– 21	63	– 189	567	112	21.84	11
12.45	10	– 2	– 20	40	– 80	160	10	21.20	21
13.45	14	–1	– 14	14	– 14	14	0	15.68	35
14.45	20	0	0	0	0	0	20	2.40	55
15.45	12	1	12	12	12	12	192	10.56	67
16.45	9	2	18	36	72	144	729	16.92	76
17.45	7	3	21	63	184	367	1792	20.16	83
18.45	4	4	16	64	256	1024	2500	15.52	87
19.45	3	5	15	75	375	1875	3888	14.64	90
Total	n = 90		11	431	365	5389	9567	155.40	

Mean $= \bar{x} = \bar{d} + 14.45 = 11/90 + 14.45 = 14.57$ sec

Mode : Modal Class is (13.95 Sec – 14.95 sec)

Mode = 14.38 sec (from Example 4.2)

Q_2 = Median = 14.45 sec

Check : $\sum (d_i+1)^4 f_i = \sum d_i^4 f_i + 4\sum d_i^3 f_i + 6\sum d_i^2 f_i + 4\sum d_i f_i + \sum f_i$

$= 5387 + 4 \times 365 + 6 \times 431 + 4 \times 11 + 90$

$= 9567$ (checked)

$$MD_{\bar{x}} = \frac{1}{n}\sum |x_i - \bar{x}| f_i = \frac{1}{n}\sum |d_i - \bar{d}| f_i = 155.40/90 = 1.73 \text{sec}$$

Variance : $m_2 = s^2 = \frac{1}{n}\sum (x_i - \bar{x})^2 f_i = \frac{1}{n}\sum (d_i - \bar{d})^2 f_i$

$$= \frac{1}{n}\sum d_i^2 f_i - \bar{d}^2 = \frac{431}{90} - (0.12)^2$$

$$= 4.77 \text{ sec}$$

Standard Deviation (S.D.)

$$= s = \sqrt{4.77} = 2.18 \text{ sec}$$

Moments : $m_3 = \frac{1}{n}\sum (x_i - \bar{x})^3 f_i = \frac{1}{n}\sum (d_i - \bar{d})^3 f_i$

$$= \frac{1}{n}\left(\sum d_i^3 f_i - 3\bar{d}\sum d_i^2 f_i + 2n\bar{d}^3\right)$$

$$= \left(365 - 3\times 0.12\times 431 + 2\times 90\times (0.12)^3\right) \div 90$$

$$= 2.34 \text{ sec}$$

$$m_4 = \frac{1}{n}\sum (x_i - \bar{x})^4 f_i = \frac{1}{n}\sum (d_i - \bar{d})^4 f_i$$

$$= \frac{1}{n}\left(\sum d_i^4 f_i - 4\bar{d}\sum d_i^3 f_i + 6\bar{d}^2\sum d_i^2 f_i - 3n\bar{d}^4\right)$$

$$= \left(5387 - 4\times 0.12\times 365 + 6\times (0.12)^2\times 431 - 3\times 90\times (0.12)^4\right) \div 90$$

$$= 58.32 \text{ sec}$$

1st Quartile :

Here n/4 = 22.5 , ∴ 1st Quartile class is (12.95 – 13.95)

$$\therefore Q_1 = 12.95 + \frac{22.5 - 21}{14}\times 1 = 13.05 \text{ sec}$$

3rd Quartile :

Here 3n/4 = 67.5, ∴ 3rd Quartile class is (15.95 – 16.95)

$$\therefore Q_3 = 15.95 + \frac{67.5 - 67}{9}\times 1$$

$$= 16.01 \text{ sec}$$

Measure of skewness based on central values :

$$Sk_1 = \frac{\bar{x} - M_0}{s}, \text{ or } Sk_1 = \frac{3(\bar{x} - M_e)}{s}$$

$$= 0.09, \text{ or } = 3(14.57 - 14.45)/2.18 = 0.16$$

Measure of skewness based on quartiles :

$$Sk_2 = \frac{Q_3 - 2Q_2 + Q_1}{(Q_3 - Q_1)} = (16.01 - 2\times 14.45 + 13.05)/(16.01 - 13.05) = 0.16/2.96$$
$$= 0.05$$

Kelly's coefficient of skewness (based on percentiles or deciles) :

$$Sk_3 = \frac{D_1 + D_9 - 2D_5}{D_9 - D_1} = \frac{P_{10} + P_{90} - 2P_{50}}{P_{90} - P_{10}}$$

Now, $P_{10} = 10.95 + \frac{90\times 0.10 - 4}{7}\times 1 = 11.66$ ($\because$ 16.95 - 17.95 is 9th decile or 90th percentile class).

Thus, $Sk_3 = (11.66 + 17.66 - 2\times 14.45)/6 = 0.7$

Measures of skewness based on moments :

$$Sk_4 = \frac{m_3}{m_2^{3/2}} = g_1 = \frac{2.34}{\sqrt{10.853}} = 0.22$$

Coefficient of kurtosis, $g_2 = \frac{m_4}{m_2^2} - 3 = 58.32/22.75 - 3 = -0.44$

Coefficient of skewness based on g_1 and g_2 :

$$Sk_3 = \frac{\sqrt{b_1}(b_2+3)}{2(5b_2 - 6b_1 - 9)}, \quad \text{where } b_1 = g_1^2 \text{ and } b_2 = g_2 + 3$$

$$= (0.22 \times 5.56) / 2\,(5 \times 2.56 - 6 \times 0.0484 - 9) = 1.2232/7.0192 = 0.17$$

Example 4.17 : For a grouped frequency distribution let x_i and f_i be the mid-value and class frequency of the i-th class, respectively and there are k such classes.
Then show that,

$$n\ MD_{\bar{x}} = s_2 - s_1 + \bar{x}(n_1 - n_2).$$

where $s_1 = \sum_{x_i<\bar{x}} x_i f_i, s_2 = \sum_{x_1>\bar{x}} x_i f_i, n = \sum_1^n f_i$

$n_1 = \sum_{i=1}^{k_1} f_i$ = sum of $f_i^{'}$ s of those classes in which $x_i < \bar{x}$ and $n_2 = \sum_{i=k_1+1}^{k} f_i$ = sum of $f_i^{'}$ s of those classes in which $x_1 > \bar{x}$

Solution :

For grouped frequency distribution, we know

$$nMD_{\bar{x}} = \sum_{i=1}^{k} |x_i - \bar{x}| f_i = \left(\sum_{x_i<\bar{x}}^{1} + \sum_{x_i>\bar{x}}^{2} \right) |x_i - \bar{x}| f_i$$

$$= \sum_{x_i<\bar{x}}^{1} (\bar{x} - x_i) f_i + \sum_{x_i>\bar{x}}^{2} (x_i - \bar{x}) f_i$$

$$= \bar{x} \sum_1 f_i - \sum_1 x_i f_i + \sum_2 x_i f_i - \bar{x} \sum_2 f_i$$

$$= \bar{x} n_1 - s_1 + s_2 - \bar{x} n_2 = s_2 - s_1 + \bar{x}(n_1 - n_2)$$

Example 4.18 : The following figures relate to frequencies of several classes in the distribution of scores of 328 students in certain examination. The classes are of equal width and are arranged in ascending order of magnitude. The arithmetic mean and standard deviation are given to be 48 and 11.8092 respecetively. Identify the classes of the frequency distribution assuming the class frequencies as 10, 18, 22, 30, 45, 57, 48, 46, 31, 14, 5, 2, respectively for 12 classes.

Solution :

Let $x_1, x_2, \ldots, x_{12}$ be the mid-values of the given 12 classes with class with w. So $x_1, x_2, \ldots, x_{12}$ are in A.P. with common difference w. Thus, $x_2 = x_1 + w$, $x_3 = x_1 + 2w$, ..., $x_{12} = x_1 + 11w$.

$$\text{The arithmetic mean} = \frac{10x_1 + 18(x_1 + w) + 22(x_1 + 2w) + \ldots + 2(x_1 + 11w)}{10 + 18 + 22 + \ldots + 2}$$

$$= x_1 + 1673\, w/328 = x_1 + 5.1\, w$$

The (standard deviation)^2

$$= \frac{1}{328}\left(10x_1^2 + 18(x_1 + w)^2 + 22(x_1 + 2w)^2 + \ldots + 2(x_1 + 11w)^2\right) - (48)^2$$

$$= x_1^2 + \left(10363\, w^2 + 3346\, x_1\, w\right)/328 - (48)^2$$

According to the given condition,

$$x_1 + 5.1\, w = 48 \qquad .....(1)$$

$$\text{and } x_1^2 + 31.59\, w^2 + 10.20\, x_1\, w - (48)^2 = (11.8092)^2 \qquad .....(2)$$

Now putting the value of x_1 (= 48 – 5.1 w) from (1) in the equation(2) we get,

$$0.04\, w^2 + 0.0004\, w - 1 = 0$$

$$w = \frac{-0.0004 \pm \sqrt{(0.0004)^2 + 4 \times 0.04}}{2 \times 0.04} = \frac{0.3996}{0.08} = 4.995 = 5$$

$\therefore$ x_1= 22.5 (from (1) and the 12 classes are, 20 –25, 25 –30, 30 –35, ..., 70 –75, 75 – 80.

Example 4.19 : Prove that for a variable assuming only two values x_1 and x_2, equal number of times posseses,

$$mean = \frac{x_1+x_2}{2},\ m_2 = \left(\frac{x_1-x_2}{4}\right)^2,\ m_3 = 0 \ and \ m_4 = \frac{(x_1-x_2)^4}{16}$$

Also, calculate x_1 and x_2, if mean and standard deviation (S.D.) are 30 and 4, respectively.

Solution :

Let each of the variable values occur m number of times.

So, mean = $(x_1 \times m + x_2 \times m) / 2m = (x_1+x_2)/2$,

$$m_2 = \frac{\left(x_1 - \frac{x_1+x_2}{2}\right)^2 \times m + \left(x_2 - \frac{x_1+x_2}{2}\right)^2 \times m}{2m}$$

$$= \frac{(x_1-x_2)^2 \frac{m}{4} + (x_2-x_1)^2 \frac{m}{4}}{2m}$$

$$= \frac{2m(x_1-x_2^2)}{2m\times 4} = \frac{(x_1-x_2)^2}{4},$$

$$m_3 = \frac{\left(x_1 - \frac{x_1+x_2}{2}\right)^3 m + \left(x_2 - \frac{x_1+x_2}{2}\right)^3 m}{2m}$$

$$= \frac{(x_1-x_2)^3 m - (x_1-x_2)^3 m}{2m} = 0$$

and $$m_4 = \frac{\left(x_1 - \frac{x_1+x_2}{2}\right)^4 m + \left(x_2 - \frac{x_1+x_2}{2}\right)^4 m}{2m}$$

$$= \frac{(x_1-x_2)^4 \frac{m}{16} + (x_2-x_1)^4 \frac{m}{16}}{2m}$$

Substituting the given values,

$$Mean = \frac{x_1+x_2}{2} = 30 \Rightarrow x_1 + x_2 = 60 \quad \text{....(1)}$$

and $$\text{S. D.} = \sqrt{m_2} = \frac{x_1-x_2}{2} = 4 \Rightarrow x_1 - x_2 = 8 \quad \text{....(2)}$$

$x_1 = 34$ and $x_2 = 26$.

Exercise and Class Projects :

4.1 What do you understand by 'central tendency' of a distribution. Describe its 'measures'.

4.2 Under what conditions median is more sutiable than other measures of central tendency. Note such conditions for other meausres also.

4.3 What are the properties of a good average? Examine these properties with reference to Arithmetic Mean, Geometric Mean and Harmonic Mean and give an example of situations in which each is more appropriate measure for computing the average. (Calicut, U. B. Sc. 1981)

4.4 What are the desiderata of a good measure of central tendency ? Compare mean, median and mode in the light of these desiderata. Why the averages are called measures of central tendency. (B. Com., Bombay, 1970)

4.5 Explain mean, median and mode as a measure of central tendency. Give a comparative study of their relative advantages and disadvantages. (C. U. B. Com. 1991)

4.6 Explain with suitable examples the term 'dispersion'. State the relative and absolute measures of dispersion and describe the merits and demerits of standard deviation. (Bombay U., B.Sc. 1979)

4.7 What are requisites of a good measure of dispersion ? In the light of these, comment on some of the well-known measures of dispersion. (I. C. W. A. 1967)

4.8 Define 'mean deviation'. How does it differ from 'standard deviation'?(C.A. 1968; H.S. 1991)

4.9 What are the various measures of dispersion ? Explain them with their uses.

4.10 Discuss briefly the relative merits and demerits of various measures of dispersion. What is the difference between 'variance' and 'variation'?

4.11 What are quartiles ? How are they used for measuring dispersion?

4.12 What is Lorenz Curve? How is it drawn ? In what way does it help in studying variation/ concentration of certain variable values ? What is the Gini's mean difference ? How Gini's mean difference has been used to find the area of concentration in Lorenz Curve ? Illustrate with the help of an example.

4.13 What is the 'coefficient of variation'? What purpose does it serve ? Also distinguish between 'variance' and 'coefficient of variation'. (B. Com Bombay 1973)

4.14 Define raw and central moments of a frequency distribution. Obtain the relation between the central moments of order and in terms of the raw moments. What are Sheppard's corrections to the central moments. (Calicut U., B.Sc. 1977, Mysore U., B.Sc. 1979)

4.15 What is skewness ? How does it differ from dispersion ? Describe the various measures of skewness. (B. Com Nagpur 1968)

4.16 Compare the different measures of skewness.

4.17 Define kurtosis. How do we measure it? How does the measure of kurtosis help in understanding the shape of a frequency distribution with known mean and standard deviation ?

4.18 Compute the mean, median and mode of the following data on the number of seeds germinated abont 10 seeds kept on a damp filter paper for 100 such filter papers.

x (No. of seeds germinated :	0	1	2	3	4	5	6	7	8	9	10
f (No. of filter papers) :	7	23	37	15	9	6	2	1	0	0	0

Check whether the empirical relation among mean, median and mode holds and comment, and also give textual representation of the given frequency distribution table.

4.19 The expenditure of 100 familes is given below with two missing frequencies.

Expenditure (Rs. '00) : 0 – 10, 10 – 20 20 – 30 30 – 40 40 – 50

No. of families : 14 – 27 – 15

Mode of the distribution is 24 Calculate missing frequencies. Calculate different measures of skewness and kurtosis. Graphically determine the partition values : Q_1, Q_2 and Q_3. Determine the number of families whose expenditure are, (a) less than Rs. 1900, (b) between Rs. 2200 and Rs. 4100 and (c) more than Rs. 3200.

4.20 If G_x is the geometric mean of n x' s and G_y is the geometric mean of n y's., then the geometric mean (G) of 2n observations will be, $G=\sqrt{G_x\, G_y}$.

4.21 Let x be a variable assuming the values 1, 2, ..., n and $F_1', F_2', ..., F_n'$ be the corresponding less than type cumulative frequencies. Show that

$$\bar{x} = \sum_{i=1}^{n} F_1' \div F_n'$$

4.22 Determine the mean deviation about mean, the mean deviation about median standard deviation, coefficient of variation, measures of skewness and kurtosis, and also draw Lorenz Curve from the following frequency distribution table.

Income Distribution of a Factory Workers

Income (Rs.)	Numbers of workers
300 – 309	9
310 – 319	20
320 – 329	24
330 – 339	38
340 – 349	48
350 – 359	27
360 – 369	17
360 – 379	6

4.23 Given below is the distribution of 140 candidates obtaining marks x or higher in a certain examination. Calculate mean, median, mode, standard deviation, all the measures of skewness and kurtosis.

x :	10	20	30	40	50	60	70	80	90	100
C. F.	140	133	118	100	75	45	25	9	2	0

4.24 In a sample of observations, $x_1, x_2, ..., x_n$ it is found that one of the values, x_1 is incorrect and its correct value was x_1'. Show that the adjustment to the variance to correct this error is

$$\frac{1}{n}(x' - x_1)\left(x_1' + x_1 - \frac{x_1' - x_1 + 2n\,\bar{x}}{n} \right)$$

4.25 Scores of two golfers for 24 rounds were as follows :

Golfer A : 74, 75, 78, 72, 78, 77, 79, 78, 81, 76, 72
77, 74, 70, 78, 79, 80, 81, 74, 80, 75, 71, 73

Golfer B : 86, 84, 80, 88, 89, 85, 86, 82, 82, 79, 86, 80
82, 76, 86, 89, 87, 83, 80, 88, 86, 81, 84, 87

Find which golfer may be considered to be a more consistent player ?

Class Projects :

Generally, we work out the numerical exercises on the basis of the data collected by others. To develop idea of variability and uncertainty existing in nature, create your own data-bank in the field of your interest. We suggest the following few cases :

1. Once while you were waiting on a railway platform for a passenger train, a goods train with oil racks passed by. You noted down tare weight of the oil racks. From the noted observations you have an idea of average tare weight of oil rack with certain variability. Represent those data graphically. Perform all types of statistical treatments, you know, on the data.
2. Suppose you want to know the average number of automobiles passing through a busy road per minute. For this purpose on 30 different days of a year for 10 minutes on each day, observations on the number of automobiles passing by are to be noted. For the selection of 30 days from 365 days in a year and the selection of 10 minutes from 1440 such units in a day, sampling technique should be used. Since sampling technique has not yet been discussed, you may use your common sense in selecting the days with the corresponding times. Present the data, so obtained, in a table, on a graph paper and perform all statistical treatments, you know so far, on the data.

Objective Questions

1. Examine the validity of the following statements :

(a) For a set of 8 observations A.M., G.M., H.M., are 5.2, 6.3 and 7.1, respectively.

(b) Geometric mean of a given number of values cannot be obtained representatively if one of the values is zero.

(c) The mean depth of water of a river is 130 cm., therefore a man with a height of 165 cm can cross the river safely.

(d) For the wholesale manufacturer interested in selecting the right type of goods for manufacturing, he has to take median of the available data in respect of different goods as the most appropriate measure.

(e) A.M. = 25, G.M. = 14, and H.M. = 9 for two positive values of a variable.

(f) The geometric mean of 2, 4, 16 and 32 is 16.

(g) The mean and median of 100 items are 50 and 52, respectively. It was later found that the value of the largest item is recorded as 100 instead of 110. So the corrected mean and median will be 50.1 and 52 respectively.

(h) The algebraic sum of the deviations of 20 observations measured from 30 is 2. So the mean of the observations is 32.

(i) One quintal of wheat was purchased at 0.8 kg per rupee and another quintal at 1.2 kg per rupee. So the average rate per rupee is 1 kg.

(j) The number of deaths among military people in the recent war was 10 out of 1000 while the number of deaths among civilians in Calcutta was 18 out of 1000 people. Hence it is safe to join military service than to live in Calcutta as a civilian.

(k) A fixed number of workers in a small scale factory produced 40000 candles and 38000 candles in the months of January '90 and February '90, respectively. So the workers will get more incentives in January.

(l) Twelve persons gambles on a certain night. Seven of them lost on a average Rs. 10.50 while the remaining five gained on an average Rs. 14.70 (assuming loss incurred by one person is the gain of others).

(m) The interest rates 3% on Rs. x, 4% on Rs. y and 5% on Rs. z yield same amounts per annum. The average interest rate will be G.M. of 3, 4 and 5.

(n) Mean deviation from mean is 3 and that from median is 4 for a given set of values.

(o) (i) For a symmetrical distribution, Median = $\frac{1}{2}(Q_1 + Q_3)$

(ii) It is known that the median of a frequency distribution is either 56 or 65. It is also known $\sum f_i|x_i - 56| = 964$ and $\sum f_i|x_i - 65| = 1064$. So the median is 56.

(p) Standard deviation of a given of values will be positive or negative according as the given values are positive or negative.

(q) The mean and standard deviation of a set of observations are 50 and 8, respectively. By adding 23 to each of those observations, we get

(i) the new mean = 52, and

(ii) the new standard deviation = 10.

(r) If $u = 5x - 10$ and variance of $x = 100$, then the standard deviation u is 50.

(s) If for a set of 50 observations x_i, i = 1, 2, ..., 50.

$$\sum_{i=1}^{50}(x_i - 90) = 10, \text{ then } \bar{x} = 90.2$$

(t) The first central moment of a set of observations is the average of those values, and it is positive for positive observations.

(u) (i) If the standard deviation of a distribution is 3 and the fourth central moment is 243, then the distribution is mesokurtic.

(ii) For a leptokurtic distribution, $\mu_4 = 3\mu_2^2$.

(iii) Coefficient of kurtosis β_2 may take both positive and negative values.

(iv) A negative coefficient of skewness implies that the mean is greater than mode.

(v) In a frequency curve the mode was found to be higher than the mean. So the distribution was negatively skewed.

(vi) Coefficient of skewness varies from – 3 to + 3.

(vii) Coefficient of kurtosis is less than 1.

Answers

(a) The statement is invalid since we know A.M. ≥ G.M. ≥ H.M. but here A.M. < G.M. < H.M. is observed from the given values.

(b) The G.M. of $x_1, x_2, ..., x_n$ with one zero value can be calculated as zero mathematically, but it cannot be a representative value statistically. Thus the statement is valid.

(c) The given statement is not valid as there may be an observation on depth which is more than 165 cm.

Note : Statistical methods are the tools for decision makers. It is to be remarked that we should not use a particular type of statistical measure everywhere as in this case use of arithmetic mean may cause the loss of life if the depth exceeds 165 cm at some point on the river surface on the path of crossing. Here other relevant statistical measures should be taken into account for making a right decision. Thus the decision 'cross the river' based on A.M. only may be dangerous.

(d) The statement is not valid. Mode is the most suitable measure in this case.

(e) It is not valid; G.M. = $\sqrt{A.M. \times H.M.}$ for two positive values. Here G.M. = $\sqrt{25 \times 9} = 15$

(f) G.M. of 2, 4, 16, 32 = $(2 \times 4 \times 16 \times 32)^{1/4} = (2^{12})^{1/4} = 2^3 = 8$. Thus the statement is not valid.

(g) Corrected mean = $50 + \dfrac{110-100}{100} = 50.1$. But median remains unchanged. So the statement is valid.

(h) According to the given question $\sum(x-30) = 2$

$\Rightarrow \sum x = 602 \Rightarrow \bar{x} = 30.1$ So the statement is not valid.

(i) Here average rate will be H.M. of 0.8 and 1.2, and it is 0.96. Hence the statement is not valid.

(j) The statement is not valid, since statistical methodology does not deal with data from abnormal situations.

(k) The statement is not valid, since the number of days in a month varies, and the number of working days is not considered.

(l) The statement is valid, since 7 × 10.50 = 5 × 14.70

(m) According to the given questions $3x = 4y = 5z = k$, say, ...(1)

Let r be the average of 3, 4, and 5.

So we can write $3x + 4y + 5z = r(x + y + 2)$

$$\therefore \quad k + k + k = r\left(\frac{k}{3} + \frac{k}{4} + \frac{k}{5}\right)$$

$$\Rightarrow \quad r = \frac{1}{\frac{1}{3}\left(\frac{1}{3} + \frac{1}{4} + \frac{1}{5}\right)}$$

$\therefore$ r is H.M. of 3, 4 and 5.

Thus the given statement is invalid.

(n) The statement is valid, since mean deviation about median is least.

(o) For a symmetric distribution,

(i) Median − $Q_1 = Q_3$ − Median

$$Median = \frac{Q_1 + Q_3}{2}$$

Hence the statement is valid.

(ii) Since $\sum f_i|x_i - 56| < \sum f_i|x_i - 65|$ and it is also known that median is either 56 or 65, we can identify the median as 56. Hence the statement is valid.

(p) Statement is not valid, since by definition, standard deviation of any set of values is positive.

(q) (i) Mean = 52 is valid, since mean is changed by origin.

(ii) Standard deviation = 10 is not valid, since standard deviation remain unchanged by the shifting of origin.

(r) $u = 5x - 10 \Rightarrow \bar{u} = 5\bar{x} - 10$

$$\text{var}\ (u) = \frac{1}{n}\sum(u-\bar{u})^2 = \frac{1}{n}\sum(5x-5\bar{x})^2 = \frac{25}{n}\sum(x-\bar{x})^2 = 25\times\text{var}(x)$$

= 25 × 100

Standard Deviation (u) = 5 x 10 = 50

Hence the statement is valid.

(s) $\sum(x_i - 90) = 10, \quad or, \frac{1}{25}\sum x_i - 90 = 10/50$

or $\bar{x} - 90 = 0.2 \Rightarrow \bar{x} = 90.2$ Hence the statement is valid.

(t) The first central moment of $x_1, x_2, \ldots, x_n$ is

$$= \frac{1}{n}\sum(x_1 - \bar{x}) = \frac{1}{n}\sum x_i - \bar{x} = \bar{x} - \bar{x} = 0$$

So the first central moment is zero for any set of values. Hence the statement is not valid.

(u) (i) A distribution is mesokurtic if $b_2 = \frac{m_4}{m_2^2} = 3$

Here, $m_4 = 243$ and $m_2^2 = s^4 = 3^4 = 81$

$\therefore\ b_2 = 243/81 = 3$. So the distribution is mesokurtic. Hence the statement is valid.

(ii) For a leptokurtic distribution $\mu > 3\mu_2^2$. So the statement is not valid.

(iii) For any set of values from a distribution, $\beta_2 = \frac{\mu_4}{\mu_2^2}$ cannot be negative, since it is a ratio of two non-negative values. Hence the statement is not valid.

(iv) For a negatively skewed distribution mean is less than mode. So the statement is not valid.

(v) The statement is valid, since for negatively skewed distribution mode is greater than mean.

(vi) The statement is not valid generally.

(vii) The statement is not valid, since there is no such limit on kurtosis.

2. Identify '*true*' or '*false*' statements with reasons, if any.

(a) The harmonic mean of n observations is the arithmetic mean of the reciprocals of the n values.

(b) The algebraic sum of the deviations of a series of observations from their mean is always zero.

(c) Mode is the suitable measure of location for the eight observations : 10, 8, 12, 6, 7, 11, 12 and x ($x < 6$).

(d) Geometric mean is the appropriate average when observations are increasing at a fixed ratio.

(e) Harmonic mean is infinite when one of the observations is zero.

(f) Mean lies between median and mode.

(g) Mean is not affected by extreme values.

(h) Mean, Median and mode of 100 positive integers is 50.5.

(i) If each value is increased by 10 then the new mean, median and mode will be 60.5

(ii) Co-efficients of skewness remain unchaged, if all the observations are decrased by 5.

(iii) If each value is multiplied by 2 and decreased by 3, (a) the new mean, median and mode will be 98, and (b) measure of kurtosis will be 2 times of the original.

(iv) If each value is a multiplied by 10 and also multiplied by 100 separately and added and the sum so obtained for each value is increased by 10, then the mean, median and mode of the new set of values will be 5065.

(i) G.M. of first five positive integers is $(120)^{1/5}$

(i) If each value is increased by 1, the new G.M. is $(120)^{1/5} + 1$

(ii) G.M. of first five even integers is $(120)^{1/5}$.

(iii) If each value is multiplied by 10 and multiplied by 100 separately and the sum so obtained for each value has the G.M. equal to 110 $(120)^{1/5}$

(f) H.M. of 1st five positive integers is 2.2

(i) If each value is increased by 3, the new H.M. will be 5.2.

(ii) H.M. of 1st five even positive integers is 4.4.

(iii) If 1st five even positive integers are added to 4 times 1st five positive integers then the new H.M. will be 12.2

Answers

(a) False; Harmonic mean is the reciprocal of the arithmetric means of the reciprocal of the given values.

(b) True : $\sum (x_i - \bar{x}) = 0$, for n values , x_1, x_2, ..., xn of x-variable.

(c) False; Median is the appropriate average. Ascending the given values, x, 6, 7, 8, 10, 11, 12, 12.

$$\text{Median} = \frac{8+10}{2} = 9$$

(d) True ; Geometric mean is related to the arithmetic mean of logarithm of given values. If the observations are increasing at a constant ratio, their logarithmic values will lie on a straight line.

(e) False : Harmonic mean is zero when one of the observations is zero.

(f) False : For any distribution either mean > median > mode or mean> median < mode.

(g) False : Mean is based on all observations.

(h) (i) True : All measures of location are affected by by change in origin and scale of the given values.

(ii) True : All measures of skewness are not affected by change in origin and scale of the given values.

(iii) (a) True : Mean, median and mode will be changed as $2 \times 50.5 - 3 = 98$.

(b) False : Measure of kurtosis remains unchanged by change in origin and scale.

(iv) False : $10 \times 50.5 + 100 \times 50.5 + 10 = 5565$.

(i) (i) False : The new G.M. is not deductible from the original G.M. if the original values are changed by origin.

(ii) True : 1st five even values are obtained by multiplying 2 to the 1st five integers. When the original values are changed by scale only, the new G.M. can be obtained by changing the original G.M. by that scale (verify).

(iii) True : Here, in fact, each value is changed by the scale 110.

(j) (i) False : The new H.M. is not deducible from the original H.M. if the original values are changed by origin.

(ii) True : Even values can be obtained from the original values by multiplying 2. The new H.M. can be obtained from the original H.M. if the original observations are changed by scale only.

(iii) True : Here, in fact, each value is changed by the scale 6.

3. Which measure of location is suitable for the following situation.

(a) Sizes of agricultural holdings.

(b) Heights of students,

(c) Sales of product at different years.

(d) Intelligence of students.

(e) Per capital incomes of different states.

(f) Sales of shirts with different collar sizes.

(g) A frequency distribution with open-end classes.

(h) The value of all items are to be taken into consideration in calculation.

(i) Averaging of the given values so that the extreme items do not influence the average.

Answers

(a) Mode, (b) Mean, (c) Mean, (d) Median, (e) Median, (f) Mode, (g) Media or Mode, (h) Mean, (i) Median or Mode.

4. Fill in the blanks

(a) Median of 100 observations is 50 and S.D. is 10. The new median and S.D. will be _____ and ___ if.

(i) 3 is added to each observation

(ii) each observation is multiplied by 5

(iii) 5 is subtracted from each obsevation and it is divided by 3.

(b) The sum of 10 items is 12 and sum of their squares is 16.9. Therefore the S.D. will be ____

(c) In any distribution standard deviation is always ____ than or ___ to the mean deviation.

(d) If 25% of the items are less than 10 and 25% are more than 40, the co-efficient of quartile deviation is ____

(e) Variance is equal to the _____ central moment.

(f) Co-efficient of variation is unit _____

(g) Mode = 3 Median - _____ Mean, for ____ Skewed distribution

(h) Standard Deviation = - (Mean Deviation) for ____ distribution.

Answers

(a) (i) 53, 10

(ii) 250, 50

(iii) 15, 3.33

(b) 0.5

(c) greater, equal

(d) $15\left(\because Q.D. = \frac{Q_3 - Q_1}{2} = \frac{40-10}{2} = 15\right)$

(e) second

(f) free

(g) 2, moderately

(h) 5/4 mesokurtic (normal)

CHAPTER 5

STATISTICAL TREATMENT OF BIVARIATE DATA

5.1. Bivariate Date

The previous chapter(s) are devoted to representation, summarisation and characterisation of data relating to a single character (variable or attribute) only on a number of individuals in a group. This type of data is called univariate data. In some investigations, however,it is required to record the data on two characters at a time on each individual in a given group of individuals. This type of data is called bivariate data. So it is clear that the bivariate data are related to the study of two characters (variables or attributes) simultaneously with the objective to find the inter-relationship (linear) between the two variables and to predict the value of one variable from a derived relationship between the two variables or to measure independence or association between the two attributes. For example,the investigation on bivariate data may relate to the age and weight of a group of school students or to the weight and blood pressure of a selected group of persons or to the rainfall and yield of jute or to the performance of higher secondary students in final examination and income level of their parents etc. The bivariate data are obtained in pairs on each individual and the data can also be grouped in the form of a frequency distribution similar to the univariate data. Two examples on presentation of bivariate data in a tabular form for both grouped or ungrouped categories are given below.

Table 5.1 - Heights and Weights of 12 College Students

Height (x inch)	63	72	70	68	66	69	74	70	63	72	65	71
Weight (y kg)	124	184	161	164	140	154	210	164	126	172	133	150

Forty college students (ages between 18 and 22 years) participated in an intelligence scoring test. These forty pairs of observations (on age and score) have been grouped and presented in the following table

Table 5.2 - Age and Score of 40 college students

Score	Age (Year)				
	18	19	20	21	22
20 - 25	3	2	-	-	-
15 - 20	-	5	4	-	-
10 - 15	-	-	7	10	-
5 - 10	-	-	-	3	2
0 - 5	-	-	-	3	1

5.2. Two-way Frequency Distribution of Variables

In case the data on two characters have been collected from a large number of individuals, it is necessary to summarise the data in the form of a two-way frequency table. Horizontal direction is assigned for classification of one-variable and the vertical direction is assigned for the other variable. If on the basis of the data collected, one variable is classified into p classes and the other is classified into *q*-classes, when the two variables are presented jointly in a two-way table, p × q cells constitute the table. These p × q cells have been provided for *n* pairs of observations. For each and every individual, and of *n* pairs of observations corresponding to *n* individuals, one finds an appropriate cell for putting ' tally mark against the corresponding pair of observations. After completion of such operation one finds a number of tally marks i.e., frequency or number of individuals falling in each cell. The frequency for the (i, j)-th cell is denoted by f_{ij}; i = 1, 2, ..., p; j = 1, 2, 3, ..., q. The manner in which the number of individuals are distributed over p × q cells is called an two-way or bivariate frequency distribution or joint frequency distribution with respect to two variables, say, x and y. The two-way table exhibiting the bivariate frequency distribution is called a bivariate or two-way or joint frequency distribution table. It is also sometimes called the correlation table since by this table it is feasible to get some idea on the relationship between the two variables taken into account (Table 5.3).

Table 5.3 - Correlation Table

	Markes in Statistics					Marginal Total
Marks In Mathematics	Below 30	30-40	40-50	50-60	Above 60	(Frequency)
Above 60				//(2)	/(1)	3
50 - 60		/(1)	//(2)	//(2)	//(2)	7
40 - 5O		///(3)	///(4)			8
30 - 40	//(2)	/// (3)	//(2)	//(2)		9
Below 30	/// (3)		/(1)			4
Marginal Total (Frequency)	5	9	9	6	3	**32**

*This is a special type of two-way frequency table for bivariate data which gives 'an idea about the simple correlation between the two variables.

From the above two-way frequency distribution table one can easily realise (on the average) that a student who is good in Mathematics (M) is also likely to be good in Statistics (S) or a student who is poor in M also likely to be poor in S i.e., there is a (positive) correlation between the two characters, namely, performance in Mathematics and performance in Statistics of an individual.In a bivariate frequency distribution table the row totals (i.e., marginal frequencies) along with the row classification gives the marginal frequency distribution of one variable and similarly, the column totals (columnwise marginal frequencies) along with the column classification gives the marginal frequency distribution of other variable. Again the (conditional) frequencies given in any column (row) along with the row classification (column classification) give the conditional frequency distribution of a variable assigned in row(column) in conjunction with a condition on the other variable assigned in column(row).

From the given two-way table as above let us Illustrate the marginal and conditional frequency distributions.

Table -5.4 Marginal Frequency Distribution of Marks In Mathematics and Marks In Statistics

Row No . (i)	Marks in Mathematics	Marginal Frequencies (f_{io})	Column No. (j)	Marks in Statistics	Marginal Frequencies (f_{oj})
1	belov 30	4	1	belov 30	5
2	30 - 40	9	2	30 - 40	9
3	40-50	9	3	40 - 50	9
4	50 - 60	7	4	50 - 60	6
5	above 60	3	5	above 60	3

Table - 5.5 Conditional Frequency Distribution of Marks in Mathematics for Marks in Statistics in the class 40 - 50

Row No.(i)	Marks in Mathematics	Frequency (f_{i3}, conditional)
1	below 30	1
2	30 - 40	2
3	40 - 50	4
4	50 - 60	2
5	above 60	0

Table - 5.6 Conditional Frequency Distribution of Marks in Statistics for Marks in Mathematics in the class 50 - 60

Column No.(j)	Marks in Statistics	Frequency (f_{4j}, conditional)
1	below 30	0
2	30 - 40	1
3	40 - 50	2
4	50 - 60	2
5	above 60	2

5.3 . Scatter Diagram

A scatter diagram is a diagrammatic representation of ungrouped bivariate data and is obtained by plotting the observed pairs of observations (x_i, y_i), i = 1, 2,.....,n with respect to two perpendicular axes of co-ordinates on a graph paper. This is actually a two-dimensional extension of the point diagram in univariate case and is, shown in figure 5.1, From a scatter diagram, we can have an idea about the correlation between the two variables x and y, provided x and y are linearly related, either strictly or approximately.

Illustration 5.1 In the following, marks in Mathematics and Accountancy of 10 students have been given.

Serial No.	1	2	3	4	5	6	7	8	9	10
Marks in Mathematics	55	35	40	60	50	40	45	55	40	30
Marks in Accountancy	50	40	50	55	55	35	40	40	45	35

Let us plot the above points (paired data like (55,50), (35,40) etc.) for each individual against two perpendicularly selected axes as below

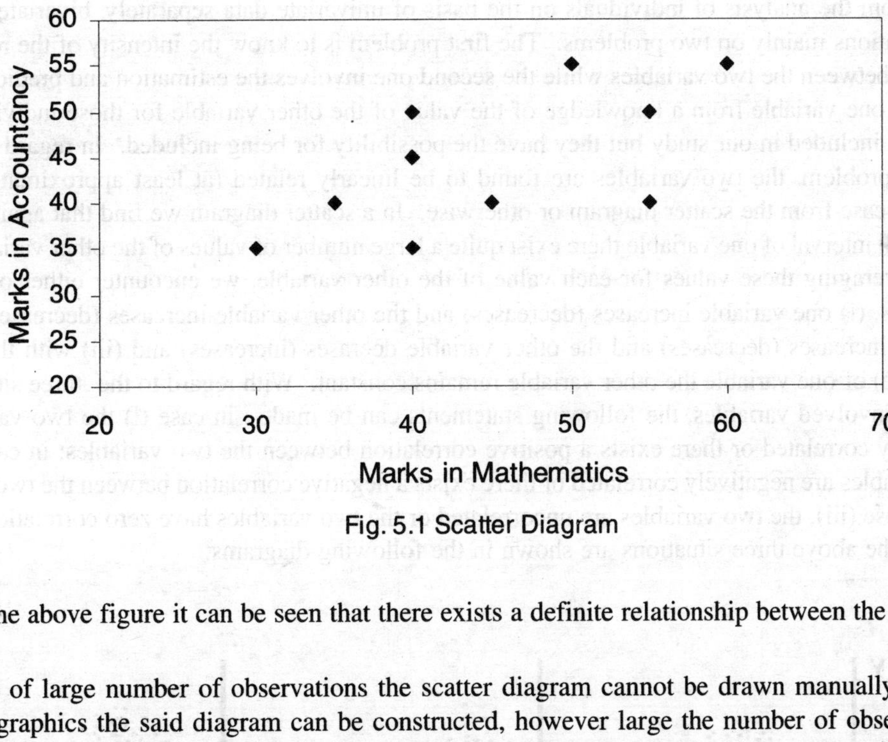

Fig. 5.1 Scatter Diagram

From the above figure it can be seen that there exists a definite relationship between the two sets of marks.

In case of large number of observations the scatter diagram cannot be drawn manually, but by computer graphics the said diagram can be constructed, however large the number of observations may be. If the number of observations is too large, the pairs are classified in two directions with respect to two variables and are presented in a two-way frequency table like Table 5.2. Now, on a graph paper, corresponding to the point,(x_i, y_i), erect a solid rectangle equivalent to the frequency of i-th class of *x*-variable and j-th class of *y* variable, f_{ij}, where x_i & y_j are the mid-values of the corresponding classes. The same procedure is follwed for the remaining (i,j) combinations. The diagram thus formed is known as stereogram which is a three dimensional analogue of histogram. With respect to the bivariate frequency distribution table (Table 5.2) the stereogram is shown below.

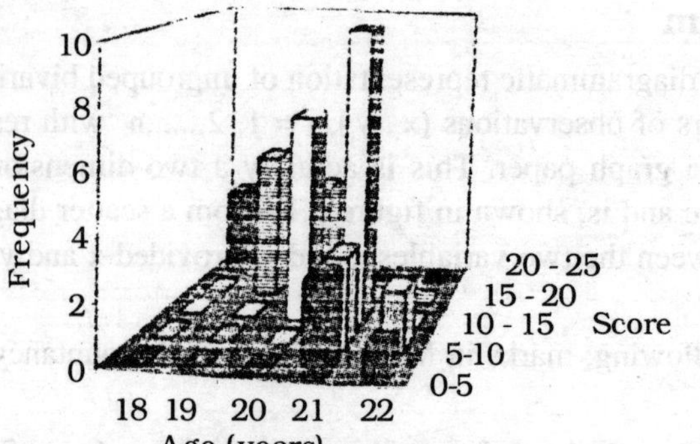

Fig. 5.2 Stereogram

5.4. Concept of Correlation and its Measure

Apart from the analysis of individuals on the basis of univariate data separately, bivariate data offer investigations mainly on two problems. The first problem is to know the intensity of the relationship existing between the two variables while the second one involves the estimation and prediction of the value of one variable from a knowledge of the value of the other variable for those individuals who were not included in our study but they have the possibility for being included. In regard to the first kind of problem, the two variables are found to be linearly related (at least approximately) in the simplest case from the scatter diagram or otherwise. In a scatter diagram we find that against a value or a small interval of one variable there exist quite a large number of values of the other variable. If we go on averaging these values for each value of the other variable, we encounter either of the three situations, (i) one variable increases (decreases) and the other variable increases (decreases), (ii) one variable increases (decreases) and the other variable decrases (increases) and (iii) with the increase (decrease) of one variable the other variable remains constant. With regard to the three situations on the two involved variables, the following statements can be made : in case (i) the two variables are positively correlated or there exists a positive correlation between the two variables; in case (ii), the two variables are negatively correlated or there exists a negative correlation between the two variables, and in case (iii), the two variables are uncorrelated or the two variables have zero correlation between them. The above three situations are shown in the following diagrams.

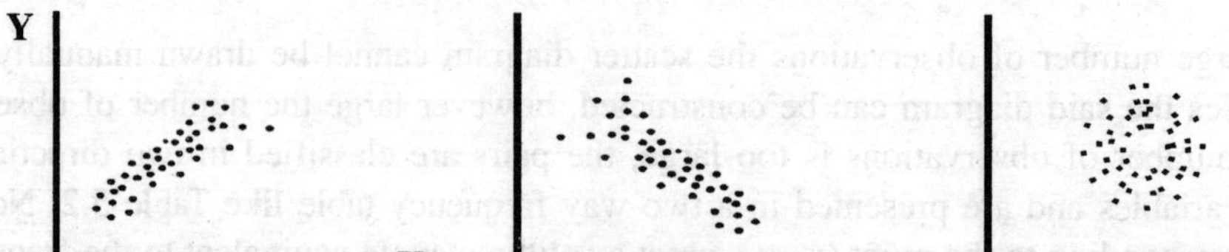

(a) : Positive Correlation Case (i)   (b) Negative Correlation Case (ii)   (c) : Zero Correlation Case (iii)

Fig. 5.3 The Schematic Scatter Diagram

In order to measure the correlation, the term correlation coefficient is used and is known as Karl Pearson's Product Moment Correlation Coefficient and is denoted by r_{xy} or r_{yx} or r, the form of which is expressed as below :

$$r_{xy}, \text{ or }, r_{yx} \text{ or, } r = \frac{\sum_{i=1}^{n}(x_i - \bar{x})(y_i - \bar{y})}{n s_x s_y}$$

where $\bar{x}, \bar{y}, s_x, s_y$, are the means and standard deviations of x and y, respectively and n is the number of pairs of obsevations. The derivations of the correlation coefficient, r is given below. Let the origin of the two axes x and y (on a scatter diagram) be shifted to $(\bar{x}, \bar{y})$. Taking $x_i' = (x_i - \bar{x})$ and $y_i' = (y_i - \bar{y})$, i = 1, 2, ..., n and thus forming a new set of coordinates x' and y' (imposed on the same graph), it can be seen (Fig. 5.4) that all the points are distributed over the four quadrants of the (x', y') plane. Further, it is also seen that the products, $x_i' y_i'$'s are positive in 1st and 3rd quardrants (because x_i''s and y_i''s are of same sign in those quadrants) and the products are negative in 2nd and 4th quadrants (because x_i''s and y_i''s are of opposite signs in those quadrants). For positive correlation most of the observations have the tendency to lie in 1st and 3rd quadrants so that the total of the positive products is more than the total of the negative products

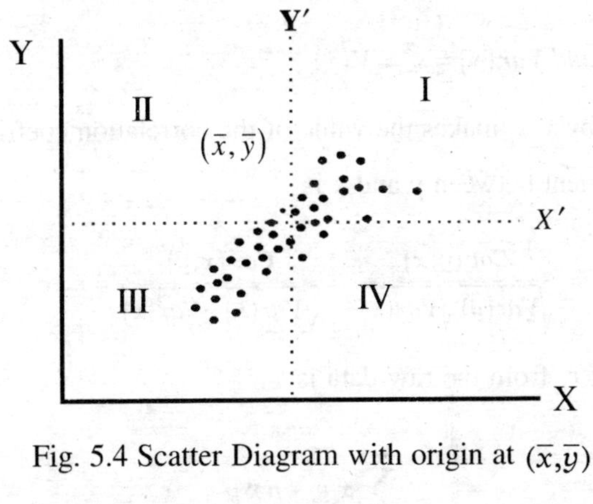

Fig. 5.4 Scatter Diagram with origin at $(\bar{x}, \bar{y})$

making the sum $\sum_{i=1}^{n} x_i' y_i'$ a positive quantity. Similarly, for negative correlation most of the observations have the tendency to lie in 2nd and 4th quadrants so that the sum $\sum_{i=1}^{n} x_i' y_i'$ becomes a negative quantity, total of the negative products being larger than the total of the positive products. Finally, in case of no correlation all observations are distributed over the four quadrants in more or less equal numbers and the sum $\sum_{i=1}^{n} x_i' y_i'$ becomes approximately zero, because the total of positive products becomes nearly equal to the total of negative products. Consequently, the sum $\sum_{i=1}^{n} x_i' y_i' = \sum_{i=1}^{n} (x_i - \bar{x})(y_i - \bar{y})$ may be considered as a measure of correlation but, indeed, this expression depends on the total number of observations (n), the units of the two variable and the variabilities of the two variables. So in order to make it free from the effect of the number of observations and also from the effect of the variabilities of *x* and *y* (i.e., s_x and s_y), we have to divide the above sum by $ns_x s_y$, and hence the ultimate formula of the Product Moment Correlation Coefficient between *x* and *y* becomes,

$$r_{xy}, \text{ or } r = \frac{\sum_{i=1}^{n} (x_i - \bar{x})(y_i - \bar{y})}{ns_x s_y} = \frac{\sum_{i=1}^{n} (x_i - \bar{x})(y_i - \bar{y}) \div n}{s_x s_y}$$

The numerat of *r* is called the covariance of x and y and is denoted by Cov (x y) which is analogus to the term variance for univariate data. In fact, the formula can also be written as,

$$r_{xy} = \frac{Cov(x, y)}{\sqrt{Var(x)}\sqrt{Var(y)}} \quad \text{.... (5.1)}$$

since, $Var(x) = s_x^2 = V(x) \text{ and } Var(y) = s_y^2 = V(y)$

The division of *Cov (x, y)* by $s_y s_x$ makes the value of the correlation coefficient unit free.

Also the correlation coefficient between y and x is

$$r_{yx} = \frac{Cov(y, x)}{\sqrt{Var(y)}\sqrt{Var(x)}} = \frac{Cov(x, y)}{\sqrt{Var(x)}\sqrt{Var(x)}} = r_{xy} \quad \text{... (5.2)}$$

The computational form of r_{xy} from the raw data is

$$r_{xy} = \frac{\sum_{i=1}^{n} x_i y_i - n\bar{x}\bar{y}}{\sqrt{\left\{\left(\sum_{i=1}^{n} x_i^2 - n\bar{x}^2\right)\left(\sum_{i=1}^{n} y_i^2 - n\bar{y}^2\right)\right\}}} \quad \text{.... (5.3)}$$

Another simple computational formula of r_{xy} when x and y values are very large has been mentioned in (5.4).

In what follows, the subscript i under $\sum$ notation will be dropped, as it is otherwise understood.

5.5. Properties of Correlation Coefficient

Correlation coefficient between x and y and correlation coefficient between y and x are equal, i.e.,

$r_{xy} = r_{yx}$

Proof : Since $Cov\,(x, y) = \frac{1}{n}\sum (x_i - \bar{x})(y_i - \bar{y}) = \frac{1}{n}\sum (y_i - \bar{y})(x_i - \bar{x})$

$$= Cov\,(y, x)$$

So, $r_{xy} = r_{yx}$

Property 2 : Correlation coefficient is a pure number , *i.e.,* r_{xy} is independent of units of which the variables are measured.

Property 3. Correlation coefficient is independent of the change of origin and scale provided the scales have been changed in the same direction (positive or negative).

Proof : Let $u_i = \frac{x_i - a}{h}$, $v_i = \frac{y_i - b}{k}$ so that $x_i = a + hu_i$ and $y_i = b + kv_i$ where *a, b h* and *k* are constanst, $i = 1, 2, ..., n$.

Now,

$$x_i - \bar{x} = (a + hu_i) - (a + h\bar{u}) = h(u_i - \bar{u})$$

$$y_i - \bar{y} = (b + kv_i) - (b + kv_i) = k(v_i - \bar{v})$$

Hence, $Cov(x, y) = \frac{1}{n}\sum (x_i - \bar{x})(y_i - \bar{y}) = \frac{hk}{n}\sum (u_i - \bar{u})(v_i - \bar{v})$

$$= hk\,Cov(u, v)$$

$$Var(x) = \frac{1}{n}\sum (x_i - \bar{x})^2 = \frac{h^2}{n}\sum (u_i - \bar{u})^2 = h^2\,Var(u)$$

$$Var(y) = \frac{1}{n}\sum (y_i - \bar{y})^2 = \frac{k^2}{n}\sum (v_i - \bar{v})^2 = k^2\,Var(v)$$

Thus, $r_{xy} = \frac{Cov(x,y)}{\sqrt{Var(x)}\sqrt{Var(y)}} = \frac{hkCov(x,y)}{|h|\,|k|\sqrt{Var(u).Var(v)}} = \frac{hk}{|h||k|}.r_{uv}$

If *h* and *k* are of same sign, $\frac{hk}{|h||k|} = 1$ and hence $r_{xy} = r_{uv}$

Note 1. If h and k are of opposite signs, $\frac{hk}{|h||k|} = -1$ and in that case $r_{xy} = -r_{uv}$, i.e., the correlation coefficient between the transformed variable is equal to the correlation coefficient between the original variables in magnitude but opposite in sign.

Note 2. If $h = k = 1$, i.e., variables are changed with respect to the origins (bases) only, then $r_{xy} = r_{uv}$.

Due to the Property (Note 2), the calculation of r_{xy} can be made more easily by the following formula,

$$r_{xy} = \frac{\sum_{i=1}^{n} d_{x_i} d_{y_i} - n d_{\bar{x}} d_{\bar{y}}}{\sqrt{\left\{\left(\sum_i d_{x_i}^2 - n d_{\bar{x}}^2\right)\left(\sum_i d_{y_i}^2 - n d_{\bar{y}}^2\right)\right\}}} \quad \text{... (5.4)}$$

where $d_{x_i} = x_i - a$ and $d_{y_i} = y_i - b$, a and b being the respective assumed means/origins $d_{\bar{x}} = \bar{x} - a$ *and* $d_{\bar{y}} = \bar{y} - b$.

Property 4 : The value of the correlation coefficient lies between –1 and + 1, i.e., $-1 \le r \le 1$.

Proof. It is known,

$$r_{xy} = \frac{\frac{1}{n}\sum (x_i - \bar{x})(y_i - \bar{y})}{s_x s_y}, \text{ where } s_x = \sqrt{Var(x)} \text{ and } s_y = \sqrt{Var(y)}$$

Now, let $x_i' = \dfrac{x_i - \bar{x}}{s_x}$ and $y_i' = \dfrac{y_i - \bar{y}}{s_y}$

then, $r_{xy} = \dfrac{1}{n}\sum x_i' y_i' = r$

and $\sum x_i'^2 = \sum y_i'^2 = n$

Now, $\dfrac{1}{n}\sum (x_i' + y_i')^2 \ge 0$

or $\dfrac{1}{n}\sum x_i'^2 + \dfrac{1}{n}\sum y_i'^2 + \dfrac{2}{n}\sum x_i' y_i' \ge 0$

or, $2(1 + r) \ge 0 \Rightarrow r \ge -1$

Again, $\dfrac{1}{n}\sum (x_i' - y_i')^2 \ge 0$

or $\dfrac{1}{n}\sum x_i'^2 + \dfrac{1}{n}\sum y_i'^2 - \dfrac{2\sum x_i' y_i'}{n} \ge 0$

or $2(1 - r) \ge 0 \Rightarrow r \le 1$

and hence the property $-1 \le r \le 1$ holds.

Note 3. We know, $r_{xy} = \frac{1}{n}\sum x_i' y_i'$ and r takes the smallest value, – 1, if for each i,

$$y_i' = -x_i' \left(\because r = \frac{1}{n}\sum x_i' y_i' = \frac{1}{n}\sum x_i'^2 = -\frac{1}{n} \times n = -1 \right)$$

which implies $y_i = \bar{y} - \frac{s_y}{s_x}(x_i - \bar{x})$ and $x_i = \bar{x} - \frac{s_x}{s_y}(y_i - \bar{y})$

and r takes the highest value, 1 if, for each i,

$$y_i' = x_i' \left(\because r = \frac{1}{n}\sum x_i' y_i' = \frac{1}{n}\sum x_i^2 = \frac{1}{n} \times n = 1 \right)$$

which implies $y_i = \bar{y} + \frac{s_y}{s_x}(x_i - \bar{x})$ and $x_i = \bar{x} + \frac{s_x}{s_y}(y_i - \bar{y})$ for each *i*.

In the two above mentioned cases the variables are said to be perfectly (negatively or positively) correlated since each variable is an exact linear function of the other. Though these lines (i.e., $y = \bar{y} - (s_y / s_x)(x_i - \bar{x})$ and $y = \bar{y} + (s_y / s_x)(x_i - \bar{x})$, come out to be equal to the regression lines (average line or predicted line) with r= – 1 (with negative slope), and r = + 1 (with positive slope) respectively, but in the above derivation these lines are obtained exactly by algebraic manipulation and not by averaging.

5.6. Linear Regression Analysis

In the discussion on bivariate data so far, besides the measurement of the extent of relationship between the two variables, it has been realised that the determination of the nature of relationship is also necessary as in many scientific investigations the variation in measurement of one variable is caused by other variable (s) to a great extent in course of experimentation. By using the data on the influencing/ independent variable (x) and the influenced/dependent variable (y) in a statistical analysis, it is possible to determine the nature of the relationship and to infer about the influenced variable, utilising this relationship, for a given value of the influencing variable. Searching for the relationship among variables is, indeed, important for prediction of one variable from the observations on the other variable, and such relationship is also used to determine the optimum value of the independent variable which produces the target value of the dependent variable. These aspects can be studied with the help of the theory of regression analysis. In short, the regression analysis is a course of statistical techniques which deal with mathematical modelling that describes the relationship between two variables for the prediction and other statistical inferences. Sir Francis Galton first used the term "regression" in 1885 under the published title "Regression Toward Mediocrity in Hereditary stature." Due to various nature of possible relationship existing among variables, a broad clssification (as given below) is, indeed required to identify the area applicable for regression analysis.

5.6.1. Deterministic Relation.

Variables are often related by a precise (exact) mathematical function recognised by universally accepted laws and any deviation in observation by using the relation is treated as an experimental error. For example, temperature in Farenheit (^{0}F) scale and temperature in Centigrade (^{0}C) scale are related as,

$$F = 32 + 1.8C$$

Take another example. If an amount of Rs. x is to be deposited by a person as a life insurance premium and if it is unpaid for n terms (years), he has to pay Rs. y at 100i % interest, then y and x are related as

$$y = x + x(1+i) + x(1+i)^2 + ... + x(1+i)^{n-1}$$

i.e., $y = \frac{x}{i}\left((1+i)^n - 1\right)$, which is a law of annuity.

These cases are not of interest in regression analysis.

5.6.2. Semideterministic Relation

In these cases an established theory provides a form relating the variables through some non-fixed constants (parameters). To get these parameters, we must perform experiments. Owing to many limitations in experimentation (e.g., limitations of measuring instruments, uncontrollable experimental conditions etc.), the data contain error that may cause some variation in the true relation. For example, to predict emission rate (or decay value, λ) of α- particles from radioactive substances (whose atomic weights are more than 206) a law has been proposed by Geiger and Nuttal as $\lambda = cR^b$, where R is the energy range of α-particles, b and c have to be estimated from the experiment. Such types of relations, originated from partial theoretical background, may not be so precise as to be acceptable universally and also may not include other uncontrollable factors. Such investigations should be conducted through statistical analysis of data relating to two or more variables.

5.6.3. Empirical Relation

Besides the above situations, there are many natural phenomena where the relation between a variable and a number of influencing or casual variables is not guided by a precise physical law. There are copious examples of such relations-yield of a cereal per plot and plot size; production of cow milk and amount of food taken; performance in completing a job and duration of training rendered for it, etc. After conducting such experiments sufficient number of times one may acquire some theoretical knowledge about the form of the mathematical relation governing the involved variables (dependent and independent) and then can use the same on future occasions. This kind of study is considered under the general theory of regression analysis.

In a simple regression problem, we begin with an experiment to determine the relationship existing between the two variables x and y, y is a predicting or dependent or response variable and x is a predictor or independent or controlled variable. The first step in regression analysis is to plot the data points (x_i, y_i), $i = 1, 2, ..., n$ on a graph paper. If it is interpreted from the scatter diagram that there exists a linear relationship (the corresponding problem being linear regression analysis), we adopt the following model.

$$y_i = \alpha + \beta x_i + e_i = Y_i + e_i, Y_i = \alpha + \beta x_i; i = 1, 2, ..., n \quad ...(5.5)$$

where y_i is the observed value of the response varialbe y related to the value x_i of the controlled variable x, α and β are the parameters to be estimated from the set of observations, e_i's are nonassisgnable/

uncontrolled elements representing the uncontrolled parts of y_i's and are called errors (variable) on the true relation. In fact, (x_i, y_i)'s constitute n pairs of observations, whereas e_i's are n unknowns under the model and constitute the error part.

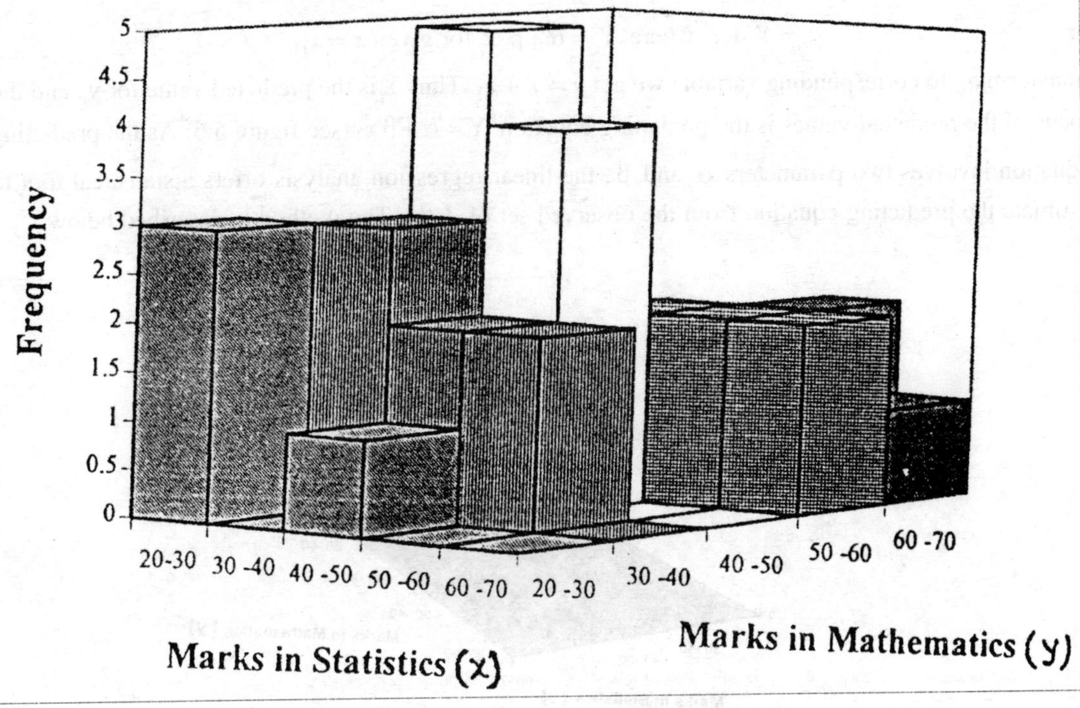

Fig. 5.5 Observed Conditional Distribution of (x_i, y_i)

For a fixed x_i, let there be f_i values of y_i, which is possible for grouped data, ungrouped discrete pairs, etc. If we draw a frequency polygon or a histogram with those f_i, frequencies which are distributed over the different classes of y - variable at x_i, we get a frequency curve with respect to y variable at that particular value of x_i (vide frequency curve). Now let us assume that the frequency curve is normal with a mean value $\alpha + \beta\ x_i$ at x_i. More specifically, it is assumed that the theoretical (conditional) distribution of the variable y at the fixed point $x = x_i$ is normal with (conditional) mean $E(y / x = x_i) = \alpha + \beta\ x_i$ and the (conditional) variance $V(y / x = x_i) = \sigma^2$, a constant.

The observed conditional distribution of y at the point $x = x_i$ (based on (x_i, y_i), $i = 1, 2, ...n$) has got a mean value y_i with a superimposed error e_i. The two distributions are shown in diagrams. 5.5 and 5.6. Thus, e_i is the difference between the observed y_i and the predicted (theorical) mean value $y_i = \alpha + \beta x_i$. Also error variables $e_i's$ are assumed to be independetly and identically normally distributed with zero mean and constant variance σ^2. It is to be noted, that the assumption of normality is not required for derivation of the regression line; it is enough to assume that the conditional mean

$E(y/x = x_i) = \alpha + \beta x_i$ of y at the point $x = x_i$ is linear in x_i, and the observed value y_i is composed of two parts, namely.

$$y_i = E(y/x = x_i) + e_i,\ i = 1, 2, ..., n$$

or $$y_i = \alpha + \beta\, x_i + e_i \qquad ..... (5.6)$$

or $$y_i = Y_i + e_i \text{ where } Y_i = \alpha + \beta\, x_i \text{ for given } x = x_i,$$

transforming to corresponding variable we get $y = Y + e$. Thus Y_i is the predicted value for y_i, and the locus of the predicted values is the predicting equation $Y = \alpha + \beta x$ (see figure 5.6) As the predicting equation involves two parameters α and β, the linear regression analysis offers a statistical tool to estimate the predicting equation from the observed set of data. The method is described below.

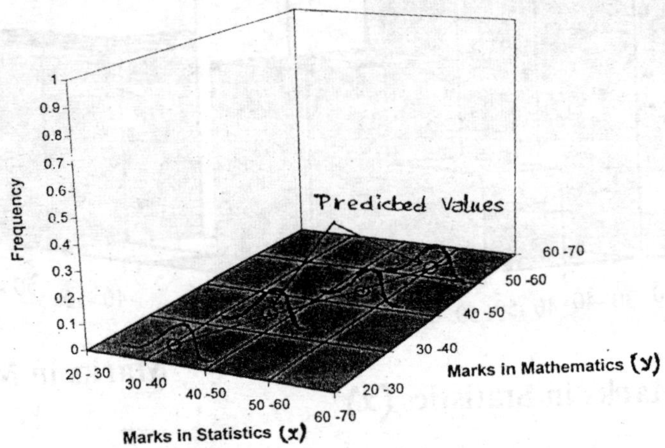

Fig. 5.6 Predicted Conditional Distribution of y under the assumption of linearity

The predicting equation is so determined (i.e., α and β to be so estimated) that the errors e_i's are small. In order to achieve the objective, minimisation of $\sum_{i=1}^{n} e_i$ with respect to α and β does not produce satisfactory estimates, as positive and negative errors may cancel each other to a minimum value in spite of large positive and negative error values, whereas minimisation of $\sum_{i=1}^{n} e_i^2$ with respect to α and β produces estimates with reasonably low individual errors (e_i's). This procedure of minimisation of sum of squared errors for estimation of parameters is known as the method of least squares.

Thus, by minimising $S = \sum_{i=1}^{n} e_i^2 = \sum (y_i - Y_i)^2 = \sum (y_i - \alpha - \beta x_i)^2$ with respect to α and β, the following normal equations (estimating equations) are obtained :

$$\frac{\partial S}{\partial \alpha}=0 \equiv \sum\left(y_i-\hat{\alpha}-\hat{\beta}\, x_i\right)=0 \Rightarrow \sum y_i=n\hat{\alpha}+\hat{\beta}\sum x_i \qquad ....(5.7)$$

and , $$\frac{\partial S}{\partial \beta}=0 \equiv \sum x_i\left(y_i-\hat{\alpha}-\hat{\beta}\, x_i\right)=0 \Rightarrow \sum x_i\, y_i=\hat{\alpha}\sum x_i+\beta^2\sum x_i^2 \qquad ...(5.8)$$

Equations (5.7) gives $\hat{\alpha}=\bar{y}-\hat{\beta}\bar{x}$ estimate of α. ...(5.9)

Substituting this value of $\hat{\alpha}$ in (5.8), we get

$$\sum x_i y_i=\left(\bar{y}-\hat{\beta}\bar{x}\right)\sum x_i+\hat{\beta}\sum x_i^2$$

$$=n\bar{y}\bar{x}-n\hat{\beta}\, \bar{x}^{2}+\hat{\beta}\sum x_i^2 \left[\because \sum x_i=n\bar{x}\right]$$

$$\Rightarrow \qquad \hat{b}=\frac{\sum (x_i y_i)-n\bar{x}\,\bar{y}}{\sum x_i^2-n\bar{x}^2}=\frac{\frac{1}{n}\sum(x_i-\bar{x})(y_i-\bar{y})}{\frac{1}{n}\sum(x_i-\bar{x})^2}=\frac{Cov(x,y)}{Var(x)}=\frac{Cov(x,y)}{\sqrt{V(x)V(y)}}\cdot\frac{\sqrt{V(y)}}{\sqrt{V(x)}}$$

$$=r_{xy}\frac{s_y}{s_x}=b, \textit{ say, estimate of } \beta \qquad ...(5.10)$$

It should be noted that $\hat{\alpha}$ and $\hat{\beta}$ are estimates of α and β respectively from the normal equations and they are variables since they involve $y_i's$ and $e_i's$ which are variables, whereas α and β are parmameters.

Substituting the estimated values of α and β we arrive at the desired predicting formula as regression equation (or regression line) of *y* on *x*. For the *i*-th pair, $\hat{Y}_i=\bar{y}+r\frac{s_y}{s_x}(x_i-\bar{x})$, and for all i, the locus of $\hat{Y}_i's$ will produce the regression equation of y on x as $\hat{Y}=\bar{y}+r\frac{s_y}{s_x}(x-\bar{x})$.

... (5.11)

with $\hat{\alpha}=a=\bar{y}-r\frac{s_y}{s_x}\cdot\bar{x}$ as the *y* intercept and $\hat{\beta}=b=r\frac{s_y}{s_x}$ as the slope of the regression line. The estimate $\hat{\beta}$ is also called regression coefficient of *y* on *x* and is denoted by b_{yx}. Thus, $Y=\bar{y}+b_{xy}$ (*x* – $\bar{x}$) gives the linear regression equation (line of *y* on *x* and is used for the prediction of responses (*y's*) for the given *x's*.

Similarly, for prediction of x values for given y values, let us take $x_i = X_i + e_i' = \alpha' + \beta' y_i + e_i'$, say (assuming linear relationship as before). i = 1, 2, ..., *n*.

Now we minimize $S' = \sum e_i^2 = \sum (x_i - X_i)^2 = \sum (x_i - \alpha' - \beta' y_i)^2$ with respect to α' and β' and produce the following normal equations.

$$\frac{\partial S'}{\partial \alpha'} = 0 = \sum \left(x_i - \hat{\alpha}' - \hat{\beta}' y_i\right) = 0 = \Rightarrow \sum x_i = n\hat{\alpha}' + \hat{\beta}' \sum y_i \qquad ...(5.12)$$

$$\frac{\partial S'}{\partial \beta'} = 0 \equiv \sum y_i \left(x_i - \hat{\alpha}' - \hat{\beta}' y_i\right) = 0 \Rightarrow \sum x_i y_i = \hat{\alpha}' \sum y_i + \hat{\beta}' \sum y_i^2 \qquad ....(5.13)$$

From the equation (5.12)

$$\hat{\alpha}' = \bar{x} - \hat{\beta}' \bar{y} = a', \text{ say} \qquad ....(5.14)$$

Using (5.14) in (5.13) we get,

$$\hat{b}' = \frac{\sum x_i y_i - n\bar{x}\,\bar{y}}{\sum y_i^2 - n\bar{y}^2} = \frac{\frac{1}{n}\sum (x_i - \bar{x})(y_i - \bar{y})}{\frac{1}{n}\sum (y_i - \bar{y})^2}$$

$$= \frac{Cov(x,y)}{Var(y)} = \frac{Cov(x,y)}{\sqrt{Var(x)}\sqrt{Var(y)}} \times \frac{\sqrt{Var(x)}}{\sqrt{Var(y)}}$$

$$\equiv r_{xy} \frac{s_x}{s_y} = b' = b_{xy} \text{ (regression coefficient of x on y)} \qquad ... (5.14a)$$

Substituting (5.14a) in (5.14), we get $\hat{\alpha}' = \bar{x} + r\frac{s_x}{s_y}\bar{y}$ (x-intercept). Thus the regression line of x on y becomes.

$$\hat{x} = \bar{x} + r\frac{s_x}{s_y}(y - \bar{y}) = \bar{x} + b_{xy}(y - \bar{y})$$

Note 1. In order to avoid notational complexity we generally express the regression line of *y* on *x* as $Y = a + bx = \bar{y} + r\frac{s_y}{s_x}(x - \bar{x})$ and that of *x* on *y* as and that of *x* on *y* as $X = a' + b'y = \bar{x} + r\frac{s_x}{s_y}(y - \bar{y})$.

In this context, it may be mentioned that sometimes in question papers or in different text books the above mentioned equations are also written as

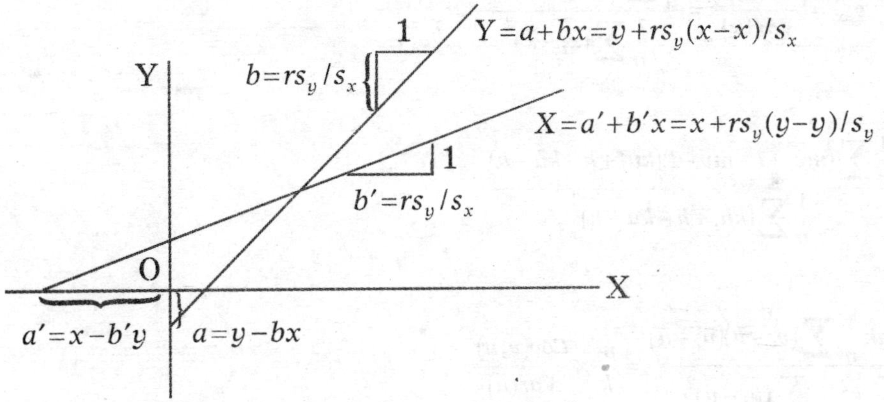

Fig. 5.7 Two regression lines showing regression coefficients as the increment of predicting variable per unit increment of predictor variable

y = a + bx and x = $a' + b'y$ respectively,

It can be noted that both the regression lines pass through the point $(\bar{x}, \bar{y})$ which gives the point of intersection of two regression lines.

5.7. Results on Regression Lines and Regression Coefficients

Consider the regression line of y on x as $Y = a + b_{yx}x$ and corresponding to a given value at $x = x_i$, $Y_i = a + b_{yx} x_i$,

where, $a = \bar{y} - b_{yx}\bar{x}$, and $b_{yx} = r\frac{s_y}{s_x} = \frac{Cov(x,y)}{Var(x)}$. Similarly, the regression line of x on y is

$$X = a' + b_{xy}\, y, \text{ where } a' = \bar{x} - b_{xy}\bar{y} \text{ and } b_{xy} = r\frac{s_x}{s_y} = \frac{Cov(x,y)}{V(y)}.$$

Property 1. Regression coefficients are affected by change in scale but not by change in origin (base).

Let x variable be changed by the scale, k and the base, h to u, and y variable be changed by the scale, m and the base, l to v, so that

$$u_i = \frac{x_i - h}{k}, i = 1.2, \ldots n, \text{ and}$$

$$v_i = \frac{y_i - l}{m}, i = 1.2 \ldots .n$$

Now, $$b_{yx} = \frac{Cov(y,x)}{Var(x)} = \frac{\frac{1}{n}\sum(y_i - \bar{y})(x_i - \bar{x})}{\frac{1}{n}\sum(x_i - \bar{x})^2}$$

$$= \frac{\frac{1}{n}\sum(mv_i + l - m\bar{v} - l)(ku_i + h - k\bar{u} - h)}{\frac{1}{n}\sum(ku_i + h - k\bar{u} - h)^2}$$

$$= \frac{mk\frac{1}{n}\sum(v_i - \bar{v})(u_i - \bar{u})}{k^2\frac{1}{n}\sum(u_i - \bar{u})^2} = \frac{m}{k} \times \frac{Cov(v,u)}{Var(u)}$$

So, $b_{yx} = \frac{m}{k} b_{vu}$ .... (5.15)

Similarly, $b_{xy} = \frac{k}{m} b_{uv}$

Hence the proposition

Note 2. If $m = k$ then $b_{yx} = b_{vu}$ and $b_{xy} = b_{uv}$, so regression coefficients are not affected by changing the original variables by equal value on the scale.

Note 3. If $|m| = |k|$ but m and k are of opposite signs then $b_{yx} = -b_{vu}$ and $b_{xy} = -b_{uv}$, then the regression coefficients are changed by signs only.

Note 4. If $m = k = 1$, i.e., variables are changed by origin only, then $b_{yx} = b_{vu}$ and $b_{xy} = b_{uv}$ i.e., the values of the regression coefficients remain unchanged.

Property 2 : The mean of the observed values, y_i, $i = 1, 2, ..., n$ is equal to the mean of the predicted values, Y_i, $i = 1\ 2\ ..., n$ from linear regression equation of y on x, i.e., $\bar{y} = \bar{Y}$.

Proof : The predicted value for given $x = x_i$ is $Y_i = \bar{y} + r\frac{s_y}{s_x}(x_i - \bar{x}), i = 1, 2, ... n$

Taking sum on both sides,

$$\sum Y_i = n\bar{y} + r\frac{s_y}{s_x}\sum(x_i - \bar{x})$$

or, $$\frac{1}{n}\sum Y_i = \bar{y} + r\frac{s_y}{s_x}\frac{1}{n}\sum(x_i - \bar{x})$$

$$\Rightarrow \bar{Y} = \bar{y} \left[\because \frac{1}{n}\sum(x_i - \bar{x}) = 0\right] \quad ... (5.16)$$

Note 5. The mean of the observed values , x_i, $i = 1, 2, ..., n$ is equal to the mean of the predicted values, X_i, $i = 1, 2, ..., n$ (from linear regression equation of x and y) i.e., $\bar{x} = \bar{X}$. ...... (5.17)

Property 3 : The square of the correlation coefficient (r^2) is the proportion of variability due to regression line (of y on x) to the total observed variability of y i.e., r^2 is the part of total observed variation of Y accounted for by the regression line.

Proof : We know, $Y_i = \bar{y} + r\frac{s_y}{s_x}(x_i - \bar{x})$

or, $$Y_i - \bar{y} = r\frac{s_y}{s_x}(x_i - \bar{x}) \qquad (\because \bar{y} = \bar{Y} \text{ from } (5.16))$$

or, $$\frac{1}{n}\sum(Y_i - \bar{Y})^2 = r^2\frac{s_y^2}{s_x^2}\frac{1}{n}\sum(x_i - \bar{x})^2$$

or, $$Var(Y) = r^2\,Var(y) \Rightarrow r^2 = \frac{Var(Y)}{Var(y)} \qquad (5.18)$$

Also, $$|r| = \frac{\sqrt{Var(Y)}}{\sqrt{Var(y)}} = \frac{\textit{Variation due to regression line}}{\textit{Total observed variation of } y}$$

Thus r^2 may be interpreted as the proportion of total observed variability of y as explained by the variability due to linear regression on x. r^2 is also called summary statistic or summary measure of the series of bivariate data.

Note 6. The square of the correlation coefficient (i.e., r^2) is the proportion of variability due to regression line (of x on y) to the total observed variability of x. i.e.,

$$r^2 = \frac{Var(X)}{Var(x)} \text{ or } |r| = \frac{\sqrt{Var(X)}}{\sqrt{Var(x)}} \qquad \text{... } (5.19)$$

Note 7 : The absolute value of the correlation coefficient , r (or the value r^2) does not reflect the nature and degree of association between the two variables in all cases. The above fact can be clearly demonstrated through the following four examples (given by F. J. Anscombe, 1973, American Statisticians) where in each of the examples the value of the correlation coefficient is around 0.82. (Figure 5.7).

Ex. 1

x :	10	8	13	9	11	14	6	4	12	7	5
y :	9.14	8.14	8.74	8.77	9.26	8.10	6.13	3.10	9.13	7.26	4.74

Ex, 2 -

x :	10	8	13	9	11	14	6	4	12	7	5
y :	7.46	6.77	12.74	7.11	7.81	8.84	6.08	5.39	8.15	6.42	5.73

Ex. 3 -

x :	8	8	8	8	8	8	8	19	8	8	8
y :	6.58	5.76	7.71	8.84	8.87	7.04	5.25	12.50	5.56	7.91	6.89

Ex. 4 -

x :	10	8	13	9	11	14	6	4	12	7	5
y :	8.04	6.95	7.58	8.81	8.33	9.96	7.24	4.26	10.84	4.82	5.68

A close look at the diagrams reveals, (i) a curvilinear fit in Ex 1, (ii) a strictly linear fit (exclusing an outlying observation) in Ex 2. (iii) absence of correlation (excluding the influencing point) between the two variables in Ex 3, and (iv) in Ex 4 there exists and approximately linear relationship between two variables with a high value of the correlation coefficient. In fact, in examples 1 and 3, though there do not exist any kind of linear relationship (curvilinear relationship in Ex 1 and absence of linear relationship in Ex -3) yet the value of the correlation coefficients show existence of high degree of correlation in both cases. Thus the values of the correlation coefficients in the above two cases mislead in matter of understanding the true nature and degree of relationship existing between the two variables.

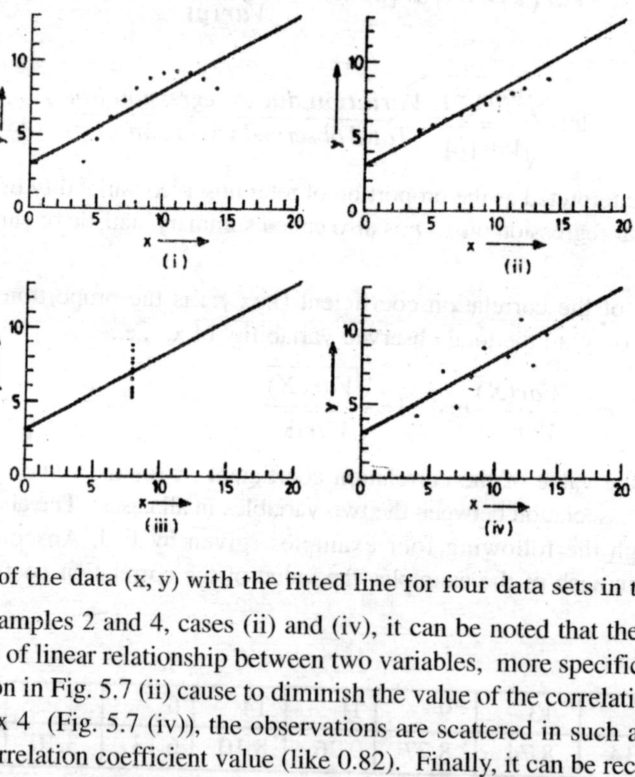

Fig. 5.8 Scatter plot of the data (x, y) with the fitted line for four data sets in the examples 1 to 4.

Recalling the examples 2 and 4, cases (ii) and (iv), it can be noted that these two situations are ideal for the existence of linear relationship between two variables, more specifically, the existence of an outlying observation in Fig. 5.7 (ii) cause to diminish the value of the correlation coefficient from 1 to 0.82, whereas in Ex 4 (Fig. 5.7 (iv)), the observations are scattered in such a way so as to justify existence of a high correlation coefficient value (like 0.82). Finally, it can be recommended that Ex 4 exposes an ideal situation for application of simple correlation and regression analysis on bivariate data.

If two variables possess a high value of correlation coefficient then existence of a causal relationship between the variables is not always evident. Actually one of the following three situations might have boosted the value of correlation coefficient.

(i) One variable might be the cause of the other variable.

(ii) One or more other causes affect both the variables.

(iii) The present value of correlation coefficient has been obtained only by chance.

Property 4. $(1-r^2)$ is the proportion of error variance or residual variance (as obtained from regression line of y on x) to the total observed variance of y.

$$\text{The residual variance} = Var(e) = \frac{1}{n}\sum e_i^2 \left[\because \frac{1}{n}\sum e_i = 0\right]$$

$$= \frac{1}{n}\sum\left(y_i - a - b_{yx}x_i\right)^2 = \frac{1}{n}\sum\{(y_i-\bar{y}) - b_{yx}(x_i-\bar{x})\}^2$$

$$= \frac{1}{n}\left\{\sum(y_i-\bar{y})^2 - 2b_{yx}\sum(y_i-\bar{y})(x_i-\bar{x}) + b_{yx}^2\sum(x_i-\bar{x})^2\right\}$$

$$= s_y^2 - 2r\frac{s_y}{s_x}.r\,s_y s_x + r^2\frac{s_y^2}{s_x^2}.\,s_x^2$$

$$= s_y^2 - r^2 s_y^2 = \left(1-r^2\right)Var(y)$$

$$\Rightarrow 1-r^2 = \frac{Var(e)}{Var(y)} = \frac{s_e^2}{s_y^2} \qquad \text{... (5.20)}$$

Hence the property.

Note 8. $(1-r^2)$ is the proportion of error variance or residual variance (as obtained from regression line of x on y) to the total observed variance of x, i.e.,

$$1-r^2 = \frac{s_{e'}^2}{s_x^2}, \text{ where } s_{e'}^2 = Var(e') \qquad \text{... (5.21)}$$

Note 9. The standard deviation of errors (s_e) is called the standard error of the estimate of y from the linear regression and generally denoted by $s_{y.x}$. So.

$$s_{y.x} = \sqrt{Var(e)} = s_y\sqrt{1-r^2} = s_e \qquad \text{... (5.22)}$$

Similarly,

$$s_{x.y} = \sqrt{Var(e')} = s_x\sqrt{1-r^2} = s_{e'} \qquad \text{... (5.23)}$$

From (5.22), we have $-1 \le r \le 1$ since, $s_y\sqrt{1-r^2} \ge 0 \Rightarrow r^2 \le 1$.

$$\Rightarrow |r| \le 1 \Rightarrow -1 \le r \le 1$$

If $r = \pm 1$ then Var (e) = 0 which implies that all the observed values (y_i) of y on the scatter diagram fall on the linear regression equation of y on x, i.e., linear regression equation of y on x is an ideal predicting equation for given values os x. And if r=0, then Var (e) = s_y^2, implying thereby that the errors of estimation (estimated errors) are as much variable as the original observations on y, and hence the linear regression equation of y on x is of no help in predicting the values of y-variable for given values of x-variable. A similar interpretation will follow from the equation (5.23)

Property 5 : The predictor variable , x and the error variable , *e* are uncorrelated, i.e., r_{xe}=*0* (from the regression line of y on x). We have

$$Cov(x,e)=\frac{1}{n}\sum(x_i-\bar{x})e_i,\ \text{since}\ \bar{e}=0$$

$$=\frac{1}{n}\sum x_i\,e_i-\bar{x}\frac{1}{n}\sum e_i=0 \qquad \text{.... (5.24)}$$

Since, $\sum x_i\,e_i=\sum x_i(y_i-\alpha-\beta x_i)=0$ (from the normal equation (5.8) and, $\frac{1}{n}\sum e_i=0$ Hence

$$r_{xe}=\frac{Cov(x,e)}{s_x\,s_e}=0 \qquad \text{...(5.25)}$$

Note 10. The predictor variable, y and the error variable, e' are uncorrelated, i.e. $r_{ye'}=0$ (from regression line of x on y).

Property 6. The magnitude of the correlation coefficient of y and x (based on *n* pairs (x_i,y_i), i = 1. 2, ..., *n*) is same as the correlation coefficient between y (observed value) and its predicted value Y, *i.e.*, $r_{yY}=|\,r_{yx}\,|$.

We know,

$$r_{yY}=\frac{Cov(y,Y)}{\sqrt{Var(y)Var(Y)}}$$

Now,

$$Cov(y,Y)=Cov(Y+e,Y)=Cov(e,Y)+Var(Y)$$
$$=Cov(e,\alpha+\beta x)+Var(Y)$$
$$=\beta Cov(e,x)+Var(Y)$$
$$=o+Var(Y)\left[\because Cov(e,x)=0, from\,(5.24)\right]$$
$$=Var(Y).$$

$$\therefore \qquad r_{yY}=\frac{Var(Y)}{\sqrt{Var(y)}\sqrt{Var(Y)}}=\frac{\sqrt{Var(Y)}}{\sqrt{Var(y)}}=|r|,\ \text{From(5.18)}$$

Hence, the property.

Note 11. The magnitude of the correlation coefficient between *x* and *y* is same as the correlation coefficient between *x* (observed value) and its predicted value, X, *i.e.*, $r_{xX}=|r_{xy}|$.

Note 12. It can be interpreted that correlation coefficient between *y*(*x*) and its recdicted value *Y*(*X*) is non-negative and numerically equal to the correlation between *x* and *y*.

Property 7. Correlation coefficient between *x* and *y* is numerically equal to the geometric mean of two regression coefficients b_{yx} and b_{xy} and $0\le(b_{yx}\times b_{xy})\le1$.

We know that the regression coefficient of y on x is $b_{yx}=r\frac{s_y}{s_x}$ and that of x on y is $b_{xy}=r\frac{s_x}{s_y}$.

So $b_{yx}\times b_{xy}=r\cdot\frac{s_y}{s_x}\times r\frac{s_x}{s_y}=r^2$

Hence, $|r|=\sqrt{b_{yx}\times b_{xy}}$

As $0\le r^2\le 1, so, 0\le b_{yx}\times b_{xy}\le 1$

Note 13. b_{yx} and b_{xy} are of same sign and the sign of r_{xy} is same as the sign of the regression coefficients.

Note 14. $r_{xy}\le\frac{b_{xy}+b_{yx}}{2}$, if r_{xy} is positive

and, $\ge\frac{b_{xy}+b_{yx}}{2}$, if r_{xy} is negative

Proof : Here, $\frac{b_{xy}+b_{yx}}{2}-r=\frac{r\frac{s_x}{s_y}+r\frac{s_y}{s_x}}{2}-r=\frac{r}{2s_xs_y}(s_x-s_y)^2$ ... (A)

Now (A) ≥ 0, if r_{xy} is positive,

and (A) < 0, if r_{xy} is negative.

Hence the inequalities hold.

Equality holds, i.e., $\frac{b_{xy}+b_{xy}}{2}=r_{xy}$, if $s_x = s_y$

The above result can also be stated as

$|r_{xy}|\le\left|\frac{b_{xy}+b_{yx}}{2}\right|=\frac{|b_{xy}|+|b_{yx}|}{2}$ (since r_{xy}, b_{xy} and b_{yx} are of same sign)

It can be proved by the relation A.M. $\ge$ G.M. on two values $|b_{xy}|$ and $|b_{yx}|$.

Property 8. The angle between the two regression lines, θ can be expressed in terms r (correlation coefficient), s_x and s_y. In fact,

$$\theta=\tan^{-1}\left\{\frac{1-r^2}{r}\cdot\frac{s_xs_y}{s_x^2+s_y^2}\right\}.$$

Equations of the regression lines y on x and of x on y are, $Y-\bar{y}=r\frac{s_y}{s_x}(x-\bar{x})$ and $X-\bar{x}=r\frac{s_x}{s_y}(y-\bar{y})$, respectively. Slopes of these two lines are rs_y/s_x and s_y/rs_x respectively. So the angle, θ between the two lines of regression is

$$\tan\theta=\pm\frac{\dfrac{s_y}{rs_x}-\dfrac{rs_y}{s_x}}{1+\dfrac{s_y}{r.s_x}.\dfrac{rs_y}{s_x}}=\pm\frac{1-r^2}{r}.\frac{s_ys_x}{s_x^2+s_y^2}$$

$$\text{Thus } \theta=\tan^{-1}\left(\frac{1-r^2}{r}\times\frac{s_ys_x}{s_x^2+s_y^2}\right) \text{ or } \theta=\tan^{-1}\left(\frac{r^2-1}{r}\times\frac{s_ys_x}{s_x^2+s_y^2}\right)$$

The angles between the two regression lines can either be acute (θ_1) or obtuse (θ_2).

$$\text{Now, } \theta_1(acute\,angle)=\tan^{-1}\left(\frac{1-r^2}{|r|}.\frac{s_ys_x}{s_x^2+s_y^2}\right)$$

$$\text{and, } \theta_2(obtuse\,angle)=\pi-\theta_1 \quad =\tan^{-1}\left(\frac{r^2-1}{|r|}.\frac{s_ys_x}{s_x^2+s_y^2}\right)$$

In the following figures two regression lines have been drawn with the corresponding typcial value of θ and r.

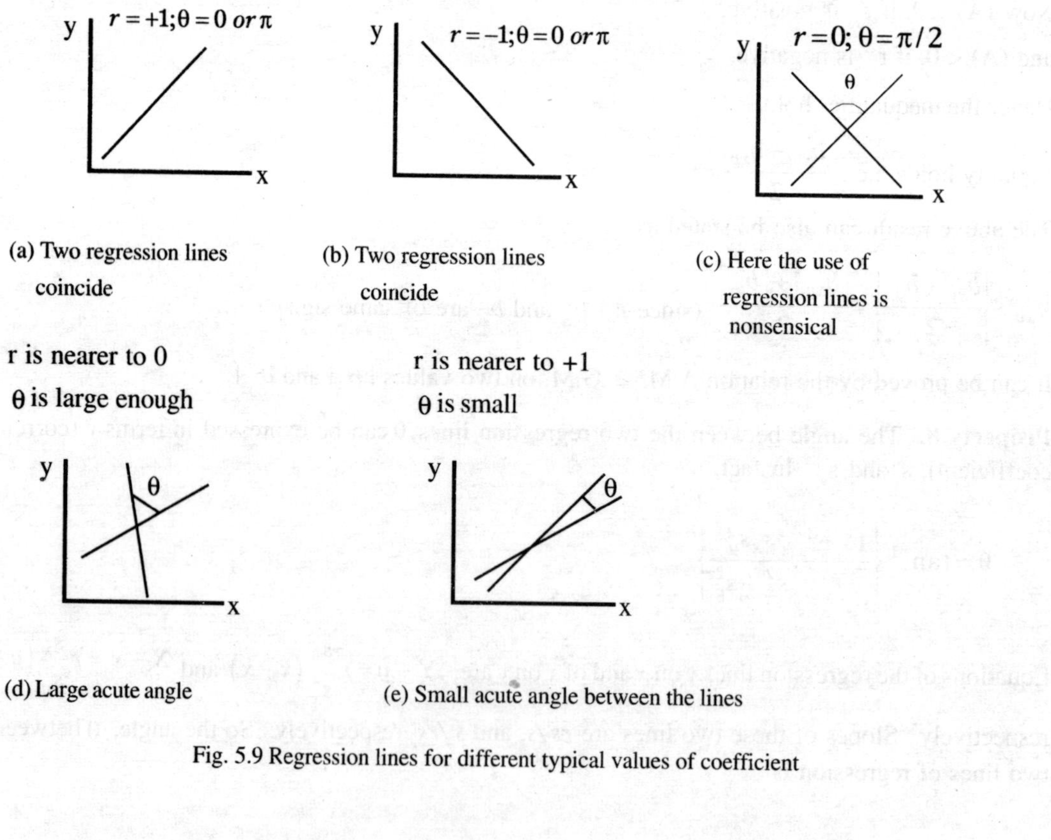

(a) Two regression lines coincide

(b) Two regression lines coincide

(c) Here the use of regression lines is nonsensical

r is nearer to 0
θ is large enough

r is nearer to +1
θ is small

(d) Large acute angle

(e) Small acute angle between the lines

Fig. 5.9 Regression lines for different typical values of coefficient

5.8 Limitations of Correlation Coefficient

Limitations of Pearson's product moment correlation coefficient exist in the underlying assumptions. First one is that the variables under study are to be linearly related. So the correlation coefficient will serve as a satisfactory measure in case of linear relationship existing between the two variables. For non-linear relationship between two variables, the value of the correlation coefficient may be approximately zero. So nearly zero value of correlation coefficient does not always mean that the variables are not related. In fact, there may exist non-linear relationship between the two variables. In the following figures non-linear relationships between the two variables have been shown.

The following hypothetical examples show the fact even if there exists perfect relationship, the correlation coefficient is zero.

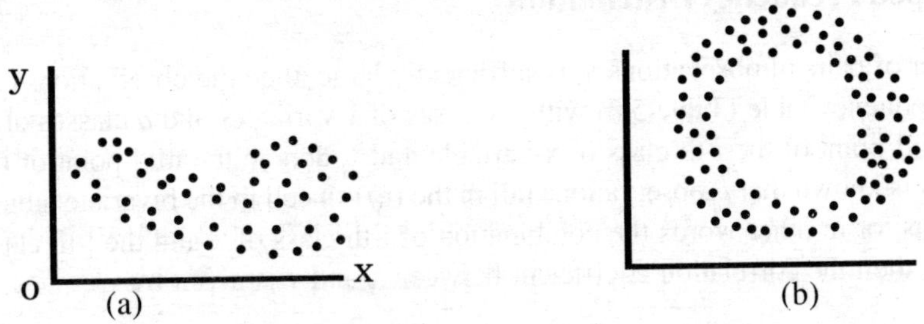

Fig. 5.10 Scatter Diagram for non-linear relations

Illustration 5.2.

x:	–5	–4	–3	–2	–1	0	1	2	3	4	5
y:	0	1	2	3	4	5	4	3	2	1	0

Here, $\bar{x} = 0, \sum x_i y_i = 0 \Rightarrow Cov(x, y) = \frac{1}{n}\sum x_i y_i - \bar{x}\bar{y} = 0$ and hence, $r_{xy} = 0$.

Here two straight lines $y = x + 5$ and $y = 5 - x$, one for x-values in the domain (–5, 0), another for x-values in the domain (1, 5), express the relationships between x and y.

Illustration 5.3

X:	–3	–2	–1	0	1	2	3
Y:	9	4	1	0	1	4	9

In each of these cases r = 0, so the data suggest that there does not exist linear type of relationship between the two variables x and y but they can exhibit other types of dependence. That is why in regression analysis we draw the scatter diagram as a first step to know whether there exists linear relationship between the variables or not.

The second assumption in determination of correlation coefficient is that we choose those variables so that the factors affecting each of the two variables are either identical or related in a causal fashion. For example, the variables relating to age, height, weight, supply, demand, price etc. are causally related, and therefore calculation of correlation coefficient is justified since regarding an individual any factor affecting height and any factor affecting its weight is either identical or interrelated. If the causes operating on the variables are entirely independent, then there cannot exist any correlation between the variables taken into account. For example, correlation between steel production in India and number of birds that immigrated to India during the last ten years or between the number of road accidents per day and the number of still births in a region cannot exist even if the numerical calculations of the correlation coefficient between such variables show very high values. Such a correlation is called spurious or non-sense or chance correlation.

Presence or absence of correlation between two variables cannot be judged by the values of the correlation coefficient alone, since spurious correlation may also rise due to non-elimination of the third factor, which may include non-homogeneity of data (blocking or grouping effect) or social norms or any other variables related to the two involved variables.

5.9 Calculation of Correlation Coefficient and Determination of Regression Lines from Grouped Frequency Distribution

In the number of pairs of obsevations n, is sufficiently large, then the observations can be presented in a bivariate frequency table (Table 5.3) with p classes of x-variables and q classes of y-variable. Let x_i denote the mid-point of the i-th class of x-variable and y_i donote the mid-point of the j-th class of y-variable. If it is known that f_{ij} observations fall in the (i,j)-th cell in the bivariate table involving p rows and q columns, or in other words the combination of i-th class of x and the j-th class of y contains f_{ij} observations, then the correlation coefficient between x and y is given by

$$r = \frac{Cov(x,y)}{\sqrt{Var(x)Var(y)}} = \frac{\sum_{i,j}(x_i-\bar{x})(y_j-\bar{y})f_{ij}}{\sqrt{\left(\sum(x_i-\bar{x})^2 f_{i0} \sum(y_j-\bar{y})^2 f_{0j}\right)}}$$

$i = 1, 2, ..., p \quad j = 1, 2, ..., q.$

where, $f_{i0} = \sum_j f_{ij}, f_{0j} = \sum_i f_{ij},$

$$n = \sum_i \sum_j f_{ij} = \sum_i f_{i0} = \sum_j f_{0j},$$

$$\bar{x} = \frac{\sum x_i f_{i0}}{n}, \text{ and } \bar{y} = \frac{\sum y_j f_{0j}}{n}$$

To make the calculation simpler the following steps are performed. Firstly, the base (origin) and scale of x and y variables are changed. It will be advantageous if midvalues of the middle or nearly middle classes in the classified ranges of x and y variables are taken as origins and widths of the class intervals of x and y variables as the corresponding scales.

Let the new varibles be, $u = \dfrac{x-h}{k}$ and $v = \dfrac{y-1}{m}$, k and m being the corresponding widths. Using the Property 3 of section 5.5, the expression of r reduces to

$$r = \frac{\sum_i \sum_j (u_i - \bar{u})(v_i - \bar{v}) f_{ij}}{\sqrt{\left(\sum (u_i - \bar{u})^2 f_{i0} \sum (v_i - \bar{v})^2 f_{0j}\right)}}$$

$$= \frac{n \sum_i \sum_j u_i v_j f_{ij} - \left(\sum u_i f_{i0}\right)\left(\sum v_j f_{0j}\right)}{\sqrt{\left\{\left(n \sum u_i^2 f_{i0} - \left(\sum u_i f_{i0}\right)^2\right)\left(n \sum v_j^2 f_{0j} - \left(\sum v_i f_{0j}\right)^2\right)\right\}}}$$

$$\left[\because \bar{u} = \frac{\sum u_i f_{io}}{n} \text{ and } \bar{v} = \frac{\sum v_j f_{0j}}{n}\right]$$

We can calculate $\sum u_i f_{i0}$, $\sum u_i^2 f_{i0}$, $\sum v_j f_{0j}$ and $\sum v_j^2 f_{0j}$ very easily from the given bivariate frequency table and we can also calculate $\sum_i \sum_i u_i v_j f_{ij}$ in both ways we can check the correctness of the calculation, since $\sum_i \sum_i u_i v_j f_{ij}$ can be expressed as,

$$\sum_i \sum_i u_i v_j f_{ij} = \sum_i u_i \left(\sum_j v_j f_{ij}\right) = \sum_i u_i V_i, \text{ where } V_i = \sum_j v_j f_{ij},$$

or

$$= \sum_j v_j \left(\sum_i u_i f_{ij}\right) = \sum_j v_j U_j \text{ where } U_j = \sum_i u_i f_{ij}$$

Again,

$$\sum_i V_i = \sum_i \sum_j v_j f_{ij} = \sum_j v_j \left(\sum_i f_{ij}\right) = \sum_j v_j f_{0j}$$

and,

$$\sum_j U_j = \sum_j \sum_i u_i f_{ij} = \sum_i u_i \left(\sum_j f_{ij}\right) = \sum_i u_i f_{io}$$

So these identities, (1) : $\sum_i u_i V_i = \sum_j v_j U_j$, (2): $\sum_i V_i = \sum_i v_j f_{0j}$ and (3) : $\sum_j U_j = \sum_i u_i f_{i0}$ are used as checks in the calculation.

Illustration 5.4

Recalling the example of Table 5.3, we can calculate the correlation coefficient by the above method. Let x denote the marks in Mathematics and y denote the marks in Statistics, then the following table shows necessary calculations for correlation coefficient (r) between x and y.

Table 5.7 – Determination of correlation coefficient from grouped data

Class mark x_i		25	35	45	55	65					
Class mark y_j	u_i $\frac{x_i - 45}{10}$ / v_i $\frac{y_j - 45}{10}$	–2	–1	0	1	2	f_{oj}	$v_j f_{oj}$	$v_j^2 f_{oj}$	U_i	$v_j U_j$
25	–2	3		1			4	–8	16	–6	12
35	–1	2	3	2	2		9	–9	9	–5	5
45	0		5	4			9	0	0	–5	0
55	1		1	2	2	2	7	7	7	5	5
65	2				2	1	3	6	12	4	8
	f_{io}	5	9	9	6	3	32	–4	44	–7	30
	$u_i f_{io}$	–10	–9	0	6	6	–7	check			
	$u_i^2 f_{io}$	20	9	0	6	12	–41	check	check		
	V_i	–8	–2	–2	4	4	–4				
	$u_i V_i$	16	2	0	4	8	30				

So using the formula the value of correlation coefficient,

$$r = \frac{n\sum u_i V_i - \left(\sum U_j\right)\left(\sum V_i\right)}{\sqrt{\left\{\left(n\sum_i u_i^2 f_{i0} - \left(\sum_j U_j\right)^2\right)\left(n\sum_j v_j^2 f_{oj} - \left(\sum_i V_i\right)^2\right)\right\}}}$$

$$= \frac{32 \times 30 - (-7)(-4)}{\sqrt{\left\{\left(32 \times 41 - (-7)^2\right)\left(32 \times 44 - (-4)^2\right)\right\}}} = 0.706$$

The value of regression coefficient of v on u,

$$b_{vu} = \frac{n\sum u_i V_i - \left(\sum U_j\right)\left(\sum V_i\right)}{n\sum u_i^2 f_{io} - \left(\sum U_j\right)^2} = \frac{32 \times 30 - (-7)(-4)}{32 \times 41 - (-7)^2} = 0.7379$$

and hence the regression coefficient of y on x, $b_{yx} = (b_{vu})$=0.7379, (since the scales of both variables are equal)

Now, $\bar{x} = 45 + 10\bar{u} = 42.81$

and $\bar{y} = 45 + 10\bar{v} = 43.75$

since $\bar{u} = \frac{-7}{32}$ and $\bar{v} = \frac{-4}{32}$

So the regression line of y on x becomes,

$$Y = \bar{y} + b_{yx}(x - \bar{x}) = 43.75 + 0.7379(x - 42.81)$$

or, $$Y = 12.16 + 0.7379x.$$

Similarly, the regression coefficient of x on y.

$$b_{xy}(= b_{uv}) = \frac{n\sum u_i V_i - \left(\sum U_j\right)\left(\sum V_i\right)}{n\sum v_j^2 f_{0j} - \left(\sum V_i\right)^2} = \frac{32 \times 30 - (-7)(-4)}{32 \times 44 - (-4)^2}$$

= 0.6695, (since the scales used to change the variables are equal)

And the regression line of x on y becomes

$$X = \bar{x} + b_{xy}\left(y - \bar{y}\right) = 42.81 + 0.6695(y - 42.75)$$

or $$X = 13.52 + 0.6695\, y.$$

5.10. Fitting of Curves to Observed Data

In a study concerning two variables, x and y, we collect n pairs of observations on two variables as (x_i, y_i), $i = 1, 2, ..., n$.

Here we treat the variable x as an independent variable (whose values are controlled and hence are not likely to be subjected to errors) and the other variable y as the dependent variable (which is subjected to errors in observations). If we plot these points on a graph paper we see apparently an irregular series of points, as if, no curve (mathematical expression) of the form $y = f(x)$ can pass through those series of points.

The general problem in curve fitting is to find the most possible approximate expression of the form, $Y = f(x; a, b, c, ...)$ representing the functional relationship between the dependent and independent variables with the parameters (unknown constants), a, b, c, ... etc. to be estimated from the observed data. So the precision of the suggested relation depends largely on the way the parameters are estimated. Here we discuss the method of least squares as a suitable method for estimating the parameters. When the assumption of linear relationship holds, $Y = a + bx$ exhibits graphically the notions of this method.

The method of least squares suggests that the straight line $Y = a + bx$ will be the best for which $\sum_{i=1}^{n} e_i^2$ is minimum and the normal equations have to be formed to get the solutions of the parameters by minimising $\sum e_i^2$ with respect to the parametes a and b respectively. In the following, we discuss the method in details:

(i) For $Y = a + bx$, we minimise $S = \sum (y_i - a - bx_i)^2$ with respect to a and b and the respective equations are as follows :

$$\frac{\partial S}{\partial a} = 0 \Rightarrow \sum y_i = na + b\sum x_i \quad .... (5.26)$$

and,

$$\frac{\partial S}{\partial b} = 0 \Rightarrow \sum x_i y_i = a\sum x_i + b\sum x_i^2 \quad ... (5.27)$$

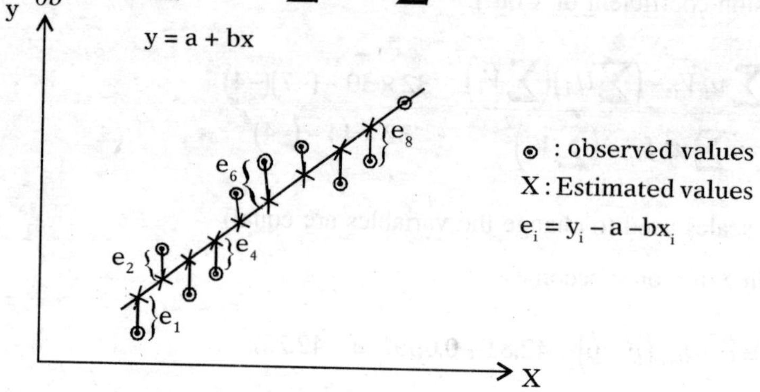

Fig. 5.11 Scatter diagrams of 10 hypothetical observations with an arbitrary straight line

The solutions of the normal equations (5.26) and (5.27) will be easier if the values of x(controlled variable) are taken at equal distances from each other and are transformed to the values t_i's, such that $\sum t_i = 0$. Then the best predicting equation becomes $Y = a' + b't$ with $\hat{a}' = \sum y_i / n$ and $\hat{b}' = \sum y_i t_i / \sum t_i^2$ (n being the number of observations), where $\hat{a}'$ and $\hat{b}'$ are the respective estimates. For example, if x_i values are 10, 20, 30, 40 and 50, the corresponding t_i values will be –2, –1, 0, 1, 2 $\left(here\ t_i = \frac{x_i - 30}{10}\right)$ and if the x_i values are 10, 20, 30, 40, 50, and 60, the corresponding t_i values will be –5, –3, –1, 1, 3, 5 $\left(here,\ t_i = \frac{x_i - 35}{5}\right)$ so that $\sum t_i = 0$ in each case.

(ii) For second degree equation (linear model)

For a second degree polynomial (predicting equation, $Y = a + bx + cx^2$, we minimise $S = \sum \left(y_i - a - bx_i - cx_i^2\right)^2$ with respect to *a, b* and c respectively and get the following normal equations (the summation being taken over n pairs of observations).

$$\frac{\partial S}{\partial a} = 0 \Rightarrow \sum y_i = na + b\sum x_i + c\sum x_i^2$$

$$\frac{\partial S}{\partial b} = 0 \Rightarrow \sum y_i x_i = a\sum x_i + b\sum x_i^2 + c\sum x_i^3 \qquad \text{.... (5.28)}$$

$$\frac{\partial S}{\partial c} = 0 \Rightarrow \sum y_i x_i^2 = a\sum x_i^2 + b\sum x_i^3 + c\sum x_i^4$$

Solving the above set of equations, we get the best estimate for *a, b* and c̲ and hence we get the best second degree predicting equations as $\hat{Y} = \hat{a} + \hat{b}x + \hat{c}x^2$, $\hat{a}, \hat{b}$ *and* $\hat{c}$ being the respective estimates.

Similarly, we get the best fitted p-th degree polynomial curve (linear model) like $Y = a_0 + a_1 x + a_2 x^2 + ... + a_p x^p$ by estimating (p+1) unknowns (parameters) $a_0, a_1, a_2, ..., a_p$ as the solution of (p+1) normal equations given as follows :

$$\frac{\partial S}{\partial a_0} = 0, \frac{\partial S}{\partial a_1} = 0, ..., \frac{\partial S}{\partial a_p} = 0 \qquad \text{..... (5.29)}$$

where $S = \sum \left(y_i - a_0 - a_1 x_i ... - a_p x_i^p\right)^2$, the summation being taken over all pairs of observations.

The greatest limitation in curve-fitting by least squares method lies in the choice of the appropriate curve to be fitted on the basis of the set of observations. The choice of curve needs skill, intelligence and experience of the investigators associated with the work. If the suggested curve is neither linear nor quadratic nor even of polynomial type, some transformation can be made on either of or both the dependent and independent variables so that we may have the above set of normal equations (for linear models) involving transformed variables for estimation for the parameters. Let us take the example of the exponential curve, $Y = ae^{bx}$.

Take 5.8 - Linear Transformation of some Linear and Non-linear Models

Function	**Linear Form**	**Identity**
$Y = ax^b$	$Y' = a' + bx'$	$Y' = \log_{10} Y,\ a' = \log_{10} a,\ x' = \log_{10} x;$
$Y = a + b\log_{10} x$	$Y = a + bx'$	$x' = \log_{10} x;$
$Y = \frac{e^{a+bx}}{1+e^{a+bx}}$	$Y' = a' + b'x$	$Y' = \log_{10}\frac{Y}{1-Y}, a' = a\log_{10} e, b' = b\log_{10} e;$
$Y = \frac{x}{ax+b}$	$Y' = a + bx'$	$Y' = \frac{1}{y}, x' = \frac{1}{x}$
$Y = \frac{1}{(a+bx)^2}$	$Y' = a + bx$	$Y' = \frac{1}{\sqrt{Y}}$
$\frac{1}{Y} = a + \frac{b}{1+x}$	$Y' = a + bx'$	$Y' = \frac{1}{y}, x' = \frac{1}{1+x}$

In order to linearise the equation, logarithms are taken on both sides and the above curve reduces to $Y' = a' + b'x$. where

$$Y' = \log_{10} Y, a' = \log_{10} a, \text{ and } b' = b\log_{10} c$$

Next, we apply the least squares method on the transformed relationship for estimating a' and b' and finally on solving the set of normal equations so obtained, the estimates of a' and b' are found.

5.11. Concept of Rank Correlation and Determination of Different Rank Correlation Coefficients

Suppose measurements on two characters, x and y representing marks obtained in admission test and final examination, for n individuals are recorded from an equispaced series ordered numbers. These measurements can be treated as the respective ranks of the individuals and these numbers are generally taken as n natural numbers. An individual is assigned rank r with respect to a character if $(r - 1)$ individuals have shown qualitatively better performance than him with respect to the said character. Ranks are assigned to individuals after judging their comparative performance. In fact, ranking is done because of the following three basic causes :

(i) when the quantitative measurements on the charactetistics of individuals are not possible,

(ii) even if the quantitative measurements are possible, existence of time and/or money and/or other resource constraints make actual measurements impossible, and

(iii) characters are measured quantitatively, but when it is necessary to get a quick idea about the degree of relationship between the two variables under study.

A ranking wherein two or more individuals are allotted the same rank is called a tie. In order to measure the degree of relationship existing between the two sets of ranks, the concept of rank correlation coefficient is visualised. The corelation coefficient between the paired ranks (r_{x_i}, r_{y_i}), i = 1, 2, ..., n is called Spearman's rank correlation coefficient; r_{x_i} and r_{y_i} are the ranks of the i-th individuals on characters (variables) x and y respectively (i.e., $r_{x_i}, r_{y_i} \in (1, 2, ..., n), \forall\ i = 1, 2, ..., n$). Thus, it is a Product Moment Correlation coefficient between two series of ranks. It's derivation is done as follows.

Case I, (Case of no tie)

It is konwn that

$$\sum_{i=1}^{n} r_{x_i} = \sum_{i=1}^{n} r_{y_i} = 1 + 2 + ... + n = \frac{n(n+1)}{2}$$

$$\therefore \text{ Mean rank for each variable } \bar{r}_x = \bar{r}_y = \frac{1}{n} \times \frac{n(n+1)}{2} = \frac{n+1}{2}$$

Again, $\sum r_{x_i}^2 = \sum r_{y_i}^2 = 1^2 + 2^2 + 3^2 + ... + n^2 = \frac{n(n+1)(2n+1)}{6}$

$\therefore$ Variances of ranks for each variable,

$$s_{r_x}^2 = s_{r_y}^2 = \frac{1}{n} \times \frac{n(n+1)(2n+1)}{6} - \left(\frac{n+1}{2}\right)^2 = \frac{n^2-1}{12}$$

Now the average of squares of rank differences of individuals is given by,

$$\frac{1}{n}\sum d_i^2 = \frac{1}{n}\sum \left(r_{x_i} - r_{y_i}\right)^2 = \frac{1}{n}\sum \left\{\left(r_{x_i} - \bar{r}_x\right) - \left(r_{y_i} - \bar{r}_y\right)\right\}^2, \left[\because \bar{r}_x = \bar{r}_y\right] \quad ... (5.30)$$

$$= s_{r_x}^2 + s_{r_y}^2 - 2Cov\left(r_x, r_y\right) = 2\left(s_{r_s}^2 - Cov\left(r_x, r_y\right)\right), \left[\because s_{r_x}^2 = s_{r_x}^2\right]$$

or $Cov\left(r_x, r_y\right) = \frac{n^2-1}{12} - \frac{1}{2n}\sum d_i^2$

Finally, Spearman's correlation coefficient is,

$$R_s = \frac{Cov\left(r_x, r_y\right)}{s_{r_x} \times s_{r_y}} = \frac{\frac{n^2-1}{12} - \frac{1}{2n}\sum d_i^2}{\left(n^2-1\right)/12} = 1 - \frac{6\sum d_i^2}{n\left(n^2-1\right)} \quad ... (5.30a)$$

In the case of perfect agreement between the two groups of ranks relating to the character under study, the ranks are equal for each individual, i.e., $r_{x_i} = r_{y_i}, \forall i$ or $d_i = 0$ for i = 1, 2, ..., n, and hence, $R_s = 1$, whereas in the case of perfect disagreement, the ranks are totally reverse for each i = 1, 2, ..., n, *i.e.*, $r_{x_i} + r_{y_i}$ = constant (=n+1) or $r_{x_i} = n + 1 - r_{y_i}$, and hence

$$\Sigma d_i^2 = \Sigma\left(r_{x_i} - r_{y_i}\right)^2 = \Sigma\left(n+1-r_{y_i} - r_{y_i}\right)^2 = \Sigma\left(n+1-2r_{y_i}\right)^2$$

$$= \Sigma(n+1)^2 - 4(n+1)\Sigma r_{y_i} + 4\Sigma r_{y_i}^2$$

$$= n(n+1)^2 - 4\frac{(n+1)n(n+1)}{2} + \frac{4n(n+1)(2n+1)}{6} = \frac{n\left(n^2-1\right)}{3}$$

Perfect agreement (ordering of the ranks is same for both the characters)

r_{x_i} : 1 2 3 4 ....n

r_{y_i} : 1 2 3 4 ....n

$r_{x_i} - r_{y_i}$: 0 0 0 0........0 $\Rightarrow r_{x_i} = r_{y_i}, \forall i$

Perfect Disagreement (ordering of the ranks is reversed for both the characters)

r_{x_i} :1	2	3	4	... ...	$n-2$	$n-1$	n	
r_{y_i} :n	$n-1$	$n-2$	$n-3$	... ...	3	2	1	
$r_{x_i}+r_{y_i}$:	$n+1$	$n+1$	$n+1$	$n+1$	... ...	$n+1$	$n+1$	$n+1$

$$\Rightarrow r_{x_i} = n+1-r_{y_i}, \forall i$$

Thus for perfect agreement rank differences ($r_{x_i} - r_{y_i}$, $i = 1, 2, \ldots n$) is constant (=0) for each individual and for perfect diagreement rank total ($r_{x_i} + r_{y_i}$, $i = 1, 2, \ldots n$) is constant (= n+ 1) for each individual.

Putting the value of $\sum d_i^2$ in R_s, the value of the correlation coefficient in cases of perfect agreement and disagreement come out as equal to + 1 and – 1 respectively.

Case II. Case of tie

If p individuals receive the same rank (w) it is said that there exists a tie of length p and also if these p individuals follow 'r' other individuals in ranking, then the said rank, w, will be equal to the average of p such ranks, $(r+1), (r+2), \ldots, (r+p)$,

$$w = \frac{(r+1)+(r+2)+\ldots+(r+p)}{p} = \frac{rp + \frac{p(p+1)}{2}}{p} = r + \frac{p+1}{2}$$

By assigning the common rank, w, the value $r + \frac{p+1}{2}$, the average rank of all individuals remains unaltered, but the variance is reduced by the amount

Var (natural ranking) – *Var* (tied ranking)

$$= \frac{1}{n}\left\{(r+1)^2 + (r+2)^2 + \ldots + (r+p)^2 - p\left(r + \frac{p+1}{2}\right)^2\right\} = \frac{p(p^2-1)}{12n}$$

(since the arithmetic means in natural ranking and tied ranking are equal)

While ranking the n individuals, if there exist tied ranks of different lengths, then the reductions (to be needed) in variance will be additive. Let in the ranks on the first character (r_x) there be m ties of lengths, $l_1, l_2, \ldots, l_m$ and in the ranks on the second character (r_y) there be k ties of lengths, $l_1', l_2', \ldots, l_k'$ then the total reduction in both the characters will be,

$$\frac{l_1(l_1^2-1)}{12n} + \frac{(l_2^2-1)}{12n} + \ldots + \frac{l_m(l_m^2-1)}{12n} = \sum_{i=1}^{m} \frac{l_i(l_i^2-1)}{12n} = T_x, \textit{ say}$$

and $\frac{l_1'(l_1'^2-1)}{12n}+\frac{l_2'(l_2'^2-1)}{12n}+\ldots+\frac{l_k'(l_k'-1)}{12n}=\sum_{j=1}^{k}\frac{l_j'(l_j'^2-1)}{12n}=T_y$, *say.*

Then,

$$Var\,(r_x) = Var\text{ (natural ranking) – total reduction}$$

$$=\frac{n^2-1}{12}-T_x\text{, and similarly,}$$

$$Var(r_y)=\frac{n^2-1}{12}-T_y$$

Now, from (5.30) we write,

$$Cov(r_x,r_y)=\frac{1}{2}\{Var(r_x)+Var(r_y)\}-\frac{1}{2n}\sum_{i=1}^{n}d_i^2 \quad =\frac{n^2-1}{12}-\frac{T_x+T_y}{2}-\frac{1}{2n}\sum d_i^2$$

Thus, $$R_s=\frac{Cov(r_x,r_y)}{\sqrt{Var(r_x)}\sqrt{Var(r_y)}}=\frac{\frac{n^2-1}{2}-\frac{T_x+T_y}{2}-\frac{1}{2n}\sum_i d_i^2}{\sqrt{\frac{n^2-1}{2}-T_x}\sqrt{\frac{n^2-1}{12}-T_y}} \quad \ldots (5.31)$$

Another measure of rank correlation coefficient is due to Kendall and it is denoted by τ or R_k. The method of its determination is as follows.

Case 1 : (Case of no tie)

Firstly, the ranks on one of the characters are arranged in ascending order and the corresponding ranks on the other character are written against the ordered arrangement of the first. In the series of ranks on the other character for any pair of individuals (i, j), i < j (= 1, 2, ..., n), we allot the score + 1 and –1 if the pair of ranks are in ascending order and in descending order, respectively. Such types of scoring is done for each of the $\binom{n}{2}$ possible pairs. Then according to Kendall, the rank correlation coefficient, τ and R_k is defined as,

$$\tau=R_k=\frac{Total\,score}{Maximum\,possible\,score}=\frac{2\times total\,score}{n(n-1)}$$

$$=2\sum_{i<j}b_{ij}\,/\,n(n-1) \quad \ldots (5.32)$$

Maximum possible score is $\binom{n}{2}$ (since score 1 for each of $\binom{n}{2}$ pairs) and lowest possible score is $-\binom{n}{2}$, (since score – 1 for each of $\binom{n}{2}$ pairs). In case of perfect agreement the value of τ is + 1 as can be shown below :

r_x	1	2	.....	n
r_y	1	2	......	n

The total score for any series of ranks as above

$$=(n-1)+(n-2)+...+1=\frac{n(n-1)}{2}$$

So, $\tau=+1$

Now $\tau=-1$, for perfect disagreement as can be seen below :

r_x	1	2	.....	n
r_y	n	n –1	......	1

the total score with respect to the series of ranks on the second character (y variable)

$$=-(n-1)-(n-2)-...-1=-\frac{n(n-1)}{2}$$

and thus $\tau=-1$

In case the ranks on the character is either arranged in natural or ascending order, the total score can be obtained by considering the ranks in the other series. If out of $\binom{n}{2}$ possible pairs formed from the ranks of the other characters, P pairs have ranks in the natural order and Q pairs have ranks in the reverse order, then the total score is $P-Q$ as there are P pairs, each receiving score + 1 and Q pairs, each receiving a score – 1. Thus

$$\tau=\frac{P-Q}{n(n-1)/2}$$

$$P+Q=\text{total number of pairs}=\binom{n}{2}=n(n-1)/2,$$

τ can be looked upon as a product moment correlation coefficient. For ranking with respect to 1st character u_i be the rank of the i-th individual. For the pair (i, j), $i<j$, $a_{ij}=1$ if $u_i<u_j$ and $a_{ij}=-1$, if $u_i>u_j$. Similarly, b_{ij} is defined on the basis of the ranks of the second character namely, v_i and v_j of the *i*-th and *j*-th individuals respectively. In fact, $b_{ij}=1$ if $v_i<v_j$ and $b_{ij}=-1$ if $v_i>v_j$.

Thus $$\tau=\frac{\sum a_{ij}b_{ij}}{\sum a_{ij}^2\sum b_{ij}^2}$$ each summation being carried over all possibles pairs (i,j), $i<j$.

Case II : (Case of tie)

In the case of tie existing in ranks on either or both of the characters, we take $a_{ij} = 0$ if $r_{x_i} = r_{x_j}$ and $b_{ij} = 0$ if $r_{yi} = r_{yj}$ and then each of $\sum\sum_{i<j} a_{ij}^2$ and $\sum\sum_{i<j} b_{ij}^2$ is reduced by the total number of pair combinations formed from each of the fixed length. Let there be m ties of lengths, $l_1, l_2, ..., l_m$, in ranking with respect to the first character (x-variable), then $\sum\sum_{i<j} a_{ij}^2$ would be reduced to

$$\binom{n}{2} - \sum_{i=1}^{m}\binom{l_i}{2} = \binom{n}{2} - T_x \text{, say.}$$

Similarly, if there are k ties of lengths $l_1', l_2', ..., l_k'$, in ranking with respect to the second character, then $\sum\sum_{i<j} b_{ij}^2$ would be reduced to $\binom{n}{2} - \sum_{j=1}^{k}\binom{l_j'}{2} = \binom{n}{2} - T_y$, say.

Thus the formula for τ in the case of tied ranks becomes

$$\tau = \frac{\textit{Total Score}}{\sqrt{\binom{n}{2} - T_x}\sqrt{\binom{n}{2} - T_y}} \quad \text{... (5.33)}$$

If we perform rank correlation coefficient even when the characters are measurable, we may lose some information on the variables since the performance difference in a pair of individuals is never reflected in their rank differences. But rank correlation is very useful in Applied Psychology & Education and Behavioural sciences where instead of marks, scores, points on ordinal scales, indices, ranks form the databases.

5.12 Two-way Frequency Distribution of Attributes - Contingency Table

We know that attribute is a non-measurable qualitative character of an individual. Blindness, health, honesty, smoking habits, cleanliness etc., are a few examples of attributes. An individual may possess two or more qualitative characters or attributes simultaneously, and sometimes we are interested in investigating whether there exists any kind of association between the two qualitative characters, for example, whether there exists any association between blindness and honesty, health and smoking habits etc., so that we can make observations like - 'blind men are honest'. 'people with smoking habits possess bad health', etc. For this purpose, the data collected by way of individual inquiries are categorised with respect to two or more classes for each of the two attributes under study and are presented in a two-way table. The classification (distribution) of all individuals (total frequency) into different categories or classes (class frequencies) form a two-way frequency distribution of attributes and the said two-way table is called a Contigency table. If a particular attribute is categorised into k classes and the other attributes is categorised into l classes, then the tabular representation of the frequencies (in different cells) in the two-way table is called a *(k x l)* contingency table containng kl cells. The simplest of such types of tables is a 2 x 2 contingency table, wherein each of the two characters is defined into two classes – presence and absence of the character (qualitative). In the following, a 2 x 2 contingency table is presented.

Table 5.9 – A 2 x 2 Contingency Table

B : Type of Character	A : Occular Quality		Total
	Blind eye (A)	Normal eye (α)	
Honesty (B)	f_{AB}	$f_{\alpha B}$	$f_B = f_{AB} + f_{\alpha B}$
Dishonesty(β)	$f_{A\beta}$	$f_{\alpha\beta}$	$f_\beta = f_{A\beta} + f_{\alpha\beta}$
Total	$f_A = f_{AB} + f_{a\beta}$	$f_\alpha = f_{\alpha B} + f_{\alpha\beta}$	$n = f_A + f_\alpha = f_B + f_\beta = f_{AB} + f_{A\beta} + f_{\alpha B} + f_{\alpha\beta}$

f_{AB} : Number of individuals, blind and honest

$f_{A\beta}$: Number of individuals, blind and dishonest

$f_{\alpha\beta}$: Number of individuals, normal and honest

$f_{\alpha\beta}$: Number of inviduals, normal and dishonest.

Clearly, the attributes need not have just two classes each. In a ($k \times l$) contingency table, as was mentioned earlier, there are k rows and l columns. Such a table is also called an association table. A 2×2 contingency table can also be presented in the following form.

Table 5.9a – A 2 × 2 Contingency Table

Attributes		A		Total
		A	α	
B	B	a	c	a+c
	β	b	d	b+d
Total		a+b	c+d	a+b+c+d=n

More examples of Attributes (with their classes) :

(i) Readability : easy reading, difficult reading.

(ii) Smoking habit: chain, moderate, occassional

In the Table 5.9, the cell frequencies, together with their grand total, give the joint frequency distribution of the attributes because the nature of their joint variation in the given group of individuals can be visualised through the two-way contngency table. The row totals (marginal frequencies) together with the grand total give the marginal (frequency) distribution of the attribute "type of character" and the column totals together with the grand total gives the marginal (frequency) distribution of the attribute "occular quality". The table identifies the following conditional distributions also. The frequencies occuring in the first/second row along with the marginal total of the first/second row give the conditional (frequency) distribution of the attributer "occular quality" for fixed value of the form, namely, honesty/dishonesty of the attribute "type of character', respectively. Similarly, the frequencies occuring in the first/second column along with the marginal total of the first/second column give the conditional (frequency) distribution of the attribute "type of character" for fixed value of the form, namely, blind eye/normal eye of the attribute "occular quality" respectively.

5.13 Definitions and Measures of Independence and Association between two attributes from 2 ×2 Contingency.

Let *A* and *B* be the two attributes under study with two classes each. Let us denote the classes as *A* and α indicating presence and absence of the character *A*, respectively; and similarly the classes *B* and β to denote the presence and absence of the character *B*, respectively. The four cell frequency may be denoted by f_{AB} (number of individuals having both the characters), $f_{A\beta}$ (number of individuals having the character *A* but not *B*), $f_{\alpha B}$ (number of individuals having the character *B* but not *A*), and $f_{\alpha\beta}$ (number of inviduals having neither of the characters). Also, the marginal frequencies for the classifications of the attribute A are denoted by f_A (number of individuals having the character *A*) and f_α (number of individuals not having the character A) ; and similarly, for the character B the corresponding marginal frequencies are denoted by f_B and f_β. From the contingency table (Table 5.9) it is clear that,

$$f_A = f_{AB} + f_{A\beta}\,;\; f_\alpha = f_{\alpha B} + f_{\alpha\beta}$$

$$f_B = f_{AB} + f_{\alpha B}\,;\; f_\beta = f_{A\beta} + f_{\alpha\beta} \text{ and the total frequency.}$$

$$n = f_{AB} + f_{A\beta} + f_{\alpha B} + f_{\alpha\beta} = f_A + f_\alpha = f_B + f_\beta$$

If the proportion of f_{AB} to f_A is same as the proportion of $f_{\alpha\beta}$ to f_α , we say that *A* and *B* are independent of each other, since the presence of the character *B* in an individual does not determine the presence or absence of the character *A* exactly. So *A* and *B* are independent, if

$$\frac{f_{AB}}{f_A} = \frac{f_{\alpha B}}{f_\alpha} = \frac{f_{AB} + f_{\alpha B}}{f_A + f_\alpha} = \frac{f_B}{n} \left(\because \frac{a}{b} = \frac{c}{d} = \frac{a+c}{b+d}\right)$$

i.e., if $f_{AB} = \dfrac{f_A f_B}{n}$

Also,

$$\frac{f_{AB}}{f_A} = \frac{f_{\alpha B}}{f_\alpha} \Rightarrow f_{\alpha B} = \frac{f_\alpha f_B}{n}$$

$$f_{A\beta} = \frac{f_A f_\beta}{n} \quad \text{and} \quad f_{\alpha\beta} = f_\alpha f_\beta / n\,.$$

Hence, A and B are said to be positively associated (or there is positive association or only association) if $f_{AB} > \dfrac{f_A f_B}{n}$; and A and B are said to be negatively associated (or there is negative association or only association), if $f_{AB} > \frac{f_A f_B}{n}$. The perfect association or disassociation is categorically defined as follows :

(i) Complete positive association - if all A's are B's and/or all B's are A's, i.e., if $f_{A\beta} = 0$ and / or $f_{\alpha B} = 0$

(ii) Complete negative association - if no A's are B's and/or no $\alpha's$ and $\beta's$, i.e., if f_{AB}=0 and /or $f_{\alpha\beta} = 0$

(iii) Absolute positive association - if all A's are B's and all B's are A's, i.e., if $f_{A\beta} = 0$ and $f_{\alpha\beta} = 0$ and

(iv) Absolute negative association - if no A's are B's and no $\alpha's$ are $\beta's$, i.e., if f_{AB} =0 and $f_{\alpha\beta} = 0$

Now to formulate a measure of association (positive or negative) in a 2 × 2 contingency table it is necessary to impose certain characteristics on this measure. These characteristics are, (a) it should be independent of total frequency, (b) its value should be zero in the case of independence, positive in the case of positive association and negative in the case of negative association, and (c) its value should preferably vary between – 1 to + 1.

So the difference, $\delta_{AB} = f_{AB} - \frac{f_A f_B}{n}$, between the actual cell frequency for both positive A and B and its expected value if A and B are independent, may be utilised as a basis of the measure of association. Since the three characteristics (a), (b) and (c) are to be satisfied, the criterion Q_{AB} (constructed from δ_{AB}).

$$Q_{AB} = \frac{n\,\delta_{AB}}{f_{AB}\, f_{\alpha\beta} + f_{AB} f_{\alpha B}} = \frac{f_{AB} f_{\alpha\beta} - f_{A\beta} f_{\alpha B}}{f_{AB}\, f_{\alpha\beta} + f_{A\beta}\, f_{\alpha B}}$$

can be regarded as a measure of association between A and B. This measure is due to Yule.

There is another measure due to Yule which has been named as **Yule's coefficient of colligation** (Y_{AB}) and it is expressed as,

$$Y_{AB} = \frac{\sqrt{f_{AB} f_{\alpha\beta}} - \sqrt{f_{A\beta} f_{\alpha B}}}{\sqrt{f_{AB} f_{\alpha\beta}} + \sqrt{f_{A\beta} f_{\alpha B}}} \quad \ldots (5.35)$$

In case of independence between the attributes, $f_{AB}\, f_{\alpha\beta} = f_{A\beta}\, f_{\alpha B}$, i.e., $\delta_{AB} = 0$ and hence $Q_{AB} = 0$ and $Y_{AB} = 0$.

In the case of perfect positive association, $f_{A\beta} = 0$ and / or $f_{\alpha B} = 0$ and hence Q_{AB}=1 and also Y_{AB}=1. In the case of perfect negative association f_{AB} = 0 and /or $f_{\alpha B} = 0$ and hence $Q_{AB} = -1$ and also $Y_{AB} = -1$. Thus both Q_{AB} and Y_{AB} follow the desiderata of a good measure of association.

Now it can be verified that $Q_{AB} = \dfrac{2Y_{AB}}{1+Y_{AB}^2}$

and hence Q_{AB} is always greater than Y_{AB} (since $\dfrac{2}{1+Y_{AB}^2} > 1$) except when $Y_{AB} = 0$ *or* ± 1.

The third measure is coefficient of absolute association and it is expressed as

$$V_{AB} = \frac{f_{AB}\, f_{\alpha\beta} - f_{A\beta}\, f_{\alpha\beta}}{\sqrt{f_A f_B f_\alpha f_\beta}} = \frac{ad - bc}{\sqrt{(a+b)(c+d)(a+c)(b+d)}} \quad \text{...(5.36)}$$

It is the most commonly used measure of association between two attributes in a 2 × 2 contingency table.

The coefficient V_{AB} satisfies the required desiderata, but, $V_{AB} = \pm 1$ when and only when there exists absolute association between the two characters (attributes) as can be seen from the following consideration.

Using the symbols, a, b, c and d,

$V_{AB} = \pm 1$ if and only if

$(ad - bc)^2 = (a+b)\ (c+d)\ (a+c)\ (b+d)$

or, $a^2 (bc + bd + cd) + b^2 (ac + ad + cd) + c^2 (ab + ad + bd) + d^2 (ac+ab+bc) + 4\,a\,b\,c\,d = 0$.

The above expression vanishes if at least two of *a, b, c* and *d* are zero. As each marginal frequency is assumed to be non-zero, the solutions, $a = b = 0$; $c = d = 0$; $a = c = 0$ and $b = d = 0$ are not acceptable. Thus $V_{AB} = \pm 1$ implies $b = c = 0$ or $a = d = 0$. For $b = c = 0$ there is an absolute positive association between *A* and *B* and for $a = d = 0$, there exists absolute negative association between *A* and *B*.

5.14 Measures of Independence Between Two Attributes from (*k* × *l*) Contingency Table

In statistical investigations it is common and natural to classify the attributes beyond the dichotomous classification. Table 5.10 presents the frequencies corresponding to the indicated classifications for a sample of 340 individuals laid out in a 3 × 4 contingency table.

Table 5.10 – A 3 × 4 Contingency Table

Education of Mothers (A)	Performance of offsprings in a test (B)				Total
	Poor	Mediocre	Good	Very good	
Graduate	13	24	65	110	212
Higher Secondary	12	23	25	36	96
School Final	6	5	6	15	32
Total	31	52	96	161	340

If the total number of individuals is n and f_{ij} of them belong to the i-th class of attribute A and j-th class of attribute B ($i = 1, 2, ..., k, j = 1, 2, ..., l$), then

$$f_{i0} = \sum_{j=1}^{l} f_{ij}, \ (i = 1, 2, ..., k)$$

represents the number of individuals belonging to the i-th class of A and

$$f_{0j} = \sum_{i=1}^{k} f_{ij}, \ (j = 1, 2, ..., l)$$

represents the number of individuals belonging to the j-th class of B and the total frequency,

$$n = \sum_i f_{i0} = \sum_j f_{0j} = \sum_i \sum_j f_{ij}$$

Recalling the 2×2 classification, and following the notion of association between two attributes developed therein, we say that A and B are independent of disassociated, if for each j

$$\frac{f_{1j}}{f_{10}} = \frac{f_{2j}}{f_{20}} = ... = \frac{f_{ij}}{f_{i0}} = ... = \frac{f_{kj}}{f_{k0}} = \frac{f_{0j}}{n} \quad ... \ (5.37)$$

(assuming $f_{i0} > 0$ for each i and $f_{oj} > 0$ for each j), or equivalently, $f_{ij} = \frac{f_{i0} f_{0j}}{n}$, for all (i, j), and A and B are said to be associated, if $f_{ij} \neq \frac{f_{i0} f_{0j}}{n}$ for any pair (i, j).

For two attributes A and B to be independent, $k \times l$ number of equations are to be satisfied as is evident from (5.37). However, out of these k × l equations actually $kl - (k + l - 1) = (k - 1)(j - 1)$ number of equations are algebraically independent. In fact, $k + l - 1$ number of equation are satisfied owing to the relationship,

$$n = \sum_i f_{i0} = \sum_j f_{0j} = \sum_i \sum_j f_{ij}$$

Here also the differences,

$$\delta_{ij} = f_{ij} - \frac{f_{i0} f_{0j}}{n}, \ 1, 2, ..., k; \ j = 1, 2, ..., 1$$

$$= O_{ij} - E_{ij}, \text{ say,}$$

between the actual (i, j)-th cell frequency (O_{ij}, say), and the corresponding expected value (E_{ij}, say) may be used as a basis in constructing a measure of association. It may be noted that for independence, $\delta_{ij} = 0$, for all (i, j).

Measures of Contingency

The following quantities serve as measures.

(i) Chi - Square measure of contingency (Square contingency)

The following coefficient (χ^2_{AB}), based on δ_{ij}'s , is used as a measure of independence.

$$\chi^2_{AB} = \sum_i \sum_j \frac{\delta^2_{ij}}{f_{i0} f_{0j}/n} = n \sum_i \sum_j \frac{f^2_{ij}}{f_{i0} f_{0j}} - n$$
$$= \sum_i \sum_j \frac{\left(O_{ij} - E_{ij}\right)^2}{E_{ij}} \qquad \text{....(5.38)}$$

The quantity χ^2_{AB} is zero if and only if, A and B are independent, and its high value indicates that A and B are highly associated. But χ^2_{AB} is not independent of n (however large it may be assumed). To remove this defect Karl Pearson proposed the following measure.

(ii) *Mean Square contingency :*

$\phi^2_{AB} = \frac{1}{n} \chi^2_{AB}$. For a 2×2 table $\phi^2_{AB} = V^2_{AB}$. ϕ^2 has the same merit as χ^2. Though its value does not depend on n, yet its value may be large.

(iii) *Karl Pearson's coefficient of contingency*

$$C_{AB} = \sqrt{\frac{\chi^2_{AB}}{n + \chi^2_{AB}}} \qquad \text{... (5.39)}$$

This has the same property as χ^2_{AB} on the problem on independence and association, but it has also a defect that its highest value is not exactly equal to unity, since $\chi^2_{AB} < n + \chi^2_{AB}$ always. That is, C_{AB} does not attain the value unity even if A and B are perfectly associated. The above can be understood from the following.

In a table having k classes for each attribute, if $f_{ij} = 0$ for $i \neq j$ and $f_{ij} > 0$ for each i, then in the $k \times k$ case maximum degree of association is obtained.

Thus for perfect association between the attributes A and B, we must have, f_{ij}=0 (for $i \neq j$) and $f_{ij} = f_{i0} = f_{0j}$ (for all i and j) and finally the value of

$$\chi^2_{AB} = n \sum_i \sum_j \frac{f^2_{ij}}{f_{i0} f_{0j}} - n$$
$$= n \sum_i \frac{f^2_{ii}}{f_{i0} f_{0i}} - n = nk - n = n(k-1)$$

Hence, $C_{AB} = \sqrt{\frac{n(k-1)}{n + n(k-1)}} = \sqrt{\frac{k-1}{k}} < 1$.

To remove this defect Tschuprow suggested the following coefficient.

(iv) Tschuprow's coefficient of contingency

$$T_{AB} = \left\{ \frac{\chi^2_{AB}}{n\sqrt{(k-1)(l-1)}} \right\}^{\frac{1}{2}}$$

As has already been seen that in the case $(k = l)$ of perfect association, $f_{ij}=0$ (for all $i \neq j$), $f_{ij} > 0$ for each i and $f_{i0} = f_{0i} = f_{0i} = f_{ii}$, for all i, j, the value of $\chi^2_{AB} = n(k-1)$. Thus for perfect association,

$$T_{AB} = \left\{ \frac{\chi^2_{AB}}{n\sqrt{(k-1)(l-1)}} \right\}^{\frac{1}{2}} = \left\{ \frac{n(k-1)}{n(k-1)} \right\}^{\frac{1}{2}} = 1$$

The values of '*T*' are zero and unity in case of independence & perfect association respectively. But the behaviour of T_{AB} is not known much in the $k \times l$ table where $k \neq 1$.

(Note that for ($k \times l$) two-way contingency table the above measures can not meaningfully serve the distinction between positive and negative association. But this distinction is necessary for the investigations as given in Table 5.10, where classes of both the attributes can be orderly arranged. For example (a 2 × 2 contingency table).

	A	α	Total
B	70	550	620
β	80	100	180
Total	150	650	800

$Q_{AB} = -0.725$, $Y_{AB} = -0.425$

$\chi^2_{AB} = 100.75$, $C_{AB} = 0.330$ *and* $T_{AB} = 0.35$

where the investigator has to judge the level of positive or negative association).

5.15 . Yates' Correction for Continuity

For 2 × 2 contingency table we may use χ^2_{AB} for the measure of association. Where expected frequency $\frac{f_{i0} f_{0j}}{n}$ is less than 5 (i = 1, 2, j = 1, 2) Yates proposed a correction for it. The method is to add $\frac{1}{2}$ to the cell frequency which is less than 5 and then adjust the remaining cell frequencies so that marginal frequencies and total frequency remain unaltered.

For a 2 × 2 contingency table (Table 5.9a) we have

$$\chi^2_{AB} = \frac{n(ad-bc)^2}{(a+c)(b+d)(a+b)(c+d)}$$

After subtracting and adding $\frac{1}{2}$ from the respective cell frequencies, the corrected value of χ^2_{AB} is given as.

$$\chi^2_{AB} = \frac{n\left[\left(a\mp\frac{1}{2}\right)\left(d\mp\frac{1}{2}\right)-\left(b\pm\frac{1}{2}\right)\left(c\pm\frac{1}{2}\right)\right]^2}{(a+c)(b+d)(a+b)(c+d)}$$

$$= \frac{n\left[(ad-bc)-\frac{1}{2}(a+b+c+d)+\frac{1}{4}-\frac{1}{4}\right]^2}{(a+c)(b+d)(a+b)(c+d)}$$

$$= \frac{n\left[|ad-bc|-\frac{n}{2}\right]^2}{(a+c)(b+d)(a+b)(c+d)}$$

Note 1. If n is large, the correction has a very little effect.

Note 2. For $k \times l$ cases if any cell frequency is less than 5, we pool the continuous classes so that every cell frequency becomes more than 5. Yates' correction is necessary for (2×2) cases, since here such pooling is meaningless.

Examples

Example 5.1. For the given pairs (x, y) of values

x	34	37	36	41	38	36	37	34	36	34
y	37	37	34	37	40	42	40	31	38	32

(i) Plot the points on a graph paper.

(ii) Draw two nearly fitted straight lines by visual inspection.

(iii) Draw a fitted line by least squares method taking y as the dependent variable. Comment on the inaccuracy of drawing lines intuitively.

Solution :

(i) & (ii) The ten pairs of observations of x and y are plotted on the graph paper in the following. Visually, two fitted lines, Line 1 and Line 2 are drawn and shown in the following figure.

(iii) The fitted line by least squares method taking y as the dependent variable is $Y-\bar{y}=b_{yx}(x-\bar{x})$, where Y is the fitted value.

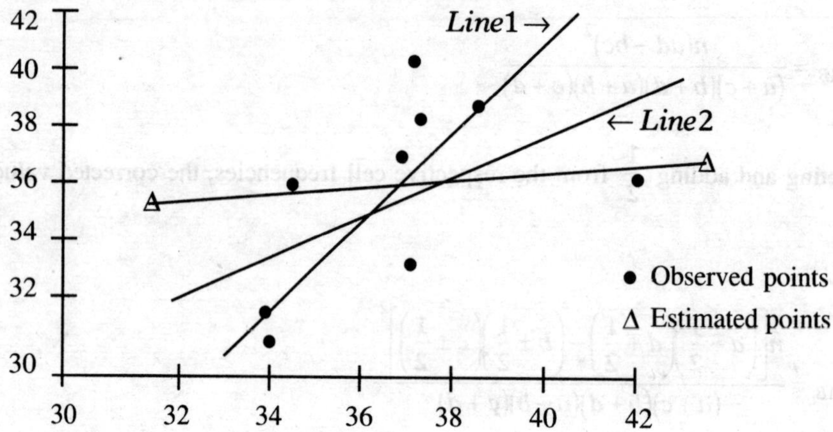

Here, $\bar{y} = 36.8, \bar{x} = 36.3$

and $b_{yx} = \dfrac{\sum xy - n\bar{x}\bar{y}}{\sum x^2 - n\bar{x}^2}$

$$= \frac{13390 - 13358.4}{13656 - 13176.9} = \frac{31.6}{479.1} = 0.0659$$

$\therefore$ Y – 36.8 = 0.659 (x – 36.3)

or, Y = 0.0659 x + 34.41 is the best fitted line and it is drawn on the above figure by joinng two arbitrary derived points from the above equation. For example, from the above equation, for x = 32 and x = 42, the values of Y are 36.5 and 37.2, respectively. The fitted line by least squares method is shown in the above figure. From the figure it is revealed that free hand curve fitting may widely differ from mathematical curve fitting.

Example 5.2 The following data relate to the stature (x cm) and sitting height (y cm) for 10 Indians.

x	17.28	166.0	164.1	168.8	164.4	165.2	170.0	163.5	169.4	159.1
y	88.8	83.6	81.3	83.9	85.4	81.1	84.9	81.1	84.9	79.6

(a) Represent the bivariate data.

(b) Compute correlation coefficient between x and y

(c) Obtain the linear regression equations of y on x and of x on y. Determine the expected sitting height of an individual if his stature is 150 cms and also the expected stature if his sitting height is 90 cms.

(d) Compute and plot the residuals in case of the above ten individuals, using two regression equations separately and comment.

Solution :

(a) To represent the data, the given ten pairs of observations are plotted on the graph paper as scatter diagram.

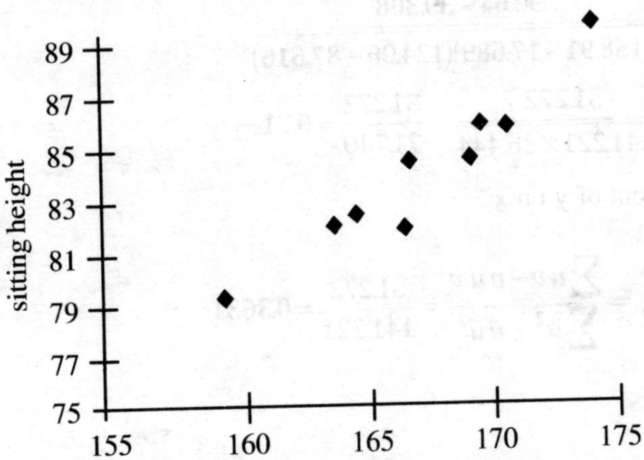

Scatter Diagram

(b) Table for the calculation of correlation coefficient and regression coefficients.

x	y	u = x – 165	v = y – 80	u^2	v^2	uv
172.8	83.8	7.8	3.8	60.84	14.44	29.64
116.0	83.6	1.0	3.6	1.0	12.96	3.6
164.1	81.3	–0.9	1.3	0.81	1.69	–1.17
168.8	83.9	3.8	3.9	14.44	15.21	14.82
164.4	85.4	–0.6	5.4	0.36	29.16	–3.24
165.2	81.1	0.2	1.1	0.04	1.21	0.22
170.0	84.9	5.0	4.9	25.00	24.01	24.5
163.5	81.1	–1.5	1.1	2.25	1.21	–1.65
169.4	84.9	4.4	4.9	19.36	24.01	21.56
159.1	79.6	–5.9	–0.4	34.81	0.16	2.36
Total		$\sum u = 13.3$	$\sum v = 29.6$	$\sum u^2 = 158.91$	$\sum v^2 = 124.06$	$\sum uv = 90.64$

Here, $n = 10, \bar{u} = 1.33, \bar{v} = 2.96$

Correlation coefficient between x and y is

$$r_{xy} = r_{uv} = \frac{\sum uv - n\bar{u}\,\bar{v}}{\sqrt{\left(\sum u^2 - n\bar{u}^2\right)\left(\sum v^2 - n\bar{v}^2\right)}}$$

$$= \frac{90.64 - 39.368}{\sqrt{(158.91 - 17.689)(124.06 - 87.616)}}$$

$$= \frac{51.272}{\sqrt{141.221 \times 36.444}} = \frac{51.272}{71.740} = 0.71$$

(c) Regression coefficient of y on x

$$b_{yx} = b_{vu} = \frac{\sum uv - n\bar{u}\,\bar{v}}{\sum u^2 - n\bar{u}^2} = \frac{51.272}{141.221} = 0.3631$$

and that of x on y is,

$$b_{xy} = b_{vu} = \frac{\sum uv - n\bar{u}\,\bar{v}}{\sum v^2 - n\bar{v}^2} = \frac{51.272}{36.444} = 1.4069$$

Now the regression equation of y on x is

$$Y - \bar{y} = b_{yx}(x - \bar{x})$$

Here $\bar{y} = \bar{v} + 80 = 82.96$, $\bar{x} = \bar{u} + 165 = 166.33$

and $b_{yx} = b_{vu} = 0.3631$

$Y = 0.3631x + 22.5656$, (1)

is the regression line of y on x.

Again, the regression equation of x on y is

$$X - \bar{x} = b_{xy}(y - \bar{y})$$

or $X - 166.33 = 1.4069\,(y - 82.96)$,

or $X = 1.4069y + 49.6135$ (2)

represents the regression equation of x on y.

Now, if x = 150 then Y = 77.03 (from (1)). So the expected sitting height of the individual is 77.03 cm if his stature is 150cm.

Also if y = 90, X = 176.23 (from (2)).

So the expected stature of the individual is 176.23 cm if his sitting height is 90cm.

(d) The residuals e and e' as defined in the text are recalled and the following tables are constructed.

Observed Sitting height, y	83.8	83.6	81.3	83.9	85.4	81.1	84.9	81.1	84.9	79.6
Estimated Sitting height, Y	85.3	82.8	82.2	83.8	82.2	82.5	84.3	81.9	84.1	80.3
Residual e=y–Y	–1.5	0.8	–0.9	0.1	3.2	–1.4	–0.6	–0.8	0.8	0.7
Observed Stature, x	172.8	166.0	164.1	164.4	168.8	165.2	170.0	163.5	169.4	159.1
Estimated Stature, X	167.2	167.2	164.0	169.8	167.6	163.7	169.0	163.7	169.0	161.6
Residual $e' = x - X$	5.6	–1.2	0.1	– 5.4	1.2	1.5	1.0	–0.2	0.4	–2.5

The residuals are plotted in the graphs given on Page Number 195.

The linear correlation and regression analysis will be effective if the residual plots are randomly depicted near the axis of the predicting variable. From the above figures it is revealed that the errors are more or less random but prediction of sitting height (*Y*), from the linear predicting equation, on stature variable (*x*) is more appropriate than that of stature (*X*) on sitting height variable (*y*).

Example 5.3 The following statistics are recorded on two variables x and y

$n = 15, \sum x = 162. \sum y = 1840.5$

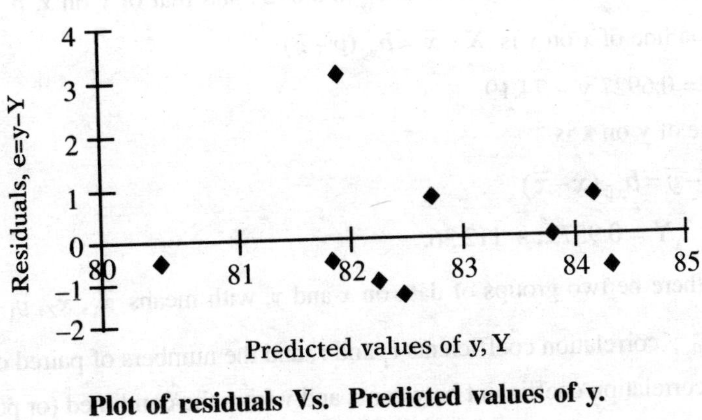

Plot of residuals Vs. Predicted values of y.

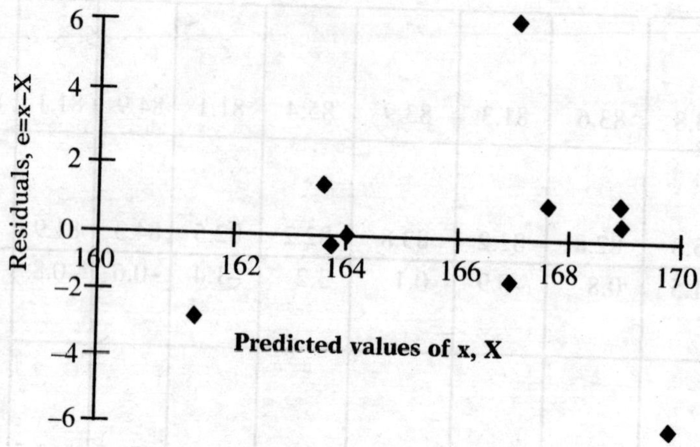

Plot of residuals Vs. Predicted values of x

$$\sum x^2 = 1820.25, \sum y^2 = 225927$$

and $$\sum xy = 19945.05$$

(a) Find regression coefficients of x on y and of y on x.

(b) Find the regression lines.

Solution :

(a) $$b_{xy} = \frac{\sum xy - n\bar{x}\,\bar{y}}{\sum y^2 - n\bar{y}^2} = \frac{66.65}{97.66} = 0.6927$$

and $$b_{yx} = \frac{\sum xy - n\bar{x}\,\bar{y}}{\sum x^2 - n\bar{x}^2} = \frac{67.65}{70.65} = 0.9573$$

So the regression coefficients of x on y, b_{xy} is 0.692., and that of y on x, b_{yx} is 0.9573.

(b) The regression line of x on y is $X - \bar{x} = b_{xy}(y - \bar{y})$

$\Rightarrow$ $$X = 0.6927\,y - 74.19.$$

The regression line of y on x is

$$Y - \bar{y} = b_{xy}(x - \bar{x})$$

$\Rightarrow$ $$Y = 0.9573x + 112.36.$$

Example 5.4 Let there be two groups of data on x and y, with means $\bar{x}_1, \bar{x}_2, \bar{y}_1$ and $\bar{y}_2$; variances, $s^2_{x_1}, s^2_{x_2}, s^2_{y_1}$ and $s^2_{y_2}$; correlation coefficients r_1 and r_2 and the numbers of paired observations, n_1 and n_2. Show that the correlation coefficient between x and y from the combined (or pooled) n (= $n_1 + n_2$) pairs of data as,

$$r = \frac{n_1 r_1 s_{x_1} s_{y_1} + n_2 r_2 s_{x_2} s_{y_2} + n_1(\bar{x}_1 - \bar{x})(\bar{y}_1 - \bar{y}) + n_2(\bar{x}_2 - \bar{x})(\bar{y}_2 - \bar{y})}{\left\{\left(n_1 s_{x_1}^2 + n_2 s_{x_2}^2 + n_1(\bar{x}_1 - \bar{x})^2 + n_2(\bar{x}_2 - \bar{x})^2\right)\left(n_1 s_{y_1}^2 + n_2 s_{y_2}^2 + n_1(\bar{y}_1 - \bar{y})^2 + n_2(\bar{y}_2 - \bar{y})^2\right)\right\}^{1/2}} \quad ...(1)$$

where $\bar{x}$ and $\bar{y}$ are the grouped means of x and y respectively.

Solution :

We obtain from the definition of the correlation coefficient.

$$r_1 = \frac{\sum(x_{1i} - \bar{x})(y_{1i} - \bar{y})}{n_1 s_{x_1} s_{y_1}} \text{ and } r_2 = \frac{\sum(x_{2j} - \bar{x}_2)(y_{2i} - \bar{y}_2)}{n_2 s_{x_2} s_{y_2}}$$

$i = 1, 2, ..., n_1 : j = 1, 2, ..., n_2,$

the combined correlation coefficient (r) between x and y is,

$$r = \frac{\sum(x_{1i} - \bar{x})(y_{1i} - \bar{y}) + \sum(x_{2i} - \bar{x}_2)(y_{2j} - \bar{y}_2)}{n s_x s_y}$$

s_x^2 and s_y^2 being the combined variances.

Again,

$$\sum(x_{1i} - \bar{x})(y_{1i} - \bar{y}) + \sum(x_{2j} - \bar{x})(y_{2j} - \bar{y})$$

$$= \sum(x_{1i} - \bar{x}_1 + \bar{x}_1 - \bar{x})(y_{1i} - \bar{y}_1 + \bar{y}_1 - \bar{y}) + \sum(x_{2j} - \bar{x}_2 + \bar{x}_2 - \bar{x})(y_{2j} - \bar{y}_2 + \bar{y}_2 - \bar{y})$$

$$= \sum(x_{1i} - \bar{x}_1)(y_{1i} - \bar{y}_1) + n_1(\bar{x}_1 - \bar{x})(\bar{y}_1 - \bar{y}) + \sum(x_{2j} - \bar{x}_2)(y_{2j} - \bar{y}_2) + n_2(\bar{x}_2 - \bar{x})(\bar{y}_2 - \bar{y})$$

$$= n_1 r_1 s_{x_1} s_{y_1} + n_2 r_2 s_{x_2} s_{y_2} + n_1(\bar{x}_1 - \bar{x})(\bar{y}_1 - \bar{y}) + n_2(\bar{x}_2 - \bar{x})(\bar{y}_2 - \bar{y}) \quad ..... (2)$$

and

$$n s_x s_y = \left(n s_x^2 \times n s_y^2\right)^{1/2}$$

$$= \left\{\left(\sum(x_{1j} - \bar{x})^2 + \sum(x_{2j} - \bar{x})^2\right)\left(\sum(y_{1i} - \bar{y})^2 + \sum(y_{2j} - \bar{y})^2\right)\right\}^{1/2}$$

$$= \left\{\left(n_1 s_{x_1}^2 + n_1(\bar{x}_1 - \bar{x})^2 + n_2(\bar{x}_2 - \bar{x})^2\right)\left(n_1 s_{y_1}^2 + n_1(\bar{y}_1 - \bar{y})^2\right) + n_2 s_{y_2}^2 + n_2(\bar{y}_2 - \bar{y})^2\right\}^{1/2} \quad ....(3)$$

Thus, we get the expression of combined correlation coefficient (r) between x and y as given in (1).

Note 1. The result can be extended for k groups of paired observations with means $\bar{x}_i$ and $\bar{y}_i$, variances $s_{x_i}^2$ and $s_{y_i}^2$, size n_i, and correlation coefficients r_i, $i = 1, 2, ..., k$ as

$$r = \frac{\sum n_i r_i s_{x_i} s_{y_i} + \sum n_i(\bar{x}_i - \bar{x})(\bar{y}_i - \bar{y})}{\left(\sum n_i s_{x_i}^2 + \sum n_i(\bar{x}_i - \bar{x})^2\right)^{1/2} \left(\sum n_i s_{y_i}^2 + \sum n_i(\bar{y}_i - \bar{y})^2\right)^{1/2}}$$

Note 2, It should also be noted that if all the group correlation coefficients are zero, i.e., $r_i = 0$ for all i, the combined correlation coefficient may become non-zero, since in that case

$$r = \frac{\sum n_i(\bar{x}_i - \bar{x})(\bar{y}_i - \bar{y})}{\left(\sum n_i s_{x_i}^2 + \sum n_i(\bar{x}_i - \bar{x})^2\right)^{1/2} \left(\sum n_i s_{y_i}^2 + \sum n_i(\bar{y}_i - \bar{y})^2\right)^{1/2}} \neq 0$$

(Unless in the situation where $\sum n_i(\bar{x}_i - \bar{x})(\bar{y}_i - \bar{y}) = 0$).

Example 5.5 Let x and y be independent variables, with standard deviations s_x and s_y. Show that the correlation coefficient between x and $x \pm y$ is $s_x / \sqrt{s_x^2 + s_y^2}$.

Solution :

$$r_{x(x\pm y)} = \frac{Cov(x, x\pm y)}{\left(Var(x)\, Var(x\pm y)\right)^{1/2}}$$

$$= \frac{\frac{1}{n}\sum(x-\bar{x})(x \pm y - \bar{x} \mp \bar{y})}{\left\{\frac{1}{n}\sum(x-\bar{x})^2 \frac{1}{n}\sum(x \pm y - \bar{x} \mp \bar{y})^2\right\}^{1/2}}$$

$$= \frac{\frac{1}{n}\sum(x-\bar{x})\left((x-\bar{x}) \pm (y-\bar{y})\right)}{\left\{\frac{1}{n}\sum(x-\bar{x})^2 \frac{1}{n}\sum\left((x-\bar{x}) \pm (y-\bar{y})\right)^2\right\}^{1/2}}$$

$$= \frac{\frac{1}{n}\sum(x-\bar{x})^2 \pm \frac{1}{n}\sum(x-\bar{x})(y-\bar{y})}{\left(\frac{1}{n}\sum(x-\bar{x})^2\right)^{1/2}\left(\frac{1}{n}\sum(x-\bar{x})^2 + \frac{1}{n}\sum(y-\bar{y})^2 \pm \frac{2}{n}\sum(x-\bar{x})(y-\bar{y})\right)^{1/2}}$$

$$= \frac{s_x}{\left(s_x^2 + s_y^2\right)^{1/2}} \qquad \left(\because \frac{1}{n}\sum(x-\bar{x})(y-\bar{y}) = 0\right)$$

Example 5.6 In a set of two variables x and y with 50 pairs of observations, the following statistics were observed.

$$\bar{x} = 10,\ \bar{y} = 6,\ s_x = 3,\ s_y = 2 \text{ and } r_{xy} = 0.3$$

On a subsequent verification it was found one pair (x, y) has been recorded as (8,11) instead of the actual value (11,8). Show how the original statistics and also the regression lines are affected.

Solution :

Corrected means :

$$\bar{x}_c = (50\times10-8+11)/50\ (\because correct\ x\ is\ 11\ not\ 8)$$

$$= 503/50 = 10.06$$

$$\bar{y}_c = (50\times6-11+8)/50 (\because correct\ y\ is\ 8\ not\ 11)$$

$$= 297/50 = 5.94$$

$$\sum x^2\ (corrected) = 50\left(s_x^2+\bar{x}^2\right)-8^2+11^2$$

$$= 5507$$

$$\sum y^2\ (corrected) = 50\left(s_x^2+\bar{y}^2\right)-11^2+8^2$$

$$= 1943$$

Corrected variances : $s_{xc}^2 = 5507/50-(10.06)^2 = 8.9364$

$\therefore$ $s_{xc} = 2.99$

and, $s_{yc} = (1943/50 - (5.94)^2)^{1/2} = 1.89$

Now, $$r = \frac{\sum xy - n\bar{x}\bar{y}}{ns_x s_y} \Rightarrow \sum xy = 50\times0.3\times3\times2+50\times10\times6 = 3090$$

$$\therefore \sum xy\,(corrected) = 3090-88+88 = 3090$$

$$r_c(corrected) = (3090-50\times10.06\times5.94)/(50\times2.99\times1.89)$$

$$= 0.36$$

Original Statistics : $\bar{x} = 10,\ \bar{y} = 6, s_x = 3,\ s_y = 2,\ r = 0.3$

$b_{xy} = 0.45,\ b_{yx} = 0.20.$

Correct Statistics : $\bar{x}_c = 10.06,\ \bar{y}_c = 5.94,\ s_{xc} = 2.99,\ s_{yc} = 1.89$

$r_c = 0.36,\ b_{xy} = 0.57\ and\ b_{yx} = 0.23$

Incorrect Regression lines :

$$y\ on\ x,\ Y-\bar{y} = b_{yx}(x-\bar{x})$$

or, $Y = 0.20x + 4$, and

x on y, $X - \bar{x} = b_{xy}(y - \bar{y})$

or $X = 0.45\,y + 7.3$.

The corrected regression line of y on x is

$$Y_c - 5.94 = 0.23\,(x - 10.06)$$

or $Y_c = 0.23\,x + 3.62$ and that of x on y is

$$X_c - 10.06 = 0.57\,(y - 5.94)$$

or, $X_c = 0.57\,y + 6.67$.

Example 5.7 x and y are two variables with variances s_x^2 and s_y^2 respectively, and r is the correlation coefficient between them. Determine the values of the constants a and b, such that $x + ay$ and $x + by$ are uncorrelated.

Solution :

The variables $x + ay$ and $x + by$ are to be uncorrelated.

So we have to find a and b such that Cov $(x + ay, x + by) = 0$

or, Var (x) + a Cov (x, y) + b Cov (x, y) + ab Var $(y) = 0$

or $s_x^2 + (a+b)\, r\, s_x\, s_y + ab\, s_y^2 = 0$

A particular solution of the above equation is obtained if we put a = – b and $b^2 = s_x^2 / s_y^2$

Thus for $\left(b = \frac{s_x}{s_y} \text{ and } a = -\frac{s_x}{s_y}\right)$

or, $\left(b = -\frac{s_x}{s_y} \text{ and } a = \frac{s_x}{s_y}\right)$ the given variables $x + ay$ and $x + by$ are uncorrelated.

Example 5.8 Given n values $x_1, x_2, \ldots, x_n$ of a variable x with weights $w_1, w_2, \ldots, w_n$ respectively, express the correlation coefficient between x and w involving the difference between the simple mean and weighted mean of x, and hence show that the weighted mean of x is > or < the simple mean of x according as r_{xw}, is > or < 0.

Solution : The correlation coefficient between x and w is

$$r_{xw} = \frac{\sum xw - n\bar{x}\bar{w}}{ns_x s_w} = \frac{\sum w\left(\frac{\sum xw}{\sum w} - \bar{x}\right)}{n s_x s_w}$$

$$= \frac{\sum w(\bar{x}_w - \bar{x})}{n\, s_x\, s_w}$$

where $\bar{x}_w$ is weighted mean and $\bar{x}$ is simple mean.

Hence, $r_{xw} > 0$ *if* $\bar{x}_w > \bar{x}$, and $r_{xw} < 0$ *if* $\bar{x}_w < \bar{x}$,

Example 5.9 Two dice are thrown, their scores being x and y. The first die is left on the table while the second is picked up and thrown again giving the score z. This process is repeated a large number of times. What is the correlation between the sums of the scores in paired operations assuming.

Var (x) = Var (y) = Var (z) and x, y, z are uncorrelated.

Solution : Hence we have to find out the correlation coefficient between

$u = x + y$ and $v = x + z$

$$\text{Now,} \quad r_{uv} = \frac{Cov(u,v)}{\sqrt{Var(u)Var(v)}} = \frac{Var(x)+Cov(x,z)+Cov(x,y)+Cov(y,z)}{\left((Var(x)+Var(y))\,(Var(x)+Var(z))\right)^{1/2}}$$

$$= \frac{Var(x)}{\sqrt{2Var(x)\times 2Var(x)}} = \frac{1}{2}$$

So the correlation between the sum of scores is 0.5.

Example 5.10 Identify the following two regression lines as y on x and x on y :

$3x + 12y = 19$, and $9x + 3y = 46$.

Solution :

Let the line, $3x + 12y = 19$ be the regression line of y on x. So,

$$y = -\frac{3}{12}x + \frac{19}{12} \Rightarrow \quad b_{yx} = -\frac{3}{12} = -\frac{1}{4}$$

The other line, $9x + 3y = 46$ is the regression line of x on y. So,

$$x = -\frac{3}{9}y + \frac{46}{9} \Rightarrow \quad b_{xy} = -\frac{3}{9} = -\frac{1}{3}$$

It is easy to check that

$$b_{xy} \times b_{yx} = \left(-\frac{1}{3}\right)\left(-\frac{1}{4}\right) = \frac{1}{12} < 1, \; \left(\because b_{xy} \times b_{yx} = r^2 < 1\right)$$

Since here $b_{xy} \times b_{yx} < 1$, our primary identification of the regression line is correct, i.e., $3x + 12y = 19$ is the regression line of y on x and $9x + 3y = 46$ is the regression line of x on y. If the value of the product of the two regression coefficients happens to be greater than one, the identification of the regression lines becomes wrong. In fact, the correct identification will be to take the reverse decision.

Now, $r_{xy} = \sqrt{b_{xy} \times b_{yx}} = \sqrt{\frac{1}{12}} = -0.29$ (since the regression coefficients are of (–) *ve* sign)

As both the given lines pass through $(\bar{x},\bar{y})$ so we can re-write the given equations as

$$3\bar{x}+12\bar{y}=19 \; and \; 9\bar{x}+3\bar{y}=46 .$$

Solving the equations we get, $\bar{x}=\bar{y}=\frac{1}{3}$.

Now, $b_{xy}=r_{xy}\frac{s_x}{s_y}\Rightarrow\frac{s_x}{s_y}=\frac{-0.25}{-0.29}=0.86 \qquad \therefore \frac{s_x^2}{s_y^2}=0.74$

Example 5.11 The combined heights of 11 pairs of father and his adult son are given below.

Pair number :	1	2	3	4	5	6	7	8	9	10	11
Combined height : (in mt.)	3.56	3.35	3.33	3.30	3.43	3.38	3.43	3.48	3.51	3.15	3.26

The following figures are also given.

	Average height (in m.)	**S.D. of height (in m.)**
Father (*x*) :	1.70	0.0508
Son (*y*) :	1.68	0.0450

Find the coefficient correlation between the height of father and that of the son. Estimate the height of a son whose father's height is 1.80 m.

Solution : Mean of the combined height is 3.33 m., i.e.,

$$\bar{x}+\bar{y}=3.38$$

Given $\qquad \bar{x}=1.70$, and hence $\bar{y}=1.68$

We know,

$$V(x+y)=V(x)+V(y)+2Cov(x,y),$$

$$V(x+y)=\frac{1}{11}\times 125.7447-(3.38)^2=0.006936 \text{ (from given data)},$$

$$V(x)=(0.0508)^2=0.002581,$$

$$V(y)=(0.0450)^2=0.002025,$$

$$Cov(x,y)=(0.006936-0.002581-0.002025)/2=0.001165,$$

$$r_{xy}=\frac{Cov(s,y)}{\sqrt{V(x)V(y)}}=\frac{0.001165}{0.0508\times0.0450}=0.51$$

We Have to estimate *y* for *x* = 1.80m.

So the regression line of *y* on *x*, $Y-\bar{y}=r_{xy}\frac{s_y}{s_x}(x-\bar{x})$

Substituting values of $\bar{x}, \bar{y}, r, s_x$ and s_y

$$Y - 1.68 = 0.51\ (0.045/0.0508)\ (x - 1.70)$$

$$\Rightarrow \qquad Y = 0.4518x + 0.9120$$

is the regression equation of y on x,

Now for $x = 1.80$, $Y = 1.72$.

So the estimated height of the son becomes 1.72 m if the height of his father is 1.80 m.

Example 5.12 A sample of 150 seed-pods is being tested for any possible correlation between length of pod and number of seeds germinating per pod and the following data were recorded.

Number of Seeds/pod	6	7	8	9	10	11
Number of Pods	8	30	70	29	10	3
Length of Pods (in cm)	5	6	7	8	9	10
Number of Pods	7	33	54	36	13	7
Total number of seeds in each category	55	251	404	305	112	85

Calculate the coefficient of correlation and the linear regression equation of the number of seeds germinating per pod on the length of the pod.

Solution:

Let y be the variable to denote the number of seeds per pod, and x be the variable to denote the length of pod (in cm).

From the given data we may form the following bivariate table to determine the correlation coefficient and the regression equation of y on x.

y_j / x_i	6	7	8	9	10	11	$*f_{i0}$	$x_i f_{i0}$	$x_i^2 f_{i0}$	$*\sum y_j f_{ij}$	$x_i \sum y_j f_{ij}$
5	f_{11}	f_{12}	f_{13}	f_{14}	f_{15}	f_{16}	07	35	175	55	275
6	.	.	.	.	.	.	33	198	1188	251	1506
7	.	.	$((f_{ij}))$	.	.	.	54	378	2646	404	2828
8	.	.	.	.	.	.	36	288	2304	305	2440
9	.	.	.	.	.	.	13	117	1053	112	1008
10	f_{61}	f_{62}	f_{63}	f_{64}	f_{65}	f_{66}	07	70	700	85	850
$*f_{0j}$	8	30	70	29	10	3	150	1086	8066		8907
$y_j f_{0j}$	48	210	560	261	100	33	1212				
$y_j^2 f_{0j}$	288	1470	4480	2349	1000	363	9950				

* column and row values are given in the problem.

Here, $\sum_i \sum_j x_i y_j f_{ij} = 8907, \sum x_i f_{i0} = 1086, \sum x_i^2 fij = 8066,$

$\bar{x} = 7.24, \bar{y} = 8.08, \sum y_i f_{0j} = 1212 \sum y_j^2 f_{0j} = 9850$

$n = \sum f_{io} = \sum f_{0j} = 150.$

The Correlation Coefficient

$$r_{xy} = \frac{n\sum\sum x_i y_i f_{ij} - \left(\sum x_i f_{i0}\right)\left(\sum y_i f_{0j}\right)}{\left(n\sum x_i^2 f_{i0} - \left(\sum x_i f_{i0}\right)^2\right)^{1/2}\left(n\sum y_j^2 f_{0j} - \left(\sum y_j f_{0j}\right)^2\right)^{1/2}}$$

= 19818/(174.65 × 153.48) = 0.7393.

The regression equation of *y* on *x* is

$$Y - \bar{y} = r_{xy}\frac{s_y}{s_x}(x - \bar{x}), or\, Y - 8.08 = 0.7393 \times \frac{153.48}{174.6}(x - 7.24)$$

or Y = 0.6503 *x* + 3.3717.

Example 5.13 For *n* pairs (*xi, yi*), *i* = 1, 2, ..., n of values on causally related variables *x* and *y*, prove the following :

(a) $V(x + y) + V(x - y) = 2(V(x) + V(y))$

(b) $r_{xy} = \frac{V(x+y) - V(x-y)}{4 s_x s_y}$

$$\frac{V(x+y) - V(x) - V(y)}{2 s_x s_y}$$

$$= \frac{V(x) - V(y) - V(x-y)}{2 s_x s_y}$$

Solution:

(a) $V(x+y) = \frac{1}{n}\Sigma(x_i + y_i - \bar{x} - \bar{y})^2 = \frac{1}{n}\Sigma((x_i - \bar{x}) + (y_i - \bar{y}))^2$

$$= \frac{1}{n}\sum(x_i - \bar{x})^2 + \frac{1}{n}\sum(y_i - \bar{y})^2 + 2\frac{1}{n}\sum(x_i - \bar{x})(y_i - \bar{y})$$

$= V(x) + V(y) + 2\,Cov(x, y)$ ... (1)

$$V(x-y)=\frac{1}{n}\sum(x_i-y_i-\bar{x}+\bar{y})^2$$

$$=\frac{1}{n}\sum((x_i-\bar{x})-(y_i-\bar{y}))^2 \qquad \ldots \qquad (2)$$

$$= V(x) + V(y) - 2\,Cov(x, y)$$

Thus $\quad V(x + y) + V(x - y) = 2\,(V(x) + V(y))$.

(b) We know, $r_{xy}=\dfrac{Cov(x,y)}{s_x\, s_y}$

Now from (1) and (2), *Cov* (*x*, *y*) can be expressed in the following alternative forms :

$$Cov(x,y)=\frac{V(x+y)-V(x-y)}{4}=\frac{V(x+y)-V(x)-V(y)}{2}$$

$$=\frac{V(x)+V(y)-V(x-y)}{2} \text{ and hence}$$

$$r_{xy}=\frac{V(x+y)-V(x-y)}{4\,s_x s_y}=\frac{V(x+y)-V(x)-V(y)}{2\,s_x s_y}=\frac{V(x)+V(y)-V(x-y)}{2\,s_x s_y}$$

Example 5.14 If two variables *x* and *y* be independent, prove that

Solution : $V(x + y) = V(x - y) = V(x) + V(y)$

$$V(x+y)=\frac{1}{n}\sum(x_i+y_i-\bar{x}-\bar{y})^2$$

$$\frac{1}{n}\sum(x_i+\bar{x})+(y_i-\bar{y}))^2$$

$$= V(x) + V(y) + 2\,Cov(x, y) = V(x) + V(y)$$

($\because Cov(x,y)=0$ for independent *x* and *y*).

Similarly, $V(x-y) = V(x) + V(y) - 2\,Cov(x, y) = V(x) + V(y)$.

Example 5.15 Obtain the correlation coefficaient between

(a) $\quad x$ and $x + c$

(b) $\quad x$ and $c + x$, c being a constant.

Solution : The correlation coefficient between *x* and $x \pm c$.

$$r_{\overline{xx\pm c}}=\frac{Cov(x,x\pm c)}{\sqrt{V(x)V(x\pm c)}} \qquad \ldots \qquad (1)$$

Now, $Cov(x, x\pm c) = \frac{1}{n}\sum (x_i - \bar{x})(x_i \pm c - \bar{x} + \bar{c}) = \frac{1}{n}\sum (x_i - \bar{x})^2 = V(x),$

$$V(x\pm c) = \frac{1}{n}\sum (x_i \pm c - \bar{x} \mp c)^2 = \frac{1}{n}\sum (x_i - \bar{x})^2 = V(x),$$

$$Con(\overline{x, c\pm x}) = \frac{1}{n}\sum (x_i - \bar{x})(c \pm x_i - c \mp \bar{x}) = \frac{1}{n}\sum (x_i - \bar{x})^2 = \pm V(x).$$

Putting these values in (.1), we get the correlation coefficients as

(a) $r(x, x \pm c) = 1$ and

(b) $r(x, c \pm x) = \pm 1$.

Example 5.16 If $V(x) = V(y)$ prove that the correlation coefficient between $(x+y)$ and $(x-y)$ is zero.

Solution :

$$Cov(x+y, x-y) = \frac{1}{n}\sum (x_i + y_i - \bar{x} - \bar{y})(x_i - y_i - \bar{x} + \bar{y})$$

$$= \frac{1}{n}\sum ((x_i - \bar{x}) + (y_i - \bar{y}))((x_i - \bar{x}) - (y_i - \bar{y}))$$

$$= \frac{1}{n}\sum (x_i - \bar{x})^2 - \frac{1}{n}\sum (x_i - \bar{x})(y_i - \bar{y}) + \frac{1}{n}\sum (x_i - \bar{x})(y_i - \bar{y}) - \frac{1}{n}\sum (y_i - \bar{y})^2$$

$= V(x) - V(y) = 0$, (according to the given condition).

Hence the correlation between $(x + y)$ and $(x - y)$ is zero.

Example 5.17 The marks secured by recruits in a selection test (x) and in the proficiency test (y) are given below :

Sl. No.	1	2	3	4	5	6	7	8	9	10
x	10	15	12	17	13	16	24	16	22	13
y	30	42	45	46	33	34	40	34	39	30

(a) Calculate product moment correlation coefficient.

(b) Calculate Spearman's rank correlation coefficient.

(c) Calculate Kendall's rank correlation coefficient.

(d) Why does rank correlation differ from product moment correlation coefficient?

Solution :

Here $n = 10, \sum x = 155, \sum y = 373,$

$$\sum x^2 = 2581, \sum y^2 = 14227, \sum xy = 5875$$

(a) $$r_{xy}=\frac{\sum xy-n\bar{x}\bar{y}}{\sqrt{\left(\sum x^2-n\bar{x}^2\right)\left(\sum y^2-n\bar{y}^2\right)}}=\frac{93.5}{\sqrt{178.5\times314.1}}=\frac{93.5}{236.78}=0.39$$

(b)

Sl. No.	1	2	3	4	5	6	7	8	9	10
rank of x (u) :	1	6	2	8	4	7	10	4	9	4
rank of y (v) :	1.5	8	9	10	3	4.5	7	4.5	6.	1.5
d = | v – u | :	0.5	2	7	2	1	2.5	3	.5	3	2.5

In the 1_{st} ranking there is only one tie of lengh 3. Thus

$$T_u=\frac{1}{10}\cdot\frac{3^3-3}{12}=0.2$$

In the 2_{nd} ranking there are two ties of length 2 each.

Here, $$T_v=\frac{1}{10}\left(\frac{2^3-2}{12}+\frac{2^3-2}{12}\right)=0.1.$$

Also, $$\frac{n^2-1}{12}=\frac{10^2-1}{12}=8.25$$

and, $$\frac{1}{n}\sum d_j^2=8.9.$$

Thus, $$r_s=\frac{8.25-\frac{0.2+0.1}{2}-\frac{8.9}{2}}{\sqrt{8.25-0.2}\sqrt{8.25-0.1}}=\frac{3.65}{\sqrt{8.05\times8.15}}=\frac{3.65}{8.10}=0.45.$$

rank of x (u) :	1	2	4	4	4	6	7	8	9	10
rank of y (v) :	1.5	9	3	4.5	1.5	8	4.5	10	6	7

The total score in this case is

8 – 6 + 5 + 4 + 5 – 2 + 3 – 2 + 1 = 16

Here, $$T_u=\frac{1}{2}\times3\times2=3$$

and, $$T'_v=\frac{1}{2}(2\times1+2\times1)=2,\frac{n(n-1)}{2}=45.$$

Thus, $$r_k = \frac{TotalScore}{\sqrt{\left(\frac{n(n-1)}{2}-T_u'\right)\left(\frac{n(n-1)}{2}-T_v'\right)}} = \frac{16}{\sqrt{(45-3)(45-2)}} = \frac{16}{42.5} 0.38$$

(*d*) The difference in calculation arises due to the fact that when ranking is used instead of the actual observation, there is always some loss of information, since rank differences outweigh the actual differences in observations.

Example 5.18 The coefficient of rank correlation between marks in Statistics and marks in Mathematics obtained by a certain group of students is 0.4606. If the sum of the squares of the difference in ranks is given to be 89, find number of students in the group.

Solution : From the formula of Spearman's rank correlation coefficient we know,

$$r_S = 1 - \frac{6\sum d_i^2}{n(n^2-1)}, n \text{ being the number of observations.}$$

Here as $r_s = 0.4606, \sum d_i^2 = 89$, we have

$$n(n^2-1) = \frac{6\sum d_i^2}{1-r_5} = \frac{6\times 89}{1-0.4606} = 989.99 = 990$$

$$= 10 \times 99 = 10\,(10^2 - 1)$$

$\Rightarrow$ n = 10.

So there are 10 students in the group.

Example 5.19 The following table shows the result of inoculation against cholera.

	Attacked	Not attacked
Inoculated	5	431
Not inoculated	9	291

Examine the effect of inoculation in controlling susceptibility to cholera.

Solution : Let" A : Inoculated, α : Not – inoculated

B : Attacked β : Not – attacked

From the given table $(AB) = 5$, $(A\beta) = 431$

$(\alpha B) = 9$, $(\alpha\beta) = 291$

The coefficient of association,

$$Q = \frac{(AB)(\alpha\beta)-(A\beta)(\alpha B)}{(AB)(\alpha\beta)-(A\beta)(\alpha B)} = \frac{5\times 291 - 431\times 9}{5\times 291 + 431\times 9} = \frac{-2424}{5334} = -0.45$$

So there is negative association between inocculation and cholera attack. Thus, inoculation is effective in controlling cholera.

Example 5.20 A census revealed the following figures of the blind and the insane persons in two age-groups in a certain population.

Age-group	15-25	25 and over
Total population	2,70,000	1,60,200
Number of blind	1000	2000
Number of insane	6000	1000
Number of insane among blind	19	9

Obtain a measure of the association between blindness and insanity for each age group and comment.

Solution :

Age-Group 15-25

	Eyed	Blind	Total
Normal	2,63,019	981	2,64,000
Insane	5981	19	6000
Total	2,69,000	1000	2,70,000

Age Group 25 and over

	Eyed	Blind	Total
Normal	157209	1991	1,59,200
Insane	991	9	1000
Total	158200	2000	1,60,200

Yule's coefficient of association for age-group 15-25 :

$$Q = \frac{2{,}63{,}019 \times 19 - 5981 \times 981}{2{,}63{,}019 \times 19 + 5981 \times 981} = \frac{870000}{10864722} = -0.08$$

and, for age-group 25 and over:

$$Q = \frac{1{,}57{,}209 \times 9 - 1991 \times 991}{1{,}57{,}209 \times 9 + 1991 \times 991} = \frac{-558200}{3387962} = -0.16$$

Comment : The low negative values of the measure of association for each age-group indicates that there is no association between insanity and blindness among the people of the two given age-groups.

Example 5.21 For the following 3×3 classification, compute a measure of association. Would it be proper to attach a sign to this measure?

	Left eyed	Ambiocular	Righteyed	Total
Left handed	48	25	52	125
Ambidextrous	32	13	25	70
Right handed	94	33	91	218
Total	174	71	168	413

Solution : Table : Showing expected cell frequencies

	Left eyed	Ambiocular	Right eyed
Left handed	52.66	21.49	50.85
Ambidextrous	29.49	12.03	28.47
Right handed	91.85	37.48	88.68

x^2 measure of association $c^2 = \sum \frac{O^2}{E} - n$

$=2.37$

Pearson's coefficient of contingency = $\sqrt{\chi^2/(n+\chi^2)}$ = 0.07

Here it is not reasonable to attach sign to the measure since attributes are classified in more than two categories as well as there does not exist any implied ranking in categories of the attributes.

Exercises and Class Projects

1. What are the coefficients of linear regression? Prove that the coefficient of correlation is the geometric mean of the coefficients of regression equations. (B.Com (H), Calcutta, 1991)
2. Define correlation coefficient in bivariate data and show that it lies between-1 and +1. (H.S.(s) 1991)
3. What is meant by correlation? How correlation between two variables is measured? Does correlation always signify a cause and effect relationship between the variables?
4. What is a scatter diagram? How does it help us in studying the correlation between two variables?
5. When can we say that the two variables are correlated? What would we infer if r_{xy} turns out to be zero?
6. What is linear regression? Why are there, in general, two regression lines? When do they coincide? Explain the use of regression equations in a decision making problem.
7. Describe the method of least squares with examples and show how it can be used to fit a linear regression.
8. What is Spearman's rank correlation coefficient? Bring out its usefulness. How does this coefficient differ from Karl Pearson's coefficient of correlation?
9. What is meant by Association of Attributes? How does it differ from the concept on correlation?
10. When are two attributes said to be

    (a) Independent, *(b)* Positively associated and

    (c) Negatively associated,

    with respect to the conditions to be satisfied by the class frequencies?
11. What are the measures of association of attributes for 2 × 2 contingency tables?
12. Generally, the term 'positive association" or 'negative association' of attributes is meaningless when the attributes are classified in more than two classes. Explain.

13. The following table gives the soil temperature and germinating time at various places.

Temperature (°F)	Germinating Time (hr)	Temp. (°F)	Germinating Time (hr)
57	10	42	27
42	26	44	19
40	30	40	18
38	41	46	19
42	29	44	31
45	27	43	29

(*a*) Draw the scatter diagram.

(*b*) Calculate correlation coefficient between time and temperature, and interpret the result.

(*c*) Derive the regression lines for both temperature and germinating time.

(*d*) What is the most likely time of germination at 50°F ?

(*e*) Plot the residuals against both the predicted values of germinating time and that of soil temperature.

(*f*) Do the basic assumptions appear to be violated?

14. In the following, diameter of trees at height 4 feet above ground and the corresponding height of the trees have been given.

x: Diameter (Inch.)	0.9	1.2	2.9	3.1	3.3	3.9	4.3	6.2	9.6	12.6	16.1	25.8
y: Height (Feet)	18	26	32	36	445	35.6	405	57.66	73	84	67	85.5

14. In the following, diameter of trees at height 4 feet above ground and the corresponding height of the trees have been given.

(*a*) Determine Pearson's product moment correlation coefficient (*r*) and Spearman's rank correlation coefficient (*R*). Mention the advantage and disadvantage of *R* over *r* with respect to this example.

(*b*) Plot the scatter diagram and determine if a straight line relation is appropriate.

(*c*) If linearity assumption is not suitable determine appropriate linearizing transformation. In particular, try for $x' = \log x$ and $y' = \log y$.

(*d*) Fit a straight line regression to the transformed data.

(*e*) What proportion of variability is explained by the fitted model?

15. The following summary appears on a survey covering 1,000 fields. Find out if the data are consistent.

Manured fields	510
Irrigated fields	490
Fields growing improved varieties	427
Fields both irrigated and manured	189
Fields both manured and growing improved varieties	140
Fields both irrigated and growing improved varieties	85

16. 88 residents of an Indian city are classified according to their smoking and tea drinking habits as follows:

	Smokers	Non-smokers
Tea Drinkers	40	33
Non-tea Drinkers	3	12

Calculate coefficient of association without and with applying Yate's correction for continuity, and comment.

17. From the following data of the marks obtained by 8 students in Accountancy and Statistics. Compute the coefficient of rank correlation.

Marks in Accountancy15	20	28	12	40	60	20	80	
Marks in Statistics	40	30	50	30	20	10	30	60

18. The coefficient of rank correlation between the marks in Economics and Statistics obtained by a group of students is 9/1 1 and the sum of squares of the differences in ranks is 30. Find the number of students in the group.

Class Project; Collect the students of your class and measure their arm reaches above head and heights. Now find the correlation coefficient between arm-reach above head and height, and also find the two regression lines. Would you expect that correlation coefficient exists among the overall students of the school?

Objective Questions

1 . Examine the validity of the following statements:

(*a*) $r_{xy} = 0 \Rightarrow x$ and y are independent.

(*b*) If $r_{xy} > 0$ then $r_{xy'} > 0$, $r_{x'y} > 0$ and $r_{x'y'} > 0$, where $x' = -x$, $y' = -y$.

(*c*) $r_{xy} > 0 \Rightarrow \Sigma\, xy > n\bar{x}\,\bar{y}$ or $E(xy) > E(x)\,E(y)$.

(*d*) If $Var(x + y) \Rightarrow Var(x) + Var(y)$, then $r_{xy} = 0$.

(*e*) If x and y are independent (uncorrelated) then $-x$ and $-y$ are independent (uncorrelated).

(*f*) $u + 3x - 5$, $2y - v = 7$ and $r_{xy} = 0.12$ then $r_{uv} = 0.12$.

(*g*) $\Sigma d^2 = 33$, $n = 10 \Rightarrow$ rank correlation coefficient =+ – 0·8.

(*h*) If $r_{xy} = 0$, then $V(x + y) = V(x - y)$.

(i) If the rank correlation coefficient of a few pairs of observations is – 0.5 with sum of squares of the differences in ranks is 126, then the number of pairs is 8.

(*j*) If $2u + 5x = 17$, $5v = 2y + 11$ and $Cov(x, y) = 3$, then $Cov(u, v) - 3$.

(*k*) Pearson's correlation coefficient is independent of origin but not of scale.

Answer

(a) Statement is invalid, since $r_{xy} = 0$ does not imply x and y are independent, we can only say that x and y are linearly independent (*i.e.*, x and y are not linearly related).

(b) $r_{xy} > 0 \Rightarrow \Sigma xy > n\bar{x}\bar{y}$ ...(1)

$\therefore r_{xy} = 0 \Rightarrow r_{xy'} > 0$ is not valid, since

$$r_{xy'} > 0 \Rightarrow -\Sigma xy > -n\bar{x}\bar{y}, (\because y' = -y)$$

or, $\Sigma xy < n\bar{x}\bar{y}$ which is contrary to the inequality (1).

Similarly, $r_{xy} > 0 \Rightarrow r_{x'y} > 0$ is not valid where $x' = -x$.

Now $r_{xy} > 0 \Rightarrow \Sigma r_{x'y'} > 0$ is valid, since

$$r_{x'y'} > 0 \Rightarrow (-x)(-y) > n(-\bar{x})(-\bar{y})$$

$$\Rightarrow \Sigma xy > n\bar{x}\bar{y} \Rightarrow r_{xy} > 0$$

(*c*) Statement is valid, since

$$r_{xy} = \frac{Cov(x,y)}{\sqrt{Var(x)Var(y)}} = \frac{\sum xy - n\bar{x}\bar{y}}{\sqrt{\left(\sum x^2 - n\bar{x}^2\right)\left(\sum y^2 - n\bar{y}^2\right)}} = \frac{E(xy) - E(x)E(y)}{\sqrt{Var(x)Var(y)}} > 0$$

$\Rightarrow Cov(x, y) > 0$ or $\Sigma xy - n\bar{x}\bar{y} > 0$ or $E(xy) - E(x)E(y) > 0$.

(*d*) $Var(x + y) = Var(x) + Var(y)$

or, $Var(x) + 2Cov(x,y) + Var(y) = Var(x) + Var(y)$.

$\Rightarrow Cov(x, y) = 0$.

$\therefore r_{xy} = \dfrac{Cov(x,y)}{\sqrt{Var(x)Var(y)}} = 0$. Thus the statement is valid.

(*e*) First part : Valid. Since x and y are independent, $-x$ and $-y$ will also be independent.

Second Part: x and y are independent $\Rightarrow r_{xy} = 0 \Rightarrow Cov(x, y) = 0.$

Now, $Cov(-x,-y) = \frac{1}{n}\sum(-x)(-y) - (-\bar{x})(-\bar{y})$

$$= \frac{1}{n}\sum xy - \bar{x}\,\bar{y} = Cov(x, y) = 0.$$

Thus $-x$ and $-y$ are also independent, so the statement is valid.
So the statement is valid.

(*f*) Here $x = \frac{5-u}{3} = \frac{u-5}{-3}$ and $y = \frac{7+y}{2} = \frac{v-(-7)}{2}$

Since the values of the variables x and y are obtained from the variables u and v by changing scales of opposite sign (*i.e.*, –3 and 2) and some origins, $r_{xy} = -r_{uv}$. So $r_{uv} = -0.12$. So the statement in not valid.

(*g*) The rank correlation coefficient $= 1 - \frac{6\sum d^2}{n(n^2-1)} = 1 - \frac{6\times 33}{10\times 99} = 0.8$

So the statement is not valid.

(*h*) $r_{xy} = 0 \Rightarrow Cov(x,y) = 0$
Now $Var\,(x + y) = Var\,(x) + Var\,(y)$
So the statement is valid and $Var\,(x - y) = Var\,(x) + Var\,(y)$, since $Cov\,(x, y) = 0$.

(*i*) Let the number of pairs be n. From the given values.

$\therefore \quad -0.5 = 1 - \frac{6\times 126}{n(n^2-1)} \Rightarrow n(n^2-1) = 504$

or, $\quad n(n^2-1) = 8\times 63 = 8(8-1)$

$\Rightarrow \quad n = 8.$

So the statement is valid.

(*j*) $2u + 5x = 17$ and $5v = 2y + 11$

or, $\quad u = \frac{x - \frac{17}{5}}{-\frac{2}{5}}$ and $v = \frac{y - \left(-\frac{17}{5}\right)}{5/2}$

Since the variables x and y are changed by scales of opposite signs, $Cov\,(u, v) = -Cov\,(x, y)$. Thus the statement is valid.

(*k*) If x and y variables are changed to u and v variables respectively with origins at a and c, scales being b and d, then

$u = \frac{x-a}{b}$ and $v = \frac{y-c}{d}$.

Obviously, $r_{uv} = \frac{|b||d|}{bx} r_{xy}$.

$\therefore\ r_{uv} = r_{xy}$, if b and d are of some sign,

$= -r_{xy}$, *if* b and d are of opposite signs.

So the correlation coefficient is independent of origin and it is also independent of scales in magnitude but not of sign. Thus the statement is partly valid.

2.(*a*) If the correlation coefficient between the variables x and y is zero then what is the correlation coefficient between x^2 and y^2 ?

(*b*) If $r_{xy} = -0.9$, then for the higher values of x, what sort of values do you expect for y?

(*c*) If $r_{xy} = 0$, what is the value of *Cov* (x, y) and how x and y are related?

(*d*) For the given values on three variables

x:	1	2	3	4	5	6	7
y:	7	8	9	10	9	8	7
z:	7	6	5	4	5	6	7

What values do you expect for

$r_{xy}, r_{xz}, r_{yz}, r_{x^2y^2}, r_{x^2z^2}$ and $r_{y^2z^2}$?

(*e*) If the correlation coefficient between the variables x and y is -1 then what will be the correlation coefficient between x^2 and y^2?

Answer:

(*a*) The correlation coefficient between x^2 and y^2 will be positive or negative low value nearly zero.

(*b*) For higher values of x, it is expected that y will retain lower values.

(*c*) *Cov* $(x, y) = 0$. Here x and y are not linearly related but there may exist a curvilinear relation.

(*d*) $r_{xy} = 0,\ r_{xz} = 0,\ r_{yz} = -1$

$r_{x^2y^2} = -0.2,\ r_{x^2z^2} = 0.2,\ r_{y^2z^2} = 1.$

(*e*) The correlation coefficient between x^2 and y^2 is also -1.

5.22 State, giving reasons, whether each ol the following statement is true or false.

(*a*) The regression coefficient of y on x is 2.3 and that of x on y is 0.7.

(*b*) Both the regression coefficients cannot exceed unity.

(*c*) $5y + x + 2 = 0$ is a regression equation of y on x and $12y - 3x - 1 = 0$ is a regression equation of x on y.

(*d*) If the regression lines are $Y = 0.25\,x$ and $X = (1/9)\,y + 1$, then $r_{xy} = 1/6$.

(*e*) In a bivariate distribution $b_{yx} = 2.2$ and $b_{yx} = 0.3$.

(*f*) The regression lines of two independent variables are parallel to each other.

Answer:

(*a*) Here $b_{yx} = 2.3$ and $b_{yx} = 0.7$.

$\therefore\ b_{yx} \times byx = 2.3 \times 0.7 = 1.61.$

Product, $b_{yx} \times b_{xy} > 1 \Rightarrow$ correlation coefficient > 1, which is impossible. Hence the statement is false.

(*b*) The statement is true, since both the regression coefficient greater than unity implies correlation coefficient greater than one which is not true.

(*c*) Since $5y + x + 2 = 0$ is a regression equation of y on x, $b_{yx} = -\ 1/5$ and

Since $12y - 3x - 1 = 0$ is a regression equation of x on y, $b_{xy} = 4$.

So the statement is false since the regression coefficients cannot be of opposite signs.

(*d*) We know $r_{xy} = \sqrt{b_{xy} \times b_{yz}}$. Here $y_{yz} = 0.25 = \dfrac{1}{4}$ and

$b_{xy} = 1/9$. So $r_{xy} = \sqrt{\dfrac{1}{4} \times 1/9} = \sqrt{1/36} = 1/6$

Thus the statement is true.

(*e*) The statement is true, since the product, $b_{yx} \times b_{xy} = 0.66 < 1$.

(*f*) The statement is false. The regression lines of two independent variables are perpendicular to each other.

CHAPTER 6

INDEX NUMBER

6.1. Introduction

An index number is an economic data figure which reflects price or quantity compared with a standard or base value. Suppose the base is taken to be the year 1990. The value of the index number in the year 2000 gives an idea of the change in price or quantity that has taken place in the next 10 years. The base usually equals 100 and the index number is usually expressed as 100 times the ratio of the base value. Suppose, a commodity costs twice as much in 1970 as it did in 1960, its index number would be 200 relative to 1960. The main uses of the index numbers are the comparison of business activity, the cost of living and the employment. Index numbers help the economists to convert huge set of business data into understandable form. In economics, index numbers generally are time series summarising movements in a group of related variables. In some cases, however, index numbers may compare geographic areas at a point of time. An example of such use of index number is the country's purchasing power parity. The best known and heavily used index number is the consumer price index number, which measures the changes of the retail prices paid by the customers. In addition, a cost-of-living index is a price index that measures the relative cost of living over time.

An index number is a statistical measure which is designed to express changes or differences in a variable or a group of related variables, usually expressed in percentage form. It is a number which is used to measure the level of a certain phenomenon at some standard period. In other words, an index number is a number which is used as a device for comparison between price, quantity or value of a group of articles in different situations, e.g., at a certain place or a period of time and that of another place or period of time.

When the comparison is done with respect to price over time, the index number is termed as price index number.

When the comparison is done with respect to quantity over time, it is termed as the index number of the quantities.

The index numbers are the useful means for the comparison of the variations arising out of the difference in situations, e.g., the change of time or the change of place.

6.2. Base Year and Current Year

As mentioned before, the index number is mainly used for comparison of the prices or quantities of the commodities at two different time points. Through index number, it can be understood that how much the price of the commodity has changed over the time frame. In order to construct the index number, we need to fix a base year, i.e. a year in which the prices or quantities of the commodities will be considered as base for comparing the prices or quantities of the future years.

Suppose, in 1980, the price of a certain commodity was Rs 30. Now, the price of the same commodity in the tear 2000 is expressed in terms of the price of the commodity in 1980. Then, 1980 is the base year whreas 2000 is the current year.

6.3. Relative Index Numbers

The index numbers are relative figures. The value of a variable in a given year (or place) divided by the value of the same variable in a specified year (or place) is called a relative and is generally expressed in percentage. Suppose, the price (P_0) of a certain commodity in the year 1970 was Rs 20 and the year 1970 had been taken as a base year. In the year 1984, the price(P_1) of the same commodity became Rs 46. Then the relative price index number for the commodity in the year 1984 was as follows -

$$\text{Price Relative Index Number} = \frac{P_1}{P_0}\times 100 = \frac{46}{20}\times 100 = \text{Rs. } 230 \qquad 1$$

The followings are the different types of relative index numbers -

Price Relative Index Number - The index number as mentioned above is an example of the price relative index number. The price (P_1) of commodity in a given year expressed as a percentage of the price (P_0) of the same commodity in a specified year (the base year) is called the price relative index number.

Suppose the price of a certain commodity in the year 2006 is P_1. The price of the same commodity in the base year 1970 was P_0. Then the Price Relative Index number for the commodity in the year 2006 can be written as follows - $(P_1 / P_0) \times 100$

Quantity Relative Index Number - The quantity of a certain commodity consumed in a given year expressed as the percentage of the quantity of the same commodity consumed in the base year is termed as quantity relative index number. Suppose Q_1 quantity of the commodity is consumed in the year 2006 whereas the Q_0 quantity of the same commodity was used in the based year 1990. Then the Quantity Relative Index Number for the commodity in the year 2006 can be written as follows -

$$\text{Quantity Relative Index Number} = \frac{Q_1}{Q_0}\times 100 \qquad 2$$

Production Relative Index Number – The amount of a certain commodity produced in a given year expressed as the percentage of the amount of the same commodity produced in the base year is termed as the production relative index number. Suppose, U_1 amount of the commodity has been produced in the year 2006 where as the U_0 amount of the same commodity produced in the base year 1990. Then the Production Relative Index Number for the commodity in the year 2006 can be written as follows –

$$\text{Production Relative Index Number} = \frac{U_1}{U_0}\times 100 \qquad 3$$

Value Relative Index Number – Suppose P_1 and Q_1 are the price and quantity respectively for a commodity used in a given year whereas P_0 and Q_0 are the specified price and quantity for the same commodity in the base year.

$$\text{Value Relative Price Index Number} = \frac{V_1}{V_0}\times 100 = \frac{P_1Q_1}{P_0Q_0}\times 100 \qquad 4$$

Value Relative Price Index Number represents variation in both price and quantity over time.

6.4. Index Number Based on Retail and Whole-sale Prices

Cost of Living Index Number – In the construction of the cost of living index number, the retail prices are used. The prices of the commodities are considered based on the amount the customers are exactly paying in order to buy it.

Wholesale Price Index Number - In the construction of the Wholesale Price Index number, the wholesale prices are used. The prices of the commodities in the wholesale market are considered in the construction of this index number.

Index Number Based on Weights

Index number can be constructed in two ways - considering different weights of each of the components and not considering the weights of each of the components used for construction of the Index Number.

Simple or Unweighted Index Number - Here all the components contribute equally in the construction of the index number.

Suppose P_{11}, P_{12}, P_{13} and P_{14} are the prices of the 4 commodities in the current year, whereas, P_{01}, P_{02}, P_{03} and P_{04} are the prices of the same commodities in the base year. Then the Simple Unweighted Index Number can be written as follows -

$$\text{Simple Unweighted Index Number} = \frac{P_{11}+P_{12}+P_{13}+P_{14}}{P_{01}+P_{02}+P_{03}+P_{04}} \qquad (5)$$

Weighted Index Number - In this type of index number, we take specific weights for the different components constructing the index number. Suppose, W_1, W_2, W_3 and W_4 are the weights of the four commodities in the construction of the Index Number.

Then the Weighted Index Number can be described as follows -

$$\text{Weighted Index Number} = \frac{W_1P_{11}+W_2P_{12}+W_3P_{13}+W_4P_{14}}{W_1P_{01}+W_2P_{02}+W_3P_{03}+W_4P_{04}} \qquad (6)$$

Characteristics of Index Number

The Chief characteristics of the index number can be described as follows -

(i) Index Number is expressed in numerical form without any unit. It represents the relative changes from the base year to the current year.

(ii) Index Number is expressed in terms of the percentage so that the base value becomes 100. However, the percentage sign is not used at the time of writing the index number.

(iii) Index Number is used to measure the changes in prices or in quantities or in values from one point of time to another point of time.

(iv) Index Number is a special case of averaging where the variables to be averaged may have different units of measurement.

(v) Changes overtime or between places or like categories are compared by the Index Numbers.

6.5. Construction of Index Number

Construction of Index Number is a complex process consisting of the following steps -

(i) Determination of the Purpose of Construction of the Index Number

(ii) Selection of the Base Year

(iii) Selection of the Commodities to be included in the Construction of the Index Number

(iv) Selection of the Representative Prices

(v) Selection of the System of Weighting

(vi) Selection of the method of averaging

(vii) Selection of the Suitable Formula for Construction of the Index Number

(i) Determination of the Purpose of the Construction of the Index Number - It is very much important to come to a definite conclusion regarding the purpose of construction of the Index Number and how it will help in the business decision making - the ultimate goal behind the construction and tracking of the index number. Suppose, the index number is to be used for the comparison of the retail prices of the different commodities in market. In that case, we will construct the index number based on the retail prices. However, if we want to have a comparative picture of the whole-sale prices of the different commodities, we shall have to construct the index number based on the whole-sale prices. So, determination of the purpose of the index number is very much important for its construction.

(ii) Selection of the Base Year - It is very much important to select an appropriate base year for the purpose of the construction of the index number. The base year should be a year where there would not be any extraordinary events like draught, flood, earthquake, etc. affecting the prices to a great extent due to scarcity. The base period of an index number is the period of time against which the comparisons are made for different commodities.

In the construction of index number, we generally encounter three types of base periods

(a) Fixed base (Single Period)

(b) Fixed base (An average of selected periods)

(c) Chain base

While selecting a base period, we have to take a specific decision whether we shall take a specific period as the base period or an average of selected periods as base period. It is also possible to take chain base for the construction of the index number.

The following things should be kept in the mind while choosing a base period for the construction of index number -

(a) The base period must be a normal period free of any sort of abnormalities or random causes such as financial crisis, floods, famines or war, etc.

(b) Reliable figures must be available in the base period.

(c) The base period should not be too distant from the current priod.

It is proper to choose the average of several periods as the base period in the case when it is difficult to choose a single period as the base period.

In case the comparisons are required on a yearly basis, the system of chain base is most appropriate for the purpose.

(iii) Selection of the Commodities to be included in the Construction of the Index Number - Selection of the commodities is an important step in the construction of the index number. It is not possible to include each and every available commodity in the construction of the index number. It is appropriate to include only some commodities in order to make the index number more effective. The following points should be considered for the selection of the commodities to be included in case of construction of index number -

(a) The commodities included should be of standard quality and there should not be any significant variation in the quality.

(b) The commodities should represent the real tastes, habits and the customs of the people.

(c) The commodities should be easily recognizable and describable.

(d) The commodities should not be non-tangible commodity, like, personal service etc.

(iv) Selection of the Representative Prices - At the time of selection of the representative prices, it is very much important to note the followings -

(a) The method of quoting the prices of the commodities,

(b) The sources from where the quoted prices may be obtained.

(v) Selection of the System of Weighing - The weighing technique is used to give variable importance to the different commodities included in the construction of the index numbers. The commodity which is needed to give more importance, the value of its weight is kept higher in the construction of the index number.

(vi) Selection of Average - In order to find the composite index, different types of averages like, arithmetic mean, geometric mean and harmonic mean etc, are used. The selection of the best average to be used in order to construct the composite index depends on the relative merits and de-merits of the different averages.

(vii) Selection of suitable formula for computation of the index number - There are several formulae for the computation of the index numbers. The different formulae are used in the different situations. We have to select the best formula based on the particular situation.

Methods of Construction of Index Numbers

There are two methods for the construction of the index numbers -

(i) Relative Method

(ii) Aggregate Method

(i) Price Relative Method - The simple average of relative price index is computed using the following procedure - (a) First, the price relative is obtained by dividing the price of each item in a given period, p_n, by its base-period price, p_0 and the result is expressed in percentage. (b) Second, The price relatives which are obtained are added together and the sum is divided by the number of items in the aggregate. Thus, the simple average of relatives is an arithmetic mean of the relatives. We have the following formula for the relative price index.

$$P_{0n} = \frac{\sum \frac{p_n}{p_0} \times 100}{n}, \quad (7)$$

where P_{on} is the relative price index for the period n with base 0.

The formula for the computation of the simple average of relative quantity index is as follows -

$$Q_{0n} = \frac{\sum \frac{q_1}{q_0}}{n} \times 100 \quad (8)$$

where Q_{on} is the relative quantity index for the period n with base 0.

A weighted price index is constructed with the purpose of measuring changes in a variety of prices over time. The most suitable weights that are used are the quantities (q) of each of them. The quantities (q) for each of the commodities either in the base year (q_0) or current year (q_n) are used as the weights.

Using the base year quantities as weights, the formula for the index number is -

$$P_{0n} = \frac{\sum \frac{q_0 p_n}{p_0}}{\sum q_0} \times 100 \quad (9)$$

This corresponds to the Laspeyres' formula.

If we use the current year quantities as weights, we get the following index number -

$$P_{0n} = \frac{\sum \frac{q_n p_n}{p_0}}{\sum q_n} \times 100 \quad (10)$$

This corresponds to the Paasche's formula.

The general formulae for construction of the index numbers through simple aggregate method are the followings -

$$P_{0n} = \frac{\sum p_n}{\sum p_0} \times 100 \quad (11)$$

$$Q_{0n} = \frac{\sum q_n}{\sum q_0} \times 100, \quad (12)$$

where P_{on} and Q_{on} are the price and quantity indices.

(ii)Price Aggregative Method - The weighted aggregative price index with base-year quantity as weight is called the Laspeyres' index (LP_{on}) and is given by the formula -

$$LP_{0n} = \frac{\sum p_n q_0}{\sum p_0 q_0} \times 100, \quad (13)$$

where p_o and p_n are commodity prices during base year and current (given) year respectively. The above is known as the Laspeyres' formula for quantity index.

Laspeyres' formula for quantity index (LQ_{on}) is as follows -

$$LQ_{0n} = \frac{\sum q_n p_0}{\sum q_0 p_0} \times 100 \tag{14}$$

Laspeyres' formula is used to compute a consumers price index, the result would measure the difference between the theoretical cost in a given year and the actual cost in the base year of maintaining the same standard of living in the base year.

Laspeyres' index answers the following question: "What is the change in aggregate value of the base year's list of goods when valued at current year prices?".

The weighted aggregative price index with current year quantity as weights is referred to as the Paasche price index. It is given by -

$$PP_{0n} = \frac{\sum p_n q_n}{\sum p_0 q_n} \times 100 \tag{15}$$

Where p_0 and p_n are the base-year and current year prices respectively and qn is the current year quantity.

The Paasche's Quantity Index is described as follows -

$$PQ_{on} = \frac{\sum q_n p_n}{\sum q_0 p_n} \times 100 \tag{16}$$

Paasche's Price Index formula used to calculate a consumer's price index, compares the cost in the given period with the theoretical cost in the base period or keeping the standard of living in the given period. In general, the Paasche's formula answers the question "What would be the value of the current year list of goods when valued at base-year prices?".

Marshall-Edgeworth formula is the combination of the Laypeyres and Paasche indices and the formula for the price is calculated as follows -

$$MP_{0n} = \frac{\sum p_n (q_0 + q_n)}{\sum p_0 (q_0 + q_n)} \times 100 \tag{17}$$

The Marshall-Edgeworth quantity index (MQ0n) is given by

$$MQ_{0n} = \frac{\sum q_n (p_0 + p_n)}{\sum q_0 (p_0 + p_n)} \times 100 \tag{18}$$

Marshall-Edgeworth formula is the half-way between Laspeyres and Paasche formula.

Fishers' Ideal Index Number for price and quantity are given as follows -

$$FP_{on} = \sqrt{\frac{\Sigma p_n q_0}{\Sigma p_0 q_0}} \times \sqrt{\frac{\Sigma p_n q_n}{\Sigma p_0 q_n}} \times 100 \tag{19}$$

$$PQ_{on} = \sqrt{\frac{\Sigma q_n p_o}{\Sigma q_o p_o}} \times \sqrt{\frac{\Sigma q_n p_n}{\Sigma q_o p_n}} \times 100 \qquad (20)$$

Test for Ideal Index Number

The index numbers should follow a few tests to be an ideal index number. These are as follows -

(i) The Time Reversal Test

(ii) The Factor Reversal Test

(iii) The Circular Test

(iv) Proportionality Test

(i) The Time Reversal Test - If we reverse the time subscripts of a price (or quantity) index, the result should be reciprocal of the original index. The meaning of the above test is

$$P_{0n} \times P_{n0} = 1 \qquad (21)$$

where P_{on} is the price index for the current year, n, with the base period 0, and P_{no} is the price index for the current year, 0, with the base period n. In case of Laspeyres' formula, we have

$$LP_{on} \times LP_{on} = \frac{\Sigma p_n q_0}{\Sigma p_0 q_0} \times \frac{\Sigma p_0 q_n}{\Sigma p_n q_n} \neq 1$$

Hence, the Laspeyres' formula of index number does not satisfy the time reversal test. In case of the Paasche's index, we have

$$PP_{on} \times PP_{no} = \frac{\Sigma p_n q_n}{\Sigma p_0 q_n} \times \frac{\Sigma p_0 q_0}{\Sigma p_n q_0} \neq 1$$

Hence Paasche's formula also does not satisfy this test.

Now, we are considering the Marshall-Edgeworth formula.

$$MP_{on} \times MP_{no} = \frac{\Sigma p_n (q_0 + q_n)}{\Sigma p_0 (q_0 + q_n)} \times \frac{\Sigma p_0 (q_n + q_0)}{\Sigma p_n (q_n + q_0)} = 1$$

Hence, the Marshall-Edgeworth index number satisfies the time reversal test.

Similarly, for Fisher's index number

$$FP_{on} \times FP_{n0} = \sqrt{\frac{\Sigma p_n q_0}{\Sigma p_0 q_0}} \times \sqrt{\frac{\Sigma p_n q_n}{\Sigma p_0 q_n}} \times \sqrt{\frac{\Sigma p_0 q_n}{\Sigma p_n q_n}} \times \sqrt{\frac{\Sigma p_0 q_0}{\Sigma p_n q_0}} = 1$$

Therefore, the Fisher's index number also satisfies the time reversal test.

(ii)The Factor Reversal Test - The test indicates that when the price index is multiplied with a quantity index, i.e., factors are reversed, it should result into a value index. Only the Fisher's Ideal Index satisfies this test. In case of Fisher's Index, we have -

$$FP_{0n} \times FQ_{0n} = \sqrt{\frac{\Sigma p_n q_0}{\Sigma p_0 q_0} \times \frac{\Sigma p_n q_n}{\Sigma p_0 q_n}} \times \sqrt{\frac{\Sigma q_n p_0}{\Sigma q_0 p_0} \times \frac{\Sigma q_n p_n}{\Sigma q_0 p_n}} = \frac{\Sigma p_n q_n}{\Sigma p_0 q_0} = V_{0n}$$

V_{0n} is the Value Index Number

(iii) The Circular Test - Circular test is an extension of the time-reversal test over several years. An index is said to meet this test in case of the following -

$P_{t_2 t_1} = P_{t_2 t_3} \times P_{t_3 t_1}$, t_3 is a time-point between t_1 and t_2. (22)

Simple Aggregative index and the Laspeyres index satisfy this test.

(iv) Proportionality Test - If all the prices change in the same proportion, then the second price index should be equal to the first price index . All index numbers studied earlier satisfy the test.

6.6. Cost of Living Index Numbers or Consumer Price Index

The 'Cost of Living Index', also known as 'consumer price index' or 'cost of living price index' is the country's principal measure in order to keep a track on the price change. The average change over time in the prices paid by the consumer of the specific baskets of goods and services is being measured by the cost of living index number. The consumer price index numbers are designed to measure the average change in the price paid by the ultimate consumers for specified quantities of goods and services over a period of time. The consumer price index number helps us in the determination of the effect of the rise and fall in prices on different classes of consumers living in different areas. Different people consume different kinds of commodities and moreover the same commodity is consumed by different people in the different proportion. The main use of the consumer price index is aiding us in the determination of the effect of rise and fall in the prices on different classes of the consumers living in different areas. The consumer price index number is important because the demand of a higher wage is based on the cost of living index and the wages and salaries in most nations are adjusted according to this index number. The cost of living index does not measure the actual cost of living or the fluctuations in the cost of living due to causes other than the change in price level but its object is to find out how much tne consumers of a particular class have to pay for a certain quantity of goods and services.

6.7. Utility of the Consumer Price Index Numbers

(i) Consumer Price Index Number is very much useful in order to measure the change in purchasing power of the currency, real income etc

(ii) Consumer Price Index number is also useful for the Government in order to take necessary policy decisions for formulating wage policy, price policy, taxation and general economic policies

(iii) Market price for a particular kind of goods and services are analysed by the consumer price index

(iv) The salaries and wages are fixed on the basis of the consumer price index. So, it is very helpful to revise wage or dearness allowance.

Assumptions for the construction of the Cost of Living Index Number

Construction of the Cost of Living Index Number is based on the following assumptions-

(a) Similar Needs - The needs of the people for which this index number is constructed are the same.

(b) Same Goods - The goods consumed in the base year and the current year are unchanged.

(c) No change in quantity of goods - It is also assumed that the quantity of goods consumed will remain the same in the base year and the current year.

(d) Price quotations are the same - It is assumed at the time of the construction of the Cost of Living Index Number that the prices at the different places are the same and they do not change frequently.

(e) True on the average - The cost of living index numbers are true on the average

(f) Representative goods - The commodities included in the cost of living index number represent the consumption of the class of people.

The steps in the Construction of the Cost of Living Index Numbers

(i) Selection of the class of people whose food and other habits should be included in the living index number.

(ii) Determination of the Scope of the Index.

(iii) Proper conduction of the family budget inquiry

(iv) Acquiring price quotation for the different commodities

(v) Framing the Index Number

Methods of Construction of the Consumer Price Index Number

The following methods are used for the construction of the Consumer Price Index Number-

(i) Aggregate Expenditure Method

(ii) Family Budget Method

(iii) Weighted Average of Price Relatives

(i) Aggregate Expenditure Method - This method is based on the Laspeyres' method, i.e., base year quantities are taken as weights (w=Q0)

$$P_{01} = \frac{\Sigma p_1 q_0}{\Sigma p_0 q_0} \times 100 \tag{23}$$

(ii) Family Budget Method - It is a method of weighted relatives where the weights are value $p_0 q_0$ in the base year which is often denoted by V.

$$P_{01} = \frac{\Sigma RV}{\Sigma V} \times 100 = \frac{\Sigma p_1 q_0}{\Sigma p_0 q_0} \times 100 \tag{24}$$

Here, we get an idea about the relative change in the current year value than the base year value.

(iii) Weighted Average of Price Relatives - It is calculated in the following way -

Suppose I=Group Index and W=Weights. Then, it is written as follows -

$$P_{01} = \frac{\sum IW}{\sum W} \tag{25}$$

6.8. Uses and Advantages of Index Numbers

(i) Index Number can be used for the comparative study

(ii) Index number can simplify the data

(iii) Index Number provides guidelines to economic policy and helps in formulating the decisions.

(iv) Index Number measures the purchasing power of money.

(v) It reveals the change in the cost of living.

(vi) It also gives an idea of the national income.

(vii) Index Number is used as control by the Government.

(viii) It reveals trends and tendencies.

(ix) Index number is very much useful for deflating.

(x) Index Number has universal utility.

6.9. Limitations of Using Index Number

The main limitations for the use of index number are the followings -

(i) Index Numbers are not perfect - These are approximate values

(ii) It is very much difficult to construct the index number. It is needed to take into account many situations like base year selection, item selections, changes in habits and average selection, etc.

(iii) Sampling errors can encroach into the constructed index number.

(iv) It is very much easy to manipulate the index numbers.

(v) Index number constructed for one purpose can not be used for other purposes.

(vi) It is very much difficult to construct index numbers as adequate and accurate data are not available.

EXAMPLES

1. Find the simple aggregative price index of the following data -

Commodity	Unit	Price in Rs. Per Unit	
		2000	2004
A	One Kg	10	15
B	One Kg	40	30
C	One Dozen	10	12
D	One Litre	5	13

Solution - Formula for Aggregative Price Index (11) is as follows -

$$P_{0n} = \frac{\sum p_n}{\sum p_0} \times 100$$

Here, $\sum p_n = 15+30+12+13=70$

Similarly, $\sum p_0 = 10+40+10+5=65$

Hence, the simple Aggregative Price Index is the following -

$$P_{0n} = \frac{\sum p_n}{\sum p_0} \times 100 = (70/65) \text{ X } 100 = 107.6923$$

This implies the price has increased 7.6923% in the year 2004 when compared to the year 2000.

2. Find out the Simple Aggregative index for the following data -

Commodity	Price in Rs. Per Quintal	
	Base Year, 2001	Current Year, 2002
Wheat	80	100
Rice	120	250
Gram	100	150
Pulses	200	300

Answer - Formula for Aggregative Price Index (11) is as follows -

$$P_{0n} = \frac{\sum p_n}{\sum p_0} \times 100$$

Here, $\sum p_n = 100+250+150+300=800$

Similarly, $\sum p_0 = 80+120+100+200=500$

Hence, the simple Aggregative Price Index is the following -

$$P_{0n} = \frac{\Sigma p_n}{\Sigma p_0} \times 100 = (800/500) \times 100 = 160$$

This implies the price has increased 60% in the year 2002 when compared to the year 2001.

3. Following data give the changes in the price and consumption (quantity) of certain major constituents of the consumption-basket of the goods of the labour class.

Commodity	Unit	1968		1978	
		Price	Consumption (Quantity)	Price (Quantity)	Consumption
Rice	Quintal	250	16	320	20
Wheat	Quintal	120	12	160	10
Cloth	Meter	8	50	10	35

Compute the following -

(i) Laspeyres' Price Index Number
(ii) Paasche's Price Index Number
(iii) Fisher' s Price Index Number
(iv) Laspeyres' Quantity Index Number
(v) Paasche's Quantity Index Number
(vi) Fisher's Quantity Index Number
(vii) Marshall-Edgeworth Price Index Number
(viii) Marshall-Edgeworth Quantity Index Number

Solution - As per the given data,

(i) Laspeyres' Price Index Number = LP68.78 = $\frac{\Sigma p_{78} q_{68}}{\Sigma p_{68} q_{68}} \times 100 = \frac{7540}{5840} \times 100 = 129.11$

(ii) Paasche's Price Index Number = PP68.78 = $\frac{\Sigma p_{78} q_{78}}{\Sigma p_{68} q_{78}} \times 100 = \frac{8350}{6480} \times 100 = 128.86$

(iii) Fisher's Price Index Number = FP68.78 = $\sqrt{LP_{68.78} \times PP_{68.78}} = \sqrt{129.11 \times 128.86} = 128.98$

(iv) Laspeyres' Quantity Index Number = LQ68.78 = $\frac{\Sigma q_{78} p_{68}}{\Sigma q_{68} p_{68}} \times 100 = \frac{6480}{5840} \times 100 = 110.96$

(v) Paasche's Quantity Index Number = PQ68.78 = $\frac{\Sigma q_{78} p_{78}}{\Sigma q_{68} p_{78}} \times 100 = \frac{8350}{7540} \times 100 = 110.74$

(vi) Fisher's Quantity Index Number = FQ68.78

$= \sqrt{LQ_{68.78} \times PQ_{68.78}} = \sqrt{110.96 \times 110.74} = 110.85$

Simarly Marshall-Edgeworth Price Index Number and Marshall-Edgeworth Quantity Index Number can be calculated by using the respective formulae.

EXERCISE

1. What is Index Number? What are its uses?

2. Write short notes on the following -

(i) Laspeyres' Index Number

(ii) Paasche's Index Number

(iii) Fisher's Index Number

(iv) Marshall-Edgeworth Index Number

3. What is the difference between price and quantity index?

4. What arc the methods of construction of Cost of Living Index Number? Write a brief note on them.

5 . Find the simple aggregative price and quantity index from the following data.

Commodity	Quantity in Kg		Price in Rs. Per Unit	
	2000	2004	2000	2004
A	15	20	110	115
B	30	36	140	130
C	20	24	210	212
D	40	50	25	43

6. Find out the Laspeyres', Paasches', Marshall Edgeworth and Fisher's Index Number for both price and quantity from the following data.

Commodity	Quantity (in 000 Kg)		Price Per Kg	
	2000	2009	2000	2009
A	60	68	5.00	6.20
B	80	86	9.25	10.75
C	100	112	12.00	14.25
D	150	176	17.75	1850
E	125	140	15.50	18.00

7. Construct an appropriate index number for the following data.

Year	Wheat		Rice		Sugar	
	Price (Rs.)	Quantity (Kg.)	Price (Rs.)	Quantity (Kg.)	Price (Rs.)	Quantity (Kg.)
1981	3.50	20	6.00	40	7	50
1982	4.00	30	7.00	60	8	30

8. For the following data calculate both the price and quantity index numbers by Paasche's method. Hence test its reliability with the help of the factor and time reversal test.

Commodity	Base Year Price	Base Year Quantity	Current Year Price	Current Year Quantity
A	10	25	15	30
B	15	30	10	20
C	20	10	25	20
D	40	25	40	26
E	60	10	50	45

9. For the following data, calculate both the price and quantity index numbers by Laspeyres' method. Hence test its reliability with the help of the factor and time reversal tests.

Commodity	Base Year Price	Base Year Quantity	Current Year Price	Current Year Quantity
X	10	25	15	30
Y	12	50	14	24
Z	20	10	25	20

10. Calculate both Laspeyres' and Paasche's aggregative price indices for the year 1960 from the following data.

Commodities	Quantities		Price Per Unit	
	1959	1960	1959	1960
A	3	5	2	2.5
B	4	6	2.5	3
C	2	3	3	2.5
D	1	2	1	0.75

CHAPTER 7

TIME SERIES

7.1. TIME SERIES - DEFINITION AND ITS NECESSITY

Time series is a set of observations taken on a variable over successive points/ periods of time, measurement of observations may be taken continuously or in discrete phases /points (mostly at equal intervals). Time series analysis constitutes an indispensable technique in the arena of management, commerce, economics, business and finance besides other areas of application.

Example: Sensex data recorded daily over a year constitute time series data where measurements of data are taken at discrete points of time (here, day-wise). Other examples are; yearly production data of different items, monthly-sale figures in a store, weekly data on bank clearings in a particular bank, daily market price of a particular vegetable at a place, annual profits in respect of an organization (bank, industry, service centre, etc) over years.

Examples of time series, where measurements may be continuously taken are, the levels of tide over time at one site, amplitudes of electric signal over time, etc. In fact, time series, where measurements are taken at discrete time points (usually equi spaced) may be considered as an accumulation of data over a period (like accumulated sale-figures over a month - monthly sale figures; rainfall accumulated over a month-monthly rainfall data, etc.). Symbolically, the value of the variable is represented by y_t, where t denotes the time point; t = 1, 2, 3, 4………, n, if there are n time points.

7.2. IMPORTANCE OF TIME SERIES

It can be seen that in a time series the observed values at different time points follow different patterns, like exhibiting trend (smooth variation: increase/decrease over time), fluctuation (up and down movement over time), or irregular/ random movement (not following any pre-defined way) or a mixture of the above mentioned three types of patterns. Thus it is necessary to analyse the given time-series data in order to know the pattern of movement (behaviour) of it over time. Further, the time series analysis assists us to get an idea of the (recent) future values of the variable i.e., to predict the future values by employing analytical methods on the past and presently available values (data).

7.3. MODELS

From the above discussion it is clear that our aim is to gain an understanding of the nature of the past behaviour of the process and to predict the future values using the past and current values. Prediction of future values require formulation of a model which is framed by identifying the explanatory variables governing the behaviour of the process (measured in terms of some variables) and to set up a model (explicit functions) by incorporating the above identified variables in it. A model is an analytic description of an objective represented in terms of a time series corresponding to a data-set to describe its features - indeed, it presents a mathematical formulation of its behaviour.

7.4. COMPONENTS OF A TIME SERIES:

The components of a time series are, Trend, Seasonal, Cyclical and Irregular. Thus y_t can be represented as, i) $y_t = T_t + S_t + C_t + I_t$ (additive model); or, ii) $y_t = T_t S_t C_t I_t$ (multiplicative model); or, iii) $yt = T_t S_t C_t + I_t$.

Additive Model : The additive model for decomposition of time series assumes that all the four components of the time series operate independently of one another. It also assumes that the behaviour of components is of an additive character. It is to be noted that only absolute values are added or deducted from the trend value to arrive at the observed value.

Multiplicative Model:: This model assumes that the observed value is obtained by multiplying the trend (T) by the rates of three other components, i,e, $Y_t = T_t x S_t \; x C_t x I_t$.

The multiplicative model assumes that the components due to the different causes are not necessarily independent and they can affect one another. It also assumes that the behaviour of the components is of multiplicative character. It may be noted that except the value of trend all the other values on the right hand side are rates or index numbers.

Most of the time series relating to economic and business phenomena conform to the multiplication model. In practice, additive model is rarely used.

Trend Component(T)

Secnlar trend is a long term upward or downward movement arisen due to the presence of factors which affect the mean of the series. Some series may remain more or less at a constant level. Thus trend is a smooth, regular and long term upward or downward (or both) movement where in sudden or frequent changes are incompatible with its concept.

Seasonal Component (S)

This measure of characteristic pattern of the time series refers to the seasonal (specific interval of time) fluctuations (occur regularly and in a periodic manner) in the period, the period being one year. Many times series in the arena of business and commerce exhibit variations in values from one season (so defined) to the other (with differing amplitudes) and such variations repeat themselves, at least approximately, year by year. A year is a time period during which a full cycle of such seasonal movement is completed. In the next year also the same pattern of the variation for the seasons within the period is followed. In fact, the seasons are the intervals of the time constituting the period. Seasonal variations occur due to the existence of either some natural events, like, rainfall, arrival of winter or some events which can be attributed to man-made conventions or practices. The sales of umbrella increases in rainy seasons, sales of winter garments boost up in winter season, price of potato falls down around October/November, etc. Other such events which occur due to man-made practices are festivals, like, Durga puja, Dushera, Christmas, ID, etc. The variations of the above nature (occur due to the existence of natural causes or man-made practices) which occur within a year, and are repeated over the subsequent years, are, in fact, seasonal variations. In fine, in such cases variations are exhibited if the data are recorded quarterly, monthly, weekly, daily, hourly, and so on.

Cyclical Component(C)

The oscillatory movement in a time series with period of oscillation greater than one year, each complete period being called as a cycle, is called the cyclical fluctuation. The cyclical component refers to the above cyclical fluctuations and it is, in fact, the slow oscillatory movements which are repeated over cycles. The length of the cycle and its intensity of fluctuation may vary from one cycle to the other. Mostly, the cycle lengths are 3-4 years, though some may be 7-10 years or longer. The business cycles are four-phased cycles comprising, prosperity, recession, depression and recovery from time to time. For short term forecasting, general trend and cycle are taken together so that it is impossible to obtain information about cycles in absence of data from several cycles. Usually, time series data contain features called trend and cycle together.

Irregular Component (I)

If in a time series the components T, S and C are accounted for, it can be seen that still some part is left which is also to be defined and this is the part, which is termed as Irregular component. It is usually a random sequence which occurs due to unforeseen and unaccountable causes/ events, such as flood, earthquake, famine, war etc. It can be noted that the characteristic movement of a time series associated with recession, increased sales during Puja, increasing demand of smaller auto mobiles and decline in death rate due to advancement in medical science, can be attributed to cyclical variation, seasonal, trend component and secular trend movement respectively.

As a distinction between seasonal and cyclical components, it can be said that the first one refers to a short term periodic movement with period being one year, where as the second one refers to an oscillatory movement with period being more than a year and the presence of cycles of varying lengths (amplitudes) are compatible with its definition. Cyclical fluctuations are not regular like seasonal fluctuations. The changes in cyclical component in a business process pass through the four phases, like, prosperity, decline, depression and recovery.

7.5. Measurement of Secular Trend

Before proceeding to measure the trend, it is necessary to eliminate the other three components from the time series. The seasonal component can be eliminated if the monthly values are added and annual values are taken for measuring the trend from the given data. For a short series, it is assumed that the trend and cyclical components are inextricably mixed up. Thus after eliminating the seasonal component following the way mentioned above, the methods of measuring trend are discussed below:

1) Free-hand Method, 2) Semi- average Method, 3) Group average Method, and 4) Mathematical curve Method.

Methods Related to Measurement of Secular Trend

7.5.1. Free hand or graphic method : Following this method, the time series values, y_t are plotted against t (time) on a graph paper. The 'n' points so obtained are then joined by free hand curve, which reveals an idea of the underlying trend in the series. This is the simplest method to find out the trend. Fitting a trend line by this method is subjective and hence arbitrary. It is needed that the trend line be a smooth curve. Owing to its subjectivity and arbitrariness this method can not be used for prediction purpose.

Example: Find out the trend values with the help of freehand curve method for the data given below:

Year	1991	1992	1993	1994	1995	1996	1997	1998	1999	2000	2001
Production (in lakh ton)	25	28	26	30	27	31	28	26	28	33	31

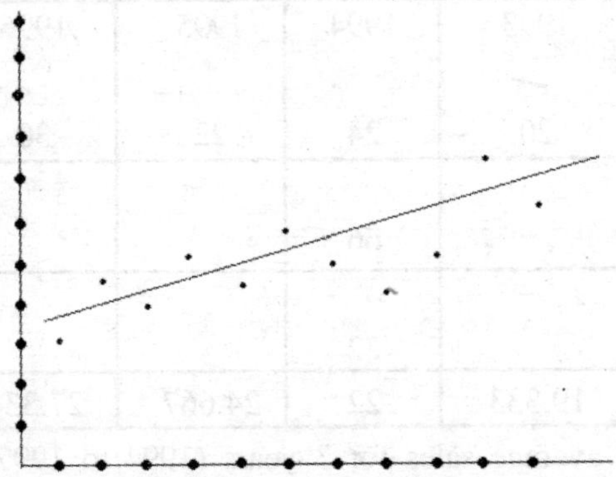

7.5.2. Semi-Average Method : The method of fitting linear trends with the help of semi average method is discussed for the following two cases:

The number of years is even :

In this case the time series data are divided into two equal parts. The total of the values corresponding to the items (years) in each part is calculated and it is then divided by the number of items (years) to obtain the arithmetic mean corresponding to each part. The averages are then plotted on the graph paper against the time interval covered under each part. The straight line joining these two points gives the trend line. The ordinate at each point gives the trend value corresponding to the point.

The number of years is odd :

When the number of years is odd, the value corresponding to the middle most year is omitted so that the time series is divided into two equal parts. Then the procedure (i) is followed.

If the actual trend is a straight line, the method of semi averages gives the values very near to the original values. If the ratios of two successive observations are equal, then, at first, the logarithms of the observations are taken. Semi-Average method is then applied to the above logarithmic data. The original trend values are obtained by taking anti-logarithms.

A trend value corresponding to any future year may be predicted by multiplying the periodic increment by the number of years (apart from the best trend value) into the future that is desired and adding the result to the best trend value listed in the series.

Merits:

i) The method is simple. ii) The trend line can be extended on either side in order to obtain past or future estimates. iii) This is an objective method, as any one applying this method gets the same trend line with that of another one.

Demerits:

i) The method of semi average assumes that a straight line relationship exists between the plotted points, regardless of the fact whether such a relationship is viable or not. ii) This method has an in-built limitation in regard to the fact that the arithmetic mean is used here. This method is not suitable in case when there exists very low or very large extreme values. Iii) There is no assurance that the influence of cycle is eliminated.

Example: Apply the method of Semi Averages for determining the trend values for the following data and estimate the trend value for year, 2000..

Year	1993	1994	1995	1996	1997	1998
Sales (thousand units)	20	24	22	30	28	32
3 Yearly Semi-Average Total		66			90	
3-yearly Semi Average Values		22			30	
Trend Values	19.333	22	24.667	27.334	30	32.667

Here the increase in average sales for 3 years (1994 to 1997) is 8 (thousand units), and annual increase is 8/3 = 2.667 (thousand units). Hence sale value for 1993 is 19.333 (= 22 – 2.667). The predicted values for 1999 and 2000 are: 35.334 (thousand units) and 38.001 (thousand units) respectively.

7.5.3. Method of Moving Averages : This method is most commonly used for smoothing out the variations in a time series and for isolating the trend. The moving average method is characterised by a constant, known as the period or extent of the moving average. This is found by plotting the time series on a graph paper and finding out the length the cycle, if any. If there is a cycle, the period of M.A. is taken as the period of the cycle. The M.A. method is applicable when the trend is linear. Following this method successive observations in the time series are averaged and more specifically, a moving average of period n is the average on n successive values of the time series. Starting from the first n observations, the averages are calculated sequentially, that is, following the method, the average of the n observations starting from the first observation, then the average of the n observations starting from the second observation, then the average of the n observations starting from the third observation, etc., until the set of the last n observations are taken into account. For a set of p values y_t, t = 1, 2, 3,........, (p-1), p, the M.A. values of period n (n is less than p) will be as follows:

1st value in the M.A series = $(y_1 + y_2 + y_3 + \ldots\ldots + y_n)/n$

2nd value in the M.A series = $(y_2 + y_3 + \ldots\ldots + y_n + y_{n+1})/n$, etc

The method of moving averages is to be described for the two cases, when the period is odd and when the period is even respectively.

Case (i): When the period of moving average is odd:

As the values of the time series are to be against the years, one has to select a period of moving average such as 3 years, 5 years, etc. Let the values of the series be, a, b, c, d, e, f, g, h, etc

Compute the moving totals according to the period of moving average. If the length of the period of moving average is 3 i,e., 3-yearly moving averages are to be calculated. The moving totals can be computed as:

a + b + c, b + c + d, c + d + e, d + e + f,........

For 5-year moving average, the moving totals are computed as:

a + b + c + d + e, b + c + d + e + f, c + d + e + f + g, , ...

The moving totals are then placed at the centre of the time span for which they are computed. Compute moving averages by dividing the moving totals by the length of the period of moving average and place them at the centre of the time span from which the moving totals are computed. These moving averages are also called the trend values.

By plotting these trend values (if desired) one can obtain the trend curve with the help of which one can judge whether the trend is increasing or decreasing. If needed, one can also compute short-term fluctuations (errors or residuals, a residual is the difference between an observed value and the corresponding MA value) by subtracting the trend values from the actual values.

Example:

The following table gives the yearly production values for the period, 1988 - 1996.

Year (t)	1988	1989	1990	1991	1992	1993	1994	1995	1996
Production (in ten ton) Yt	21	22	23	25	24	22	25	27	26

Obtain 3-yearly Moving Average values. Also find the short term fluctuations (errors).

Table showing the calculation of trend values

Year	Production (ten ton)	3 -yearly moving totals (Mov. totals)	3 -yearly MA Y_a (Trend values) (ten ton)	Short term fluctuations Error $(Y_t - Y_a)$
1988	21	-	-	-
1989	22	66	22.00	0
1990	23	70	23.33	- 0.33
1991	25	72	24.00	1.00
1992	24	71	23.67	0.33
1993	22	71	23.67	- 1.67
1994	25	73	24.33	0.67
1995	27	79	26.33	0.67
1996	26	-	-	-

Case (ii): When the period of moving average is even:

When the period of moving average is even (i.e., 4 years, 6 years, etc), the MA values are computed through the following steps:

The time series values are written down.

Select the length of the period of moving average. Let the length of the moving averages period be 4-years (usually the selection is done by plotting the time series and examining its period of the cycle, if any, the period of the cycle is taken as the period of the moving average).

Compute the 4 yearly moving totals and place them at the centre of the time span. The four - yearly moving totals are computed as follows:

a + b + c + d, b + c + d + e, c + d + e + f,

Compute the 4 - yearly moving average value for each period and place the same at the centre of each time span. It is to be noted that in this case the moving average values so obtained and placed do not coincide with original time period setting.

Thus subsequently, the two-item moving average values (based on of the 4-yearly moving average values) are computed and placed at the middle of the new time period settings. This process is called centring of the moving average values.

Case (ii): When the period of the Moving Average is even

Example - 1: Calculate 4-Yearly Moving Average values from the table given below and hence find the trend values.

Year	1958	1959	1960	1961	1962	1963	1964	1965	1966	1967
Value	50.0	36.5	43.0	44.5	38.9	38.1	32.6	41.7	41.1	33.8

To obtain the four yearly trend values by MA method we have to find the 4-yearly Moving Average values at the first stage and then apply 2-item MA on the 4-yearly MA values already obtained. The MA values obtained at the second stage give the trend values (obtained by adopting 4-Yearly MA procedure) which are centred with respect to the original data.

Table showing the calculation of Trend values by MA method (even number of period)

Serial no. time)	Actual year	Value y_t	4-year moving total	2- Items centered Moving Total	4-Years Moving Average (y_c)	Residual y_t-y_c
1	1958	50	-	-	-	-
2	1959	36.5	-	-	-	-
			174	-	-	-
3	1960	43		336.9	42.11	0.89
			162.9	-	-	-
4	1961	44.5		327.4	40.92	3.98
			164.5	-		
5	1962	38.9		318.6	39.82	-0.92
			154.1	-		-
6	1963	38.1		305.4	38.17	-0.07
			151.1	-		
7	1964	32.6		304.8	38.1	-5.5
			153.5		-	-
8	1965	41.7	-	302.7	37.83	3.87
			149.2	-	-	
9	1966	41.1				
10	1967	33.8	-	-	-	-

The MA values obtained in the 6th column give the trend values. It can be seen that the first MA value corresponds to the year, 1960, the second one corresponds to the year 1961, etc..

Example 2: The ruling price of standard gold in Rs. Per 10 gm are given below for the first 15 days of a recent month. Use the method of 4-day period moving average and estimate the trend in gold price.

4350 4360 4370 4350 4360 4370 4370 4360 4350

4340 4330 4350 4350 4360 4360 [ICWA (Intermediate), June, 1999]

Table showing calculation of 4 days Moving Average values

Day	Ruling Price (Rs. Per 10 gm)	4-Day Moving Total	2-Item Moving Total	4-Day MA values (Trend Values) (Rs. Per 10 gm)
1	4350	-		-
2	4360			
		17430		
3	4370		34870	4358.75
		17440		
			34890	4361.25
4	4350	17450		
			34890	4362.50
5	4360			
		17450		
			34900	4362.50
			3490	4363.75
6	4370			
		17460		
			34910	4363.75
7	4370			
		17450		
			34870	4358.75
8	4360			
		17420		
			34900	4360.50
9	4350			
		17380		
			34750	4343.75
10	4340			
11	4330	17370		
			34760	4345.00
12	4350			
		17390		
			34810	4351.25
13	4350			
		17420		
14	4360			
15	4360			

Example 3: For the following series of observations, verify that the 6-year centred moving average is equivalent to a 7-year weighted moving average with weights 1,2,2,2,2,2,1 respectively.

Year	Sales (in '0000 Rs.)
1990	2
1991	4
1992	3
1993	6
1994	7
1995	9
1996	4
1997	6
1998	7
1999	8
2000	10

Firstly, we have to find out 6-year centred moving averages

This will be done in two stages. In the first stage 6-year average values are computed and secondly, 2 item moving average values are computed to obtain the 6-year centred MA values.

Table showing 6-Year centred Moving Average values.

Serial number	Years	Sales	6-Years moving Total	6-year centered moving Total	6-Year centred MA values (Trend values)
1	1990	2			
2	1991	4			
3	1992	3			
			31		
4	1993	6		64	5.33
			33		
5	1994	7		68	5.67
			35		
6	1995	9		74	6.17
			39		
7	1996	4		80	6.67
			41		
8	1997	6		85	7.01
			44		
9	1998	7			
10	1999	8			
11	2000	10			

Secondly, we compute 7-year MA values with weights 1,2,2,2,2,2,1 respectively.

7-Year Weighted Moving average

Serial number	Years	Sales	7-Year Weighted Moving Average	7-Year MA values (Trend values)
1	1990	2		
2	1991	4		
3	1992	3		
4	1993	6	64	5.33
5	1994	7	68	5.67
6	1995	9	74	6.17
7	1996	4	80	6.67
8	1997	6	85	7.01
9	1998	7		
10	1999	8		
11	2000	10		

Corresponding to 1993, the 7-year weighted MA value is

$(2\times1+4\times2+3\times2+6\times2+7\times2+9\times2+4\times1)/(1+2\times5+1) = 64/12 = 5.33$

Comparing the 6 year MA values and 7 year MA values, we find that the MA values found out by applying the two methods are equal for each year.

7.5.3.1. Merits of the method of moving averages:

This method is simple and thereby most popular and extensively used.

This method is objective as it is based on averaging so that subjectivity (personal bias) is avoided. This method can be used for determining seasonal, cyclical variations also.

This method is flexible enough as an inclusion of additional data does not change the M. A. values obtained earlier.

If, in the data, the period of moving averages coincides with the period of the cyclical fluctuations, cyclical fluctuations are automatically eliminated.

7.5.3.2. Disadvantages:

There is no functional relationship between the observations and the time points. Thus, this method is not helpful in forecasting and predicting the future values beyond the time period for which the MA values are obtainable.

There are no trend values for some years in the beginning and some years at the end. For example, in case of 5 - yearly moving average there will be no trend values for the first two years and for the last two years.

In case of non - linear trend, the values obtained by this method are biased either in one direction or in the other direction.

The selection of the period of moving average is a difficult task. Therefore great care has to be taken in selecting the period of the moving average, particularly in the absence of any business cycle in the data set during the time period.

In case of weighted moving averages, weighted arithmetic means are computed using the given weights.

7.5.4. The Method of Fitting of Mathematical Equations and the Principle of Least Squares

The method of least squares is an analytical tool using which any mathematical function (i.e., the coefficients of the function are so estimated) can be fitted to the given data by minimising the sum of squared errors with respect to the parameters of the equation. This method is more accurate and precise and can be used even for forecasting future values of the regressand also. Any mathematical function (linear or curvilinear trends) can be fitted and the best function is chosen for which the sum of squared errors is minimum. For the pair of variables, (Y, X), Y, X represent the dependent and independent variables respectively. Let y be the observed values of the variable, Y and y_c be the corresponding expected value of y when the observed value of X is x.

7.5.4.1. Linear Trend

Here, $y_c = a + bx$ is a straight line to be fitted as trend, and $y = y_c + e$. The values of the coefficients, 'a' and 'b' are to be found in such a way that the sum of squares of differences between the observed and expected values of y is least, i.e., $S = \Sigma(y - y_c)^2$ is least, the least value is, however, to be obtained by applying the procedure of differentiation on S with respect to a and b respectively (Firstly, optimum values of 'a' and 'b' are obtained by solving the normal equations and then substituting the optimum values of these coefficients in the regression equation). The above procedure is called the method of least squares. The line obtained by applying the method of least squares is called 'the line of best fit.'

For a given time series data, to find a linear trend, the values of a, and b are obtained by solving the following two normal equations;

$$\Sigma y = Na + b\Sigma x \quad (1)$$

$$\Sigma xy = a\Sigma x + b\Sigma x^2, \quad (2)$$

where, N is the number of pairs in the given data.

Here a, is the intercept of the line on the Y - axis and b is the slope of the line, b is also known as growth rate (if $b > 0$) or decline rate (if $b<0$), b gives the change in the value of y against per unit change in the value of x.

Direct method

i) Convert the years into natural numbers (1, 2, 3...) and denote by x. Arrange x values correspond is to years.
ii) Find the squares of x values and obtain Σx^2.
iii) Multiply the x values with the corresponding y values and obtain the value of Σxy.
iv) Add the values of y to obtain Σy.
v) Put these values in the two normal equations, (1) and (2) and solve for 'a', and 'b'.
vi) Substitute these values of 'a', and 'b' in $y = a + bx$ to obtain the trend values for various values of x, i.e., $y_c = a_1 + b_1x_1$ where a_1 and b_1 are the estimated values of a and b respectively.

Alternative Method

Put $x_1 = x - x_0$, where x_0 is the mid-point in the X series of values. Thus $\Sigma x_1 = 0$, and then normal equations (taking Y and x_1) reduce to:

$\Sigma y = Na$, or, $a = \frac{\Sigma y}{N}$, and

$\Sigma x_1 y = b \Sigma x_1^2$, or, $b = (\Sigma x_1 y)/(\Sigma x_1^2)$

The regression equation of Y on x_1 is $y_c = a + b\, x_1$, 'a', and 'b' are the solutions of the above two equations.

7.5.4.2. Non - linear Trend

Fitting of a Non-linear function by the Method of Least Squares.

When the time series data do not conform with the linear trend then we try other forms of trend equations, like, parabolic or exponential.

An equation of the form (for n pairs of values on (Y, X)),

$Y = a + bx + ex^2 + dx^3 + \dots + kx^n$ is known as a polynomial of degree n in x, $k \neq 0$.

Let the parabolic curve to be fitted be a second degree polynomial, then

$Y = a + bx + cx^2$, with usual notations. The values of a, b, and c can be determined by solving the following normal equations :

$$\Sigma y = Na + b\Sigma x + c\Sigma x^2 \quad (1)$$
$$\Sigma xy = a\Sigma x + b\Sigma x^2 + c\Sigma x^3 \quad (2)$$
$$\Sigma x^2 y = a\Sigma x^2 + b\Sigma x^3 + c\Sigma x^4 \quad (3)$$

If x_1 is defined as before (i.e., $\Sigma x_1 = 0$), then the normal equations reduce to:

$$\Sigma y = Na + c\Sigma x_1^2 \quad (4)$$
$$\Sigma x_1 y = b\Sigma x_1^2 \quad (5)$$
$$\Sigma x_1^2 y = a\Sigma x_1^2 + c\Sigma x_1^4 \quad (6)$$

The constants a, b, and c are to be determined from the equations (4), (5) and (6) respectively.

Fitting of Exponential Trend

The exponential curve is given by the equation, $Y = a\, b^t$, ……(7), where a, and b are constants.

Taking logarithm on both sides of (7), we get, Log y = log a + t log b, or, $y_1 = A + B\,t$, where $y_1 = \log y$; A = loga and B = logb.

Again, the normal equations are, $\Sigma y_1 = na + B\Sigma t$. (8), and $\Sigma t y_1 = A\,\Sigma t + B\,\Sigma t^2$….(9)

The values of A and B can be obtained from (8) and (9) respectively. From the logarithmic form of equations we can get back to the original form of equations (i.e., the exponential equation between y and t)

Fitting by Growth Curve Model

Most of the time series relating to business and economic phenomena over long periods of time don't exhibit constant growth rate in a particular direction, but a variable growth rate over the entire time period. It is observed that the variable growth rate is initially slow, but after a while the growth rate picks up (i.e., becomes faster and gets accelerated), and it remains stable for some time and after wards it shows retardation. The following are some of the important growth curves which are generally used to describe the movements in a time series - Modified Exponential Curve; Gompertz Curve, and Logistic Curve,

Modified Exponential Curve:

$Y = a + b*c^t$; $a > 0$, a, b and c being the parameters of the curve, ...(10).

Gompertz Curve:

$y = a*b^{ct}$, ..(11), a, b and c being the parameters of the curve. Here, the left hand side 'y' is equal to 'a' multiplied by k (='b' to the power 'c' to the power 't') [or, y =a*k]. Taking log on both sides in the above equation (11), we get, log y = log a + log K, or

Taking log on both sides in the above equation (11), we get, log y = log a + log K, or log y = log a + c^t log b, or, $Y_1 = A + Bc^t$, ...(12), where A= log a, B = log b and Y_1 = log y. The model (12) is nothing but a modified exponential curve. Thus the logarithm of the given values in (12) Gompertz model follows modified exponential law.

Logistic Curve or Pearl-Read Curve:

$y = k/(1+e^{(a+bt)})$, $b<0$ ....(12)

where, k, a and b are the parameters of the curve and k = max (y) i.e., the maximum value which the variable can take over all values of time t.

Taking the reciprocal in the above equation we get,

$1/y = 1/(k/[1+e^{a+bt}]) = (1/(k)*[1+e^a.e^{bt}]) = (1/k) + (1/k)\ e^a * (e^b)^t)$, Putting, A=1/k, $B= e^a/k$ and $C = e^b$.

Thus we can write, $1/y = A + B*C^t$ ... (13),

which is the equation of the modified exponential curve. Hence, if the given observations follow Logistic law of growth, then their reciprocals follow modified exponential law.

7.5.4.3. Merits and Limitations of Trend Fitting by the Method of Least Squares

Merits: The method of least squares is the most popular and widely used method of fitting mathematical functions to a given set of observations. It has the following advantages:

Because of its analytical or mathematical character, this method completely eliminates the element of subjective judgement or personal bias on the part of the investigator.

Unlike the method of moving averages, this method enables us to compute the trend values for all the given time points in the series.

The trend equation can also be used to estimate or predict the values of the variable 'Y', with respect to any time point t in future or even in the intermediate time points (not given in the observational set) lying within the time interval used. The predicted and forecast values (y_c) so obtained are also quite reliable and precise (i.e., near to the observed values, y).

The curve fitting by the principle of least squares is the only technique which enables us to obtain the rate of growth per annum, for yearly data, if a linear trend is fitted. If we fit the linear trend y = a + b x, where x is obtained from t by change of origin such that $\Sigma x = 0$, then for the yearly data, the annual rate of growth is b or 2b according as the number of years is even or odd respectively.

Limitations : The most serious limitation of the method is the determination of the type of the model to represent the trend to be fitted, viz, whether a linear or a parabolic trend or some other more complicated trend equations to be fitted.

The addition of even a single new observation necessitates all the calculations to be done afresh which is not so in the case of moving average method.

This method requires more calculations and hence it is quite involved and time consuming as compared with other methods. It is rather difficult for a layman (not aware of the mathematical principles) to understand the philosophy and use of this method.

Future predictions of forecasts based on this method relate to long-term variations only, i.e., trend and thus ignore the cyclical, seasonal and irregular fluctuations.

It cannot be directly used to fit growth curves which are the found as the best models followed by the economic and business time series.

We shall now discuss some numerical examples to illustrate the technique of curve fitting by the principle of least squares.

Example 1: (a) Fit a straight line trend for the following data : ICWA (intermediate), Dec.2001

Year	Funds of the Bank (in Rs. Crores)
1991	147
1992	191
1993	213
1994	254
1995	296
1996	393
1997	418

Use it to estimate the working funds during 1998.

(b) For the following table: Bangalore University, 1996

i) Fit a straight line trend by the least squares.

ii) Calculate the trend values:

Year	Production
1990	12
1991	10
1992	14
1993	11
1994	13
1995	15
1996	16

1 (a) Here n = 7 (odd), hence we shift the origin to the middle of the time period i.e. 1994, 'n' being the total number of observations, y =Fund in the Bank Rs. In crores), and t = year.

Let x = t-1994 and the trend line of y (funds) on x be : y = a + b x..........(i)

Table showing the computation of straight line trend

Year (t)	Funds of the Bank (in Rs. Crores) (y)	x = t - 1994	x^2	x*y	Trend Values (y_e = 273.14 + 46.4 x)	y - y_e
1991	147	-3	9	-441	133.90	13.10
1992	191	-2	4	-382	180.14	10.86
1993	213	-1	1	-213	226.74	- 13.74
1994	254	0	0	0	273.14	- 19.14
1995	296	1	1	296	319.54	- 23.54
1996	393	2	4	786	365.94	27.06
1997	418	3	9	1254	412.34	5.66
Total	1912	$\Sigma x = 0$	$\Sigma x^2 = 28$	$\Sigma xy = 1300$		

The normal equations for estimating 'a' and 'b' with respect to the regression equation i) are:

$\Sigma y = na + b\Sigma x$ ...(ii) and $\Sigma xy = a\Sigma x + b*\Sigma x^2$....(iii)

or, 1912 =7a + b*0 or, 1285 = 273.14*0 + b*28

or, a =1912/7 or, b = 1285/28

or, a = 273.14 or, b = 46.4

Substituting the values of 'a' and 'b' in i), the straight line trend equation is given by:

$$y_c = 273.14 + 46.4x......iv)$$

Putting x =- 3, -2, -1, 0, 1, 2, 3 in (iv), we get, the trend values, where y_c = 273.14 + 46.4x

The estimated working fund during 1998 are obtained by putting t =1998 in x = t -1994 or, x = 1998 -1994 = 4. Substituting x = 4 in (iv) we get the estimated working fund for the year 1998 as below:

$$y_c \text{ (1998)} = 273.14 + 46.4*4 = 458.74$$

1 (b) Here n = 7 (odd), hence we shift the origin to the middle of the time period i.e. 1993 Let x = t -1993 and the trend line of y (production) on x be: y = a + b x..................i)

Table showing computation of straight line trend

Year(t)	Production(y)	X = t -1993	x^2	xy	Trend value y_e=13+0.75x
1990	12	-3	9	-36	10.75
1991	10	-2	4	-20	11.50
1992	14	-1	1	-14	12.25
1993	11	0	0	0	13.00
1994	13	1	1	13	13.75
1995	15	2	4	30	14.50
1996	16	3	9	48	15.25
Total	91	0	28	21	

The normal equations for estimating 'a' and 'b' are:

$\Sigma y = n\,a + b\Sigma x$ ...(ii) and $\Sigma xy = a\Sigma x + b\,\Sigma x^2$.................. (iii)

or, a =13 or, b = 0.75

Substituting the values of 'a' and 'b' in i), the straight line trend equation is given by, $y_c = 13 + 0.75x$. Putting x = -3, -2, -1, 0, 1, 2, 3 in $y_c = 13 + 0.75x$, we get the trend values.

Example 2 (a): Using the method of least squares, fit a straight line to the following data and find the trend value and short- term fluctuations. ICWA (Intermediate), June, 2002

Year	Values
1990	232
1991	226
1992	220
1993	180
1994	190
1995	168
1996	162
1997	152
1998	144

Use the fitted trend to estimate the value corresponding to the year 1999 and 2000 respectively.

Example 2 (b): Fit a straight line trend by the method of least squares to the following data:

Year	Production of TV sets in ('000)
1986	17
1987	20
1988	19
1989	26
1990	24
1991	40
1992	35
1993	55
1994	50
1995	74
1996	69

Estimate the production for the year 2001. Delhi University, B.Com (Hons)1997

2 (a) Here n = 9 (odd), hence we shift the origin to the middle of the time period i.e. 1994

Let x = t -1994 and the trend line of y (values) on x be : y = a + b x.........i)

Table showing computation of the straight line trend

Year(t)	Values(y)	x = t - 1994	x^2	x*y	Trend values y_e =186 -11.7 x	y- y_e
1990	232	-4	16	-928	234	-2
1991	226	-3	9	-678	222	4
1992	220	-2	4	-440	210	6
1993	180	-1	1	-180	198	-18
1994	190	0	0	0	186	4
1995	168	1	1	168	174	- 6
1996	162	2	4	324	162	0
1997	152	3	9	456	150	2
1998	144	4	16	576	138	6
Total	1674	0	60	-702		

The normal equations for estimating 'a' and 'b' in ii) are:

$\Sigma y = na + b\Sigma x$ and $\Sigma x.y = a \Sigma x + b \Sigma x^2$

a = 186 b = -11.7

Substituting in i), the straight line trend equation is given by, y_c = 186 -11.7 x............ii)

Putting x = -4, -3, -2, -1, 0, 1, 2, 3, 4 in ii) y_c = 186 -117x, we get the trend values.

The estimated value in1999 is obtained by putting t =1999 in x = t -1999, or, x = 1999 - 1994 = 5. Substituting x = 5 in ii) we get the estimated working fund for the year 1999 as,

y_c(1999) = 186-11.7*5 = 127.5

The estimated value during 2000 is obtained by putting t = 2000 in t - 2000 or, x = 2000 - 1994 = 6. Substituting x = 6 in iii) we get the estimated working funds for the year 1999 as

y_c (2000) = 186-11.7*6 = 115.8.

2 b) Here n =11 (odd), hence we shift the origin to the middle of the time period i.e., 1991

Let x = t -1991 and the trend line of y (production) on x be: y = a + b x...........(i)

Table showing the computation of straight line trend

Year(t)	Production of TV sets in ('000)(y)	x = t-1991	x^2	xy
1986	17	-5	25	-85
1987	20	-4	16	-80
1988	19	-3	9	-57
1989	26	-2	4	-52
1990	24	-1	1	-24
1991	40	0	0	0
1992	35	1	1	35
1993	55	2	4	110
1994	50	3	9	150
1995	74	4	16	296
1996	69	5	25	345
Total	429	0	110	638

The normal equations for estimating a and b in ii) are:

$\Sigma y = na + b\Sigma x$ and $\Sigma xy = a\ \Sigma x + b\ \Sigma x^2$

$a = 39 \qquad b = 5.8$

Therefore, the straight line trend equation is $y = 39 + 5.8x$ ....(ii)

The estimated value for 2001 is obtained by putting t = 2001 in x = t - 1991 or, x = 2001 -1991 =10. Substituting x = 10 in (ii) we get the estimated value for the year 2001 as $y_c(2001) = 39 + 5.8*10 = 97.0$ (thousand).

Example 3: Fit a straight line trend by the method of least squares to the following data and estimate the value for 1999.

Year	Value
1990	380
1991	400
1992	650
1993	720
1994	690
1995	600
1996	870
1997	930

Here n=8 (even), hence we shift the origin to the middle of the time period i.e., 1993 and 1994. Let x = [t - (1993+1994/2)]/ (1/2) = 2 (t - 1993.5) and the trend line of y on x be y = a + b x ......(ii)

Year(t)	Value(y)	x=2 (t-1993.5)	x2	x*y	Trend Values $y_c = 655+35.8x$
1990	380	-7	49	-2660	404.19
1991	400	-5	25	-2000	475.85
1992	650	-3	9	-1950	547.51
1993	720	-1	1	-720	619.17
1994	690	1	1	690	690.83
1995	600	3	9	1800	762.49
1996	870	5	25	4350	834.15
1997	930	7	49	6510	905.81
Total	5240	0	168	6320	

The normal equations for estimating 'a' and 'b' in ii) are:

$\Sigma y = na + b\Sigma x$ and $\Sigma xy = a\Sigma x + b\Sigma x^2$

Or, $a = 655 \qquad b = 35.8$

Therefore, the straight line trend equation is: $y = 655+35.8x$................(ii)

Putting x = -7, -5, -3, -1, 1, 3, 5, 7 in (ii) we get the trend values (Cloumn 6).

The estimated value for 1999 obtained by putting t =1999 in x = t -1999 or, x = 2 (1999-1993.5) =11. Substituting x=11 in ii) we get the estimated value for the year 1999 as

$$y_c(1999) = 655 + 35.8*11 = 1048.8$$

Example 4: Fit a straight line trend by the method of least squares to the following data;

Year	Size of item
1	110
2	125
3	115
4	135
5	150
6	165
7	155
8	175
9	180
10	200

Here n=10 (even), hence we shift the origin to the middle of the time period i.e. 5 and 6

Let x = [t - (5+6) /2)]/(1/2) = 2 (t - 5.5)

Let the trend line of y (size) on x be y = a + b x.........(i)

Table showing the calculation of trend values

Year(t)	Size of item(y)	X = 2 (t - 5.5)	x^2	xy	Trend value y^c = 151+ 4.75x
1	110	-9	81	-990	108.18
2	125	-7	49	-875	117.70
3	115	-5	25	-575	127.21
4	135	-3	9	-405	136.23
5	150	-1	1	-150	146.24
6	165	1	1	165	155.76
7	155	3	9	465	165.27
8	175	5	25	875	174.79
9	180	7	49	1260	184.31
10	200	9	81	1800	193.82
Total	1510	0	330	1570	

The normal equations for estimating 'a' and 'b' in (i) are:

$\Sigma y = na + b\Sigma x$ and $\Sigma xy = a\ \Sigma x + b\ \Sigma x^2$

or, a = 151 or, b = 4.75

Finally, the equation of the straight line trend is : y = 151+4.75x.......................ii)

Putting x = -9, -7, -5, -3, -1, 1, 3, 5, 7, 9 in ii), we get the trend values.

Example 5; Fit a second degree parabola to the following data

X	1	2	3	4	5
Y	1090	1220	1390	1625	1915

Table showing the calculation for Parabolic Trend

X	Y	U=X-3	V = (Y-1450)/5	U^2	U^3	U^4	UV	U^2V
1	1090	-2	-72	4	-8	16	144	-288
2	1220	-1	-46	1	-1	1	46	-46
3	1390	0	-12	0	0	0	0	0
4	1625	1	35	1	1	1	35	35
5	1915	2	93	4	8	16	186	372
Total		0	-2	10	0	34	411	73

Let the parabola of V on U be: $V = a + bU + cU^2$......i), where U = X-3 and V = (Y-1450)/5 then the normal equations for estimating a, b, c are:

$\Sigma V = na + b\Sigma U + c\Sigma U^2$ or $-2 = 5a + 10c$

$\Sigma UV = a\Sigma U + b\Sigma U^2 + c\Sigma U^3$ or $411 = 10b$

$\Sigma U^2V = a\Sigma U^2 + b\Sigma U^3 + c\Sigma U^4$ or $73 = 10a + 34c$

or, a = -11.4, b = 41.1, c = 5.5

Substituting the values of a, b, c in i) we get the fitted parabola as, $V = -11.4+41.1U+5.5\ U^2$

Puting U = X -3, and V = (Y – 1450) / 5 we get the corresponding quadratic trend equation of Y on X.

Example 6: Fit a trend function $y = A.\ B^x$ to the following data

x	1	2	3	4	5
y	1.6	4.5	13.8	40.2	125.0

Here we have to fit the exponential curve, $y = A.B^x$...(i). Taking logarithm on both sides of (i), we get,

log y = log A + x log B, or Y = a + bx...ii), where Y = log y, a = log A, and b = log B.......iii)

Equation ii) represents a straight line between the variables Y and x and hence the normal equations for estimating a and b are:

$\Sigma Y = na + b\Sigma x$ and $\Sigma xY = a\Sigma x + b\Sigma x^2$...............iv)

Table showing the computation of Exponential Trend

x	y	Y = log y	xY	x^2	Trend values (y_e)
1	1.6	0.2041	0.2041	1	1.6
42	4.5	0.6532	1.3064	4	4.6
3	13.8	1.1399	3.4197	9	13.8
4	40.2	1.6042	6.4168	16	41.1
5	125.0	2.0969	10.4845	25	122.3
$\Sigma x=15$		$\Sigma Y=5.6983$	$\Sigma xY=21.8315$	$\Sigma x^2=55$	

Substituting in iv) the normal equation for estimating a and b become:

5.6983 =5 a +15 b ....v) and 21.8315 =15 a + 55 b...........vi)

vi) - 3 * v) gives: 21.8315-3*5.6983 = 55 b - 45b, or, 10 b = 4.7366, or, b = 0.4737.

Substituting the value of b in v), we get

a = 1/5 (5.6983 -15*0.4737) = 1/5 (5.6983 - 7.1055) = -1/5*1.4072 = -0.2814

Hence using iii), we get B = antilog (b) = antilog (0.4737) = 2.977

A= Antilog (a) = Antilog (-0.2814)=Antilog (1'.7186) = 0.5231

Substituting the values A and B in (i), the required trend function is: $y = 5231*(2.977)^x$....(vii)

Putting x = 1,2,3,4 and 5 in vii), we get the trend values (shown in the last column).

Example 7: Fit a straight line trend to the following series of production data. Y values being the average production in thousand tons, what is the monthly trend increment? Find the monthly trend values from the fitted equation for January, March and December, 1963.

Year	Y
1960	37
1961	38
1962	37
1963	40
1964	41
1965	45
1966	50

Let y = a +b x ...(i) be the equation of the straight line fitted to the given yearly data (origin 1963; x unit =1 year); n = 7, x = year – 1963

The normal equations for finding the constants a and b are, $\Sigma y = a n + b\Sigma x$........i), $\Sigma xy = a\Sigma x + b \Sigma x^2$......ii)

Table showing the calculation of trend values

Year	Y	x	x^2	xy
1960	37	-3	9	-111
1961	38	-2	4	-76
1962	37	-1	1	-37
1963	40	0	0	0
1964	41	1	1	41
1965	45	2	4	90
1966	50	3	9	150
Total	288	0	28	57

The normal equations for estimating 'a' and 'b' in i) and ii) are:

288 = a*7 7 = b*28

a = 41.1 b = 2.03

Therefore, the trend equation fitted to the yearly values is, y = 41.1 + 2.03 x.....(iii)

Since y represents the monthly average value for each year and the unit of x is 12 months, we see that the trend equation for the monthly value, y = 41.1 + (2.03/12) x (unit of x = 1 month and y = monthly value) or, y = 41.1 + 0.16x .....(iv), where the origin is the 30th June 1963 (middle-most day of the year); x unit = 1 month). This equation (iv) refers to the time point, 30th June, 1963. This is to be noted that any monthly value refers to the middle of the month, i.e., 15th day of the month. So adjustment (centring on the 15th day or, indeed, half a month), is to be done after obtaining the centred yearly value in (iv).

If July 15, 1963 is chosen as origin i.e., the origin is shifted half a month later, then x should be replaced by (x + 1/2). The monthly trend equation is then, y_c = 41.1 + (x+1/2)* 0.16, or.

y_c = 41.1+0.16 x + 0.08 or, y_c = 41.18 + 0.16 x.........(v). Here for x = 0, we get the trend value for July 15, 1963 (or, for the month July in 1963)

[Note: If June 15, 1963 is chosen as origin i.e., the origin is shifted half a month earlier, then x should be replaced by (x - 1/2). The monthly trend equation is then, y_c = 41.10 + (x - 1/2)* 0.16, or, y_c = 41.10 + .16 x - 0.08 or, y_c = 41.10 + 0.16 x......(vi). Here for x = 0, we get the trend value for June 15, 1963 (or, for the month June in 1963, per say), Putting x = -5 (since January is 5 months behind June) in the equation (vi), we get the value for January 15, 1963. Thus, the value for January 15, 1963 is, y = 41.10 - .16*5 = 41.02 - .80 = 40.22 (thousand tons) ...(vii).]

The trend value for January 15, 1963 is obtained in the following way:

Since January 15, 1963 is 6 months earlier than the origin i.e., July, 15, 1963, therefore, putting x = - 6 in the trend equation (v), we get, y = 41.18 + 0.16*(-6), or, y = 40.22. Hence, the value for Jan 15, 1963 is 40.22 (thousand tons) ...(viii)

Note that the values obtained in (vii) and (viii) are the same as both refer to the January 15, 1963 value.

The trend value for March 15, 1963 is obtained as follows:

Since, March 15, 1963 is 4 months earlier than the origin, i.e., July 15, 1963

Therefore putting x = -4 in the trend equation (vi), i.e., y = 41.1 + 0.16*(-4), or, y = 40.46 (thousand tons).

The trend value for December 15, 1963 is obtained in the following way:

Since December15, 1963 is 5 moths later than the origin i.e. July 15, 1963

Therefore putting x = 5 in the trend equation, y = 41.1 + 0.16*5, we get, y = 41.9 (thousand tons).

Example 8: The trend equation for certain production is given by y = 3600 + 288 t, where

y = Annual production in thousand tons; t = Time with origin, the year 1990 and unit =1 year. Estimate the trend value for the production for September 1994. [ICWA, Intermediate, June 2000].

For the given equation, y = 3600 + 288....(i), the trend equation for monthly averages is, y = 3600/12 + 288/12 t ....(ii), (since here the given values, y' s, refer to the annual production, thus to get monthly 'y' values, the yearly 'y' values are to be converted to monthly values, i.e., yearly y values are to be divided by 12, and the impact of dividing by 12 is that we divide both 'a' and 'b' by 12 in the equation for the annual values), i.e. we write, y = 300 + 24 t (unit of y is monthly value but the unit of time, 't' is 12). So to get the monthly equation the unit of t is to be changed to monthly unit. Hence, the final equation is, y = 300 + (24/12) t, or, y = 300 +2 t .....(iii) (unit of y is month and the unit of time is 1 month). Note that the origin in this equation is at the middle of the year-range, i.e., on the 30th June 1990; t unit = 1 month)

Now the origin is shifted to July 15, 1990. Hence, y = 300 + 2*(t+1/2) or, y = 301 + 2t .....(iv) (Origin: 15th July 1990; time unit = 1 month). More specifically, y = monthly production

t = time with origin at July 15, 1990 and time unit is 1 month).

The trend value for the production for September 1994 (i.e. September 15, 1990)

Now September 1994 is 50 months (i.e. 50 units) later than the origin, July 1990, hence putting t = 50 in (iv) we get the required trend valuue, y = 301 + 2*50, or, y =401 (thousand tons).

The estimated production for 1994 is 401 thousand tons.

Example 9: You are given the following trend equation y = 480 + 36x

Origin: 1988; x = unit of one year; y = Number of units sold per year

i) Convert the above trend equation into a monthly trend equation ; and

ii) Estimate the sale for the month of Oct. 1994[[Delhi University, B. Com (Hons.) 1994]

The trend equation for monthly trend is, y =480/12 + 36/12*x, ...(i) or, y = 40 + 3*x (unit of x = 1 year)

Now if the unit of x is changed to 1 month, then, y = 40 + (3/12) x (Origin: 30th June 1988; x unit = 1 month), y = 40 + 0.25 x. Now the origin is shifted to July 15, 1988. Hence, y= 40 + 0.25* (x + 1/2), or, y = 40 + 0.25x + 0.125.... (Origin: 15th July 1988; x unit= 1 month), y = 40.125 + 0.25 x......(ii) (here, y = monthly production, x = time with origin at July 15 1988 unit being 1 month.

Now Oct. 15, 1994 is 75 months (i.e. 75 units) later than the origin July 15, 1988. Hence putting x = 75, y = 40.125 + 0.25*75 = 58.875.

Example 10: The annual production of a commodity is given as follows:

Year	Production
1990	70
1991	80
1992	90
1993	95
1994	102
1995	110
1996	115

i) Fit a straight line trend by the method of least squares.

ii) Convert the annual trend equation into monthly trend equation. [CA (Foundation, Nov. 1999)

I)Here n=7 (odd) hence we shift the origin to the middle of the time period i.e., 1993.

Let x = t -1993...i) and the trend line of y (values) on x be y = a + b x.................ii)

Computation of straight line trend

Year(t)	Production(y)	X = t -1993	x^2	xy
1990	70	-3	9	-210
1991	80	-2	4	-160
1992	90	-1	1	-90
1993	95	0	0	0
1994	102	1	1	102
1995	110	2	4	220
1996	115	3	9	345
Total	662	0	28	207

The normal equations for estimating a and b in ii) are:

$\Sigma y = n\,a + b\Sigma x$ and $\Sigma xy = a\ \Sigma x + b\ \Sigma x^2$

or, $662 = 7*a$ or, $207 = b*28$

or, $a = 94.57$ or, $b = 7.39$

Therefore, the straight line trend equation is, y = 94.57 + 7.39 x.....(iii); y = annual value of production (in thousand tons), unit of x = 1 year; origin = June 30, 1993

ii)The trend equation for monthly averages is: y = 94.57/12+(7.39/12*12), or, y =7.88 + 0.05 x.

7.6. Measurement of Seasonal Variation

As has already been pointed out that by seasonal variations in a time series we mean the variations which are attributable to forces operating in a regular and periodic manner, the period, however, is less than one year. Such variations in the form of cycles are widely exhibited in many economic and business time series. In fact, businessmen and sales managers use such series for planning future production, scheduling purchases, controlling inventories, determining personnel requirements, setting up selling and advertising programmes, and maintaining stock of goods to be made ready for sale at the right time. In what follows the effects of seasonal swings and also methods for quantifying the effects of such variations in the context of any business activity (phenomenon) are discussed. In the process of determining the seasonal effects, (i.e., quantifying and eliminating those from the data generated from business-related or such kinds of activities), we are left with residual (remaining part) free of seasonal effects and the whole process narrated in the above-mentioned lines is called de-seasonalisation of this component from the data..

From the above deliberation it is clear that for the study of seasonal variations, time series data must be available in the form of ' parts' of a year, like., monthly, quarterly, weekly, daily or hourly. Seasonal effects appear independently at specific times in a year and affect the time series data. The determination of the seasonal effects is done using two types of models, namely, additive or multiplicative models. Such seasonal variations are measured as relative effects in terms of ratios or percentages assuming multiplicative model and occasionally as absolute changes assuming additive model. The methods of measuring seasonal variations are as follows:

i) Methods of "simple average "

ii) "Ratio to Trend" method

iii) "Ratio to Moving averages" method, and

iv) "Link Relative" method

Method of Simple Averages

The method of measuring seasonal variations by the 'Method of Moving Averages' in a time series data is simple and is described below:

Firstly the time series data are arranged in a two-way table where the rows correspond to years and columns correspond to the twelve months. More specifically time series data given for a span of 5 years over 12 months in a year will be placed in a 6 x 13 two way table in which the first row contains the names of the 12 months and the first column contains the representation of 5 years. The Arithmetic Means, x_i's are then computed for the ith month; i = 1, 2,……,12, the average is done over years (here 5 years). Thus, $x'_1, x'_2,....x'_{12}$ are the average values with respect to the months, January, February,……,December respectively.

Secondly, the grand mean (obtained by averaging over12 monthly means) is computed as below:

$$x' = (x'_1 + x'_2 + ... + x'_{12})/12$$

Thirdly, seasonal indices (S.I.'s) for different months are obtained by dividing each monthly average by the grand average and then multiplying the result by 100.

Thus, the Seasonal Index (SI_i) for the ith month = (ith Monthly Average/Grand Average) *100 = (x'_i / x') *100, for I = 1, 2, ….,12. In particular, Seasonal Index for January = (x'_1/ x')*100, and Seasonal index for December = (x'_{12}/ x')*100.

Limitations: The method of simple averages, though very simple to apply, is not an effective method as it does not take care of the presence of trend and cyclical variations in the given time series on which it is applied. In fact, it assumes that the two above-mentioned components are not present in the time series, which is not a valid assumption (rather a limitation) in the context of business and financial time-series. Accordingly, the seasonal indices obtained by this method fail to represent the true seasonal swings in the data because they include the influence of trend and cyclic variations also. This method eliminates the random or irregular component to some extent by averaging the monthly values over different years. Thus in order to obtain sensitive seasonal indices the trend component is to be removed at the first stage and then the seasonal indices are to be obtained from the de-trended time series data.

7.6.1. Ratio to Trend Method :

In this method the trend component (T) is first obtained by applying the curve (appropriately chosen) fitting method (using the principle of least squares) on a given time-series data. After obtaining the trend values with respect to each month and for all years considered, a multiplicative model is assumed. Each value (Y) in the observation-set is then divided by the corresponding trend value and the ratio is multiplied by 100 (so that trend eliminated values, expressed in percentages, are obtained). The trend eliminated values (ratios in percentages) are then averaged over years to obtain the seasonal indices for the 12 months. As the sum of the seasonal indices may not be equal to 1200, an adjustment factor 1200/T (T being the total of the monthly seasonal indices) is calculated and each monthly seasonal index is multiplied by 1200/T in order to make the sum of the seasonal indices 1200. For quarterly data, the adjustment factor is, 400/ (Sum of Quarterly Indices). These adjusted Seasonal Indices provide estimates of the seasonal fluctuation present in a given time series data. Using the Adjustment factor is nothing but to expressing the preliminary seasonal indices as percentage of their arithmetic mean...

This method is an improvement over the "simple average" method as the seasonal indices are computed after eliminating the trend component from the given time-series data. As the multiplicative model is assumed and the trend is eliminated by dividing the given time series values for each season (month or quarter) by the corresponding trend values and multiplying by 100; in fact, the de-trended data (Trend eliminated values) can written symbolically as, .

Trend eliminated values = Y/T * 100= TSCI/T*100 = SCI* 100.

These percentages will include seasonal, cyclical and irregular fluctuations. Averaging the trend eliminated values over years (as mentioned above) eliminates the Irregular component, the cyclical component, however, remains mixed up with the seasonal indices to a little extent.

7.6.1.1. Merits and Demerits: As mentioned earlier, this method determines the indices of seasonal variations after eliminating the trend component, thus it definitely portrays the effects of the seasonal swings in a more representative way as compared with the 'simple average' method. However, the averaging process over different years does not completely eliminate the cyclical effects particularly, if the cyclical swings are present and pronounced in the given series. Accordingly, the indices of seasonal variations obtained by this method are mingled with cyclical effects also. In the absence of cyclical variations (swings) the seasonal indices based on "Ratio to Moving average" method, reflect the seasonal variations better than any other method. However, as compared with 'Ratio to Moving average' this method has the distinct advantage that the trend value can be obtained for each month for which original time series data are available, whereas there is loss of information of certain trend values in the 'Ratio to Moving Average' method.

Note: If the monthly figures for different years are given, then the fitting of trend equation to monthly data involves lot of calculation, and is quite involved and time consuming also. In such a situation, the calculations are simplified to a great extent by first fitting the trend equation to annual totals or average monthly or quarterly values and then obtain the monthly trend equation from the trend equation for annual totals.

7.6.2. Ratio to Moving Aveage Method :

In this method the centred 12-month moving average values for the given time series data are calculated. As the seasonal variations recur after a span of 12 months for any given monthly time series data, 12-month moving average values are supposed to completely eliminate the seasonal variations when such variations are of constant pattern and intensity. Consequently, the 12-month moving average values (after centring because of the fact that the length of the MA period is even) may be regarded to contain the trend and cyclical components only (i.e., T*C in case of a multiplicative model). It is evident that 12 month moving average values are supposed to eliminate the irregular component partially..

After obtaining the centred moving average values for all months and years (except for the first 6 months and 6 months at the end) corresponding to the table of original time series values (for all months and years) in sequence, the original values are then divided by the corresponding MA values and the ratios (which are obtained) are multiplied by 100 (briefly speaking, the original values are expressed as percentages of the centred MA values).

Subsequently, as in the cases of "simple average" and "ratio to trend" methods, the percentages (ratio to trend (MA) values) mentioned in the earlier paragraph are arranged in the form of a two-way table whose rows are years and whose columns are months Preliminary seasonal indices are obtained for each month on eliminating the irregular component by averaging (using arithmetic mean or median) the above percentages over years. The above mentioned preliminary seasonal indices are then adjusted by multiplying each by the factor (1200/T), (where 'T' is the sum total of the preliminary seasonal indices) to make the sum of the final indices as 1200. For the quarterly data, the adjustment factor is 400/T (T is to be so defined).

This method (Ratio to MA method) seems to be somewhat better than the "Ratio to Trend" method, as it tries to eliminate the cyclical variations (when the period of the cycle is one year or more) which are mixed up with the seasonal indices obtained by using the "Ratio to Trend" method.

7.6.2.1. Merits and Demerits

"Ratio to Moving Average" is considered to be the most satisfactory and widely used method for estimating the seasonal fluctuations in a time series as it removes both trend and cyclical components (being regular in periodicity and amplitude) from the indices of seasonal variations. An obvious drawback of this method is that there is loss of some trend values in the beginning and at the end and accordingly trend (MA) ratios for first six months of the first year and last six months of the last year cannot be obtained.

Additive Model: If we express the time series by an additive model, then the method of moving averages for computing seasonal indices involves the following steps:

i) Obtain 12-month moving average values. These will contain trend and cyclical components, i.e. they will represent (T+C).
ii) Trend (MA) eliminated values are obtained on subtracting the above moving average values from the given time series values to give: y - M.A. values = (T+S+C+I) - (T+C) = S+I.
iii) Irregular component is eliminated on averaging these (S+I) values for each month over different years and we get the preliminary seasonal indices for each month.
iv) Sum of these seasonal indices should be zero. In case, it is not so, the preliminary indices obtained in step ii) are adjusted to a total of zero by subtracting from each of them a constant factor.

K = 1/12 [Sum of the monthly seasonal indices]

7.6.3. Method of Link Relatives

7.6.3.1. For Monthly Data

Link relative (L.R.) is the value of a given phenomenon in any season (here either month/ quarter/week in accordance with the data given in terms of month/ quarter/ week), expressed as a percentage of its value in the preceding season, i.e., $(y_t / y_{(t-1)}) *100$, ...(i) y_t being the value of the series at the t-th point of time..In what follows, the method is discussed in case of monthly data and the same can be modified accordingly for quarterly or weekly data. Thus Link Relative corresponding to any month = [(Current month's value)/ (Previous months value)]*100.

The construction of indices of seasonal variation by Link Relatives method (also known as Pearson's method) involves the following steps;

i) The link relatives are obtained by using the formula (i) given above.

ii) The link relatives for all 12 months are then averaged (over years). Instead of AM, Median can also be used as Median is not influenced by extreme observations (which are not basically due to seasonal swings) present in the series .

iii) From the link relatives (L.R.) corresponding to any month, chain relative (C.R.) corresponding to that month is obtained by using the formula,: C.R. for any month = (L.R. of that month* C.R. of the preceding month)/100

iv) Take CR for January = 100

CR. for February = (LR for February*CR for January)/100

CR for March = (LR for March*CR for February)/100

....

CR for December = (LR for December* CR for November)/100

New CR for January = (LR for January* CR for December)/100

It is generally observed that CR for January, so obtained, is not equal to 100 due to the effect of long term secular trend and accordingly the chain indices are to adjusted for the effect of trend. The adjustment due to the effect of trend is equal to

d = (1/12) (New CR for January - 100) and this adjustment factor d is to be subtracted from the chain relatives. In case of straight line trend the adjustment factors for February, March, April...........December are d, 2d, 3d, 4d,....,11d respectively and these are to be subtracted from the respective monthly Chain Relatives. After adjustments these CR's are called monthly corrected chain relatives. Finally, the seasonal indices are obtained by multiplying each monthly corrected chain relative by a factor k, which is equal to, k = 1200 / (Sum of corrected monthly chain relatives). The impact for the above is that for a multiplicative model the sum of the seasonal indices equals 1200 (each corrected chain relative is expressed as percentage of the arithmetic mean)

7.6.3.2. For Quarterly Data,

d = (1/4)*(New CR for 1st Quarter - 100)

k= 400 / (Sum of the corrected Quarterly CR's)

7.6.3.3. Merits and Demerits

i) The averaged link relatives include both the cyclical and trend components Irregular component is eliminated by averaging the link relatives. Though the trend is subsequently eliminated by applying the correction, the indices obtained are truly representative, only if the data exhibit a straight line trend though in most of the economic and business time series the assumption of linearity is rarely found to be true.

ii) In case of Link Relative method there is loss of only one link relative i.e., the Link Relative for the first while in case of moving average method some of the values in the beginning and at the end are lost.

7.6.4. Deseasonalisation of Data

With reference to a time series data-set our objective are (i) to obtain the estimates of seasonal components by calculating seasonal indices and (ii) to eliminate them from the given time series. Elimination of the seasonal effects from the given time series is termed as deseasonalisation of data. Deseasonalised Data are obtained by dividing (in case of multiplicative models) the original time series data by the seasonal indices, or, in symbols, Deseasonalised data are = y/S = T*S*C*I/S = TCSI/S = TCI.

In case of additive model, Deseasonalised data are of the following type: y - S = T + S + C + I - S = T + C + I.

Deseasonalisation is specially needed for the study of cyclical component. It also helps businessmen and management executives for planning the future production programmes and for executing managerial control activities. It also helps in proper interpretation of the data.

Example 11: Obtain the seasonal indices by the link relative method, for the following data:

AVERAGE QUARTERLY PRICE OF A COMMODITY

Quarter	Years				
	1996	1997	1998	1999	2000
I	30	35	31	31	34
II	26	28	29	31	36
III	22	22	28	25	26
IV	31	36	32	35	33

Table showing the computation of seasonal indices by link relative method

	Link Relatives				
Year	1st Quarter	2nd quarter	3rd Quarter	4th Quarter	Total
1996	-	86.6	84.6	140.9	
1997	112.9	80	78.5	163.6	
1998	86.11	93.5	96.5	114.2	
1999	96.8	100	80.6	140	
2000	97.1	105.8	72.2	126.9	
Total of L.R.'s	392.9	465.9	412.4	685.6	
Average L.R.	98.2	93.18	82.4	137.12	
Chain Relatives	100	93.18	76.7	105.17	
Adjusted C.R.'s	100	93.18-0.817=92.36	76.7-1.634=75.06	105.17-2.451=102.71	370.13
Seasonal indices	108	99.74	81.06	110.92	399.72

The new C.R. for 1st Quarter is:

L.R. of 1st Qrt.* C.R. of last(4th) Qrt/100 = 98.2*105.17/100 =103.27

or, d = 1/4* [New C.R. of 1st Qrt - 100] = 1/4*(103.27-100) = 0.817, or, 2d = 2*0.817 =1.634

3d = 3*0.817= 2.451........

The adjusted C.R. for the 2nd , 3rd , 4th quarter is calculated by subtracting d, 2d, 3d from the corresponding C.R.'s.

Indices of seasonal variations are obtained on adjusting these adjusted C.R.'s to a total of 400 by multiplying each one of them with a constant factor

K = 400/ sum of the adjusted C.R.'s = 400/370.13 =1.08

The seasonal indices are given in the last row.

Example 12: The number of letters posted in a certain city on each day in a typical period of five weeks was as follows:

Week	Sunday	Monday	Tuesday	Wednesday	Thursday	Friday	Saturday
1st	18	161	170	164	153	181	76
2nd	18	165	179	157	168	195	85
3rd	21	162	169	153	139	185	82
4th	24	171	182	170	162	179	95
5th	27	162	186	170	170	182	120

Obtain the indices of seasonal variations within the week by the "simple average" method.

Table showing the computation of Seasonal Indices

Week	Sunday	Monday	Tuesday	Wednesday	Thursday	Friday	Saturday
1st	18	161	170	164	153	181	76
2nd	18	165	179	157	168	195	85
3rd	21	162	169	153	139	185	82
4th	24	171	182	170	162	179	95
5th	27	162	186	170	170	182	120
Total	108	821	886	814	792	922	458
Average	21.6	164.2	177.2	162.8	158.4	184.4	91.6
Seasonal Indices	15.8	119.7	129.18	118.7	115.5	133.4	66.8

The average of the averages is: x' = (21.6+164.2+177.2+162.8+158.4+184.4+91.6)/7= 137.17

7.7. Measurement of Cyclical variation

When we adopt the "residual Method" we first estimate trend (T) and seasonal (S) components and then eliminate their effects from the given time series. If the multiplicative model with respect to the given time-series is followed, the residual can be written as (in symbols), Y/T*S = TSCI / TS = C*I......(i) (i.e., a product of Cyclical and Irregular fluctuations) (here S should be taken in fraction form and not in percentage form).

In case, the random or irregular variation is not pronounced (and for the practical purpose we can ignore the said as its effect is not very significant), then the value obtained in (i) may be taken to reflect the cyclic variations.

Steps in the computation of cyclical variations by the "residual method" may be summarised below:

Compute trend values and the seasonal indices (preferably by the moving average method) [S should be expressed in fraction form and not in percentage form].

The given time series, Y is then divided by 'T' and 'S' successively to get CI...(iii)

In case the component I becomes insignificant after removing trend and seasonality, the quotient in (iii) gives the estimate of cyclical component.

Example13: Obtain the estimate of the cyclical variations for the following data:

Year	Quarter	Y	4-Quarterly Moving average
1992	I	75	
	II	60	
	II	54	63.37
	IV	59	65.37
1993	I	86	67.12
	II	65	70.87
	III	63	74.00
	IV	80	75.37
1994	I	90	76.62
	II	72	77.62
	III	66	79.50
	IV	85	81.50
1995	I	100	83.00
	II	78	84.75
	III	72	
	IV	93	

From the above table the adjusted seasonal indices for the Quarters, I, II, III and IV are computed as, 22.36, 92.42, 84.69 and 100.51 respectively. Using the above values as estimates of seasonal component, values (=Y/S*100) in the 4th column are computed. The technique of Moving Average with 4-quarter (centred) period is applied on the values in 4th column (values are given in the 5th column) to get the estimates of the fluctuations due to trend. The values in the 6th column (procedure of calculation shown in the table) give the estimates of the cyclical fluctuation (assuming fluctuation due to Irregular component to be already insignificant due to the earlier operations made on Y to eliminate the Seasonal and Trend components).

Column Year	1 Quarter	2 Original values Y	3 Seasonal Index (S)	4 Y/S*100 =TCI	5 Trend (4-Quarter-centred M.A.)	6 Col (6) = Col.(4)/Col.(5)* 100 = 100* CI
1992	I	75	122.36	61.30		
	II	60	92.42	64.92		

	III	54	84.69	63.77	63.40	100.60
	IV	59	100.51	58.73	65.41	89.77
1993	I	86	122.36	70.30	67.16	104.72
	II	65	92.42	70.34	70.88	99.24
	III	63	84.69	74.40	74.01	100.55
	IV	80	100.51	79.60	75.39	105.61
1994	I	90	122.36	73.56	76.63	96.02
	II	72	92.42	77.92	77.61	100.38
	III	66	84.69	77.92	79.51	98.05
	IV	85	100.51	84.56	81.52	103.76
1995	I	100	122.36	81.73	83.01	98.46
	II	78	92.42	84.42	84.76	99.61
	III	72	84.69	85.01	-	
	IV	93	100.51	92.52	-	

7.8. Measurement of Irregular Variations

It is difficult to obtain accurate estimates of fluctuations due to irregular components. As a crude method, the estimates of Trend (T), Seasonal (S), Cyclical (C) components are obtained from the given time-series data and the irregular component is obtained as a residual (as quotient in a multiplicative model). Symbolically, Y/TSC = TSCI/TSC = I, where, S and C are in fractional forms and not in percentage forms.

EXERCISE

Q) 1) Fit a straight line trend to the following data by least squares method after summing the given quarterly data to yearly data: Find out the trend equation for Quarters.

Export of Cotton Textiles(million Rs.)

Year	I Quarter	II Quarter	III Quarter	IV Quarter
1993	10	13	14	12
1994	12	14	15	13
1995	13	15	18	14
1996	15	18	21	18
1997	15	22	23	20

Also find out short-period fluctuations for the given years, using additive method.

Q) 2) The following table shows the consumption of butter in a district in different years. Obtain the trend values by the method of least squares.

Year	1989	1990	1991	1992	1993	1994
Consumption('000 kgs)	60	80	90	120	145	170

Q) 3) Convert the following into annual trend equation;

Y = 350 + 3X ; Origin = I - II Quarter 1986; X unit = 1 quarter; Y unit = Quarterly Production

Q) 4) Below are given the annual production (in thousand tons) of a fertilizer factory :

Year	1987	1988	1989	1990	1991	1992	1993
Production	70	75	90	91	95	98	100

i) Fit a straight line trend by the method of least squares and tabulate the trend values.

ii) Convert the annual trend equation into a monthly trend equation.

Q) 5) compute the quadratic trend of the form Y = a + bX+ cX2 for the data showing the production of an item, over a 7 years period:

Year	1985	1986	1987	1988	1989	1990	1991
Production(in '000 nos.)	10	11	12	9	10	13	11

Q) 6) Fit a parabolic trend $y = a + bx + c\,x^2$ to the following data, where y denotes the output (in thousand units):

Year(X)	1981	1982	1983	1984	1985	1986	1987	1988	1989
Y	2	6	7	8	10	11	11	11	9

Also compute the trend values. Estimate the value of 1990.

Q) 7) For the following series of observations, verify that the 4- year centered moving average is equivalent to a 5-year weighted moving average with weights 1,2,2,2,1 respectively..

Year(t)	1989	1990	1991	1992	1993	1994	1995	1996	1997	1998	1999
Annual Sales(Rs '000)	2	6	1	5	3	7	2	6	4	8	3

Q) 8) Assume a four-year cycle, calculate the trend by the method of moving averages from the following data relating to production of tea in a certain tea estate:

Year	Production	Year	Production
1961	464	1966	540
1962	515	1967	557
1963	518	1968	571
1964	467	1969	586
1965	502	1970	612

Q) 9) Assuming no trend in the series, calculate seasonal indices for the following data:

	Quarter(in units)			
Year	I	II	III	IV
1994	78	66	84	80
1995	76	74	82	78
1996	72	68	80	70
1997	74	70	84	74
1998	76	74	86	82

Q) 10) Explain 'ratio to trend' method of measuring seasonal variations and discuss its relative merits and demerits.

Find seasonal variations by the 'ratio to trend method from the data given below:

Year	1st Quarter	2nd Quarter	3rd Quarter	4th Quarter
1993	30	40	36	34
1994	34	52	50	44
1995	40	58	54	48
1996	54	76	68	62
1997	80	92	86	82

Q) 11) Apply the link relative method to the following data and calculate the seasonal indices:

Quarterly figures					
Quarter	1991	1992	1993	1994	1995
I	-	80	88	80	83
II	120	117	129	125	117
III	133	113	111	115	120
IV	83	89	93	96	79

Q)12) Calculate the link relatives by link relatives for he following data:

Quarter	Quarterly Figures for five years				
1983	1984	1985	1986	1987	
I	45	48	49	52	60
II	54	56	63	65	70
III	72	63	70	75	83
IV	60	56	65	72	86

Q)13) a) Explain the meaning of deseasonalising the data. What purpose does it serve?

b) A large company estimates its average monthly sales in a particular year to be Rs. 2,00,000. The seasonal indices of the sales data are given below. Find out the monthly sales.

Month	Seasonal index	Month	Seasonal index
January	76	July	101
February	77	August	104
March	98	September	100
April	128	October	102
May	137	November	82
June	122	December	73

Q) 14) a) What do you understand by deseasonalisation of data? Explain its uses.

January	106	July	93
February	105	August	89
March	101	September	92
April	104	October	102
May	98	November	106
June	96	December	108

The above data give the values of the seasonal indices. Calculate the adjusted seasonal indices.

Q)15) Calculate the seasonal index numbers from the following data:

RATIO OF OBSERVED VALUES TO TREND VALUES (%)

Year	I	II	III	IV
1991	108	130	107	93
1992	86	120	110	91
1993	92	118	104	88
1994	78	100	94	78
1995	82	110	98	86
1996	106	118	105	98

Q)16) a) The quarterly seasonal indices of the sales of a popular brand of colour television of a company, in Delhi are given below:

Quarter	I	II	III	IV
Seasonal Index	130	90	75	105

It is the total sales for the first quarter of 1997 is Rs. 6,50,000, estimate the worth of televisions to be kept in store to meet the demand in other quarters. Assume that there is no trend.

Ans:) We are given the total sales of a popular brand of colour televisions as Rs. 6,50,000, under the assumption that there is no trend, the estimated sales for the company for different quarters can be computed as follows :

Quarter	I	II	III	IV
Seasonal Index	130	90	75	105
Expected sales	6,50,000	4,50,000	3,75,000	5,25,000

ADDENDUM - INDEX NUMBER

EXAMPLES

1. Given below are the data on prices of some consumer goods and the weights attached to the various items. Compute price index numbers for the year 1969 (Base: 1968 = 100), using (i) simple average, and (ii) weighted average of price relatives.

Item	Unit	Price (Rs)		Weight
		1968	1969	
Wheat	Kg.	0.50	0.75	2
Milk	Litre	0.60	0.75	5
Egg	Dozen	2.00	2.40	4
Sugar	Kg.	1.80	2.10	8
Shoes	Pair	8.00	10.00	1

Solution: Calculation for the determination of the Price Relative Index

Item	p_0 1968	p_n 1969	Price Relative $I=(p_n/p_0)\times 100$	Weight (w)	Iw
Wheat	0.50	0.75	150	2	300
Milk	0.60	0.75	125	5	625
Egg	2.00	2.40	120	4	480
Sugar	1.80	2.10	117	8	936
Shoes	8.00	10.00	125	1	125
Total	_	_	637	20	2466

$$\text{Simple Average of P.R. Index} = \frac{\Sigma(\text{Price Relative})}{No.of\ Items} = \frac{637}{5} = 127.4$$

$$\text{Weighted Average of P.R. Index} = \frac{\Sigma(\text{ Price Relative} \times \text{Weight })}{\Sigma Weight} = \frac{2466}{20} = 123.3$$

2. On the basis of the following data, compute the whole price index number for the 5 groups combined:

Group	Weight (W)	Index Number for Week Ending 27-9-69 (Base Period: 1952-53=100) – I
Food Articles	50	241
Liquor and tobacco	2	221
Fuel, power, light and lubricants	3	204
Industrial raw materials	16	256
Manufactured commodities	29	179

Solution: Using the weighted arithmetic mean of group indices as the method of combination, the general index is given by the formulaGeneral Index = $\frac{\Sigma IW}{\Sigma W}$,

where I=Group Index, and W=Group Weight

Group	Weight (W)	Group Index (I)	IW
Food Articles	50	241	12.05
Liquor and tobacco	2	221	442
Fuel, power, light and lubricants	3	204	612
Industrial raw materials	16	256	4096
Manufactured commodities	29	179	5191
Total	100		22391

Therefore, the index number of Wholesale Prices = $\frac{22391}{100} = 223.91$

3. Prepare price and quantity index numbers for 1972 with 1961 as base from the following data by using (i) Laspeyres', (ii) Paasche's and (iii) Fisher's method.

Commodity	Unit	1961		1972	
		Quantity	Price (Rs)	Quantity	Price (Rs)
A	Kg	5	2.00	7	4.50
B	Quintal	7	2.50	10	3.20
C	Dozen	6	8.00	6	4.50
D	Kg	2	1.00	9	1.80

Solution:

Commodity	q_0	p_0	q_n	p_n	q_0p_0	q_0p_n	q_np_0	q_np_n
A	5	2.00	7	4.50	10.00	22.50	14.00	31.50
B	7	2.50	10	3.20	17.50	22.40	25.00	32.00
C	6	8.00	6	4.50	48.00	27.00	48.00	27.00
D	2	1.00	9	1.80	2.00	3.60	9.00	16.20
Total	—	—	—	—	77.50	75.50	96.00	106.70

(a) Price Index Numbers:

$$\text{Laspeyres' Formula} = \frac{\Sigma P_n q_0}{\Sigma P_0 q_0} \times 100 = \frac{75.50}{77.50} \times 100 = 97$$

$$\text{Paasche's Formula} = \frac{\sum p_n q_n}{\sum p_0 q_n} \times 100 = \frac{106.70}{96.00} \times 100 = 111$$

$$\text{Fisher's Formula} = \sqrt{Laspeyres' \times Paasche's} = \sqrt{97 \times 111} = 104$$

(b) Quantity Index Numbers:

$$\text{Laspeyres' formula} = \frac{\sum q_n p_0}{\sum q_0 p_0} \times 100 = \frac{96.00}{77.50} \times 100 = 124$$

$$\text{Paasche's formula} = \frac{\sum q_n p_0}{\sum q_0 p_n} \times 100 = \frac{106.7}{75.50} \times 100 = 141$$

$$\text{Fisher's formula} = \sqrt{Laspeyres' \times Paasche's} = \sqrt{124 \times 141} = 132$$

4. Annual production (in million tons) of four commodities are given below –

Commodity	1950	1954	1955	Weights
A	160	200	216	20
B	24	42	45	30
C	50	72	68	13
D	120	168	156	17

Calculate quantity index numbers for the years 1954 and 1955 with 1950 as base year, using (i) simple arithmetic mean, and (ii) weighted arithmetic mean of the relatives.

Solution: Quantity Relatives for 1954 (Base 1950)

$$\frac{200}{150} \times 100 = 125, \frac{42}{24} \times 100 = 175, \frac{72}{50} \times 100 = 144, \frac{168}{120} \times 100 = 140$$

Quantity Relatives for 1955 (Base 1950)

$$\frac{216}{160} \times 100 = 135, \frac{45}{24} \times 100 = 187.5, \frac{68}{50} \times 100 = 136, \frac{156}{120} \times 100 = 130$$

Commodity	Quantity 1954	Relatives 1955	Weight	(Q.R. for 1954)x(Weight)	(Q.R for 1955) x (Weight)
A	125	135	20	2500	2700
B	175	187.5	30	5250	5625
C	144	136	13	1872	1768
D	140	130	17	2380	2210
Total	584	588.5	80	12002	12303

Using simple arithmetic mean of quantity–relatives,

Index Number for 1954 = 584/4 = 146

Index Number for 1955 = 588.5/4 = 147

Using weighted arithmetic mean of quantity relatives,

Index Number for 1954 = 12002/80 = 150

Index Number for 1955 = 12303/80 = 154

5. Construct a suitable productivity index with base 1970 from the following information and also interpret your result.

Item	Units produced in 1970	Man-hours per unit	
	1970	1970	1975
A	20,000	0.5	0.4
B	2,000	5.0	4.5
C	1,000	2.0	1.75
D	500	1.5	1.5

Solution

Item	Units produced in	1970	1975	Total Manhours in 1970	Total Manhours in 1975
A	20,000	0.5	0.4	10000	8000
B	2,000	5	4.5	10000	9000
C	1,000	2	1.75	2000	1750
D	500	1.5	1.5	750	750
				22750	19500

$$\text{Productivity Index} = \frac{Total\ Manhours\ in\ 1975}{Total\ Manhours\ in\ 1970} \times 100 = 85.41729$$

6. Between 1980 and 1990, the price of a commodity is increased by 60% while the production of the commodity is decreased by 30%. By what percentage did the value of production of the commodity change with respect to its value in 1980?

Solution

Year	Price of commodity	Production of commodity	Price Relative	Quantity Relative
1980	100	100	160/100	70/100
1990	160	70		

$$\text{Therefore sales relative (1990)} = \text{Price Relative X Quantity Relative} = \frac{160}{100} \times \frac{70}{100} = \frac{112}{100}$$

or, the value of production increased from Rs. 100 to Rs. 112 in 1990.

Therefore, the value of production has increased by 12%.

7. The price-relatives and weights of a set of commodities are given in the following table:

Commodity	C_1	C_2	C_3	C_4
Price Relatives	120	127	125	119
Weight	$2w_1$	w_2	w_1	w_2+3

If the index for the set is 122 and the sum of the weights is 40, find w_1 and w_2.

Solution:

From the data given in the above problem, we get,

$3w_1+2w_2+3=40, 365w_1+246w_2=4523$

Solving for both w_1 and w_2, we get, $w_1=7$, and $w_2=8$.

8. Calculate index number of prices for 1995 on the basis of 1990 from the data given below:

Commodity	Weight	Price per unit 1990 (Rs.)	Price per unit 1995 (Rs.)
A	40	16	20
B	25	40	50
C	20	12	15
D	15	2	3

If the weights of the commodities A, B, C and D are increased in the ratio 1:2:3:4 what will be the increase in index number?

Solution : As per the given data, the following table is constructed –

Commodity	Weight (W)	Price per unit in Rupees		Price Relative	WP	Increased Weight $(W_1)^*$	W_1P
		1990 (P_0)	1995 (P_1)	$P = \frac{P_1}{P_0} \times 100$			
A	40	16	20	125	5000	$40 + \frac{40}{10} = 44$	5500
B	25	40	50	125	3125	$25 + \frac{2\times25}{10} = 30$	3750
C	20	12	15	125	2500	$20 + \frac{3\times20}{10} = 26$	3250
D	15	2	5	150	2250	$15 + \frac{4\times15}{10} = 21$	3150
	$\Sigma W = 100$				$\Sigma WP =$ 12875	$\Sigma W_1 = 121$	$\Sigma W_1P =$ 15650

Since weights of the commodities are increased in the ratio 1:2:3:4, (Total =10), the increase in weights are:

(A) $\frac{1}{10}\times40 = 4$, (B) $\frac{2}{10}\times25 = 5$, (C) $\frac{3}{10}\times20 = 6$, (D) $\frac{4}{10}\times15 = 6$

$$\text{Original Index Number (I)} = \frac{\sum WP}{\sum W} = \frac{12875}{100} = 128.75$$

$$\text{New Index Number } (I_1) = \frac{\sum W_1P}{\sum W_1} = \frac{15650}{121} = 129.34$$

There, the increase in index number $= I_1 - I = 0.59$.

ADDENDUM - TIME SERIES

Example 1: The sales of a company in million of rupees for the years 1994-2001 are given below:

Year	1994	1995	1996	1997	1998	1999	2000	2001
Sales	550	560	555	585	540	525	545	585

i) Find out the linear trend equation.

ii) Estimate the sales for the year 1993

iii) Find the slope of the straight line trend

iv) Do the figures show a rising trend or a falling trend?

As n = 8, the middlemost years are two i.e., 1997 and 1998;.The origin is taken as, x = [(1997+1998)/2]/(½ of the time interval) = 2 (t - 1997.5) = 2t - 3995....i)

Thus taking, t = 1997 we get x = 3994 - 3995 = -1; for t = 1996, we get x = 3992 -3995 = - 3, and so on. Let the linear trend equation between y and x be given by,:

Y = a + b x, ,,,,,,,(ii), where x = 2 (t -1997.5),

and, x unit = ½ year, y = Annual sales in million Rs.

Table showing the computation of Linear Trend

Year	Sales	x = 2 (t-1997.5)	x*y	x^2	Trend values y_e = 555.63 +0.21x
1994	550	-7	-3850	49	554.16
1995	560	-5	-2800	25	554.58
1996	555	-3	-1665	9	555.00
1997	585	-1	-585	1	555.42
1998	540	1	540	1	555.84
1999	525	3	1575	9	556.26
2000	545	5	2725	25	556.68
2001	585	9	4095	49	557.10
Total	4445	0	35	168	

The normal equations for the estimating a and b in ii) are,

$\Sigma y = na + b\Sigma x$ and $\Sigma xy = a\ \Sigma x + b\Sigma x^2$

or, 4445= 8a + 0   35 = a* 0 + 168b

or, a = 555.63   or, b = 0.21

Substituting the values in ii) the straight line trend is given by the equation:

y_e = 555.63 + 0.21x ....(iii). Putting x= -7, -5, -3, -1, 1, 3, 5 , 7 in iii) we get the trend values of sales for the years 1994 to 2001. The trend values are shown in the sixth column of the table.

ii) The estimated sales for 1993 are obtained by putting t = 1993 in (iii): Thus,

$$x = 2\,(t - 1997.5) = 2\,(1993 - 1997.5) = -9$$

Substituting x = -9 in iii) the estimated sale-value for 1993 is:

$$(y_e)_{1993} = 555.63 + 0.21*(-9) = 555.63 - 1.89 = 555.74 \text{ million Rs.}$$

iii) The slope of straight line trend iii) is given by b = 0.21

iv) The slope of the trend line represents the rate of growth per unit time i.e., here annually. Since the slope b = 0.21 is positive, the trend equation exhibits a rising trend.

Example 2); The weights of a calf taken at weekly intervals are given below. Fit a straight line using the method of least squares and calculate the average rate of growth per week.

Age (t)	1	2	3	4	5	6	7	8	9	10
Weight (kg)	52.5	58.7	65.0	70.2	75.4	81.1	87.2	95.5	102.2	108.4

Also estimate the weight of the calf when ages are, i) 12, ii) 15 weeks.

Using the procedure laid out in Example 1, we get the trend equation as,

$y_e = 79.62 + 3.08$ x ; where, Y = a +b X, 'a' intercept and 'b' = slope; and x = 2 (t - 5.5) = 2t – 11(computation below).

Table showing the computation of straight line trend

t	y	x = t - 5.5/1/2	x^2	xy	Trend values $y_e = 79.62 + 3.08x$
1	52.5	-9	81	-472.5	51.90
2	58.7	-7	49	-410.9	58.06
3	65.0	-5	25	-325.0	64.22
4	70.2	-3	9	-210.6	70.38
5	75.4	-1	1	-75.4	76.54
6	81.1	1	1	81.1	82.70
7	87.2	3	9	261.6	88.86
8	95.5	5	25	477.5	95.02
9	102.2	7	49	715.4	101.18
10	108.4	9	81	975.6	107.34
Total	796.2	0	330	1016.8	

Putting x = -9, -7, -5, -3, -1, 1, 3, 5, 7, 9, we get the trend values which are given in the sixth column of the table.

The weights of the calf, at the 12th week is; $(y_e)_{12} = 45.74 + 6.16*12 = 119.66$ and at the 15th week is $(y_e)_{15} = 45.74 + 6.16 * 15 = 138.14$

Example 3: Fit an equation of the form $Y = a + bX + cX^2$ to the data given below:

X	1	2	3	4	5
Y	25	28	33	39	46

Here n = 5 (odd), therefore, the transformed variable is, t = X - (middle value of X = 3), or, t = X – 3, so that the values of t corresponding to X = 1, 2, 3, 4 and 5 are -2, -1, 0, 1 and 2 respectively.

Let the second degree trend equation between Y and t be: $Y = a + bt + ct^2$ ....(i), where t = X-3

Table showing the calculation of the second degree trend

X	Y	t=X-3	t^2	t^3	t^4	tY	t^2Y
1	25	-2	4	-8	16	-50	100
2	28	-1	1	-1	1	-28	28
3	33	0	0	0	0	0	0
4	39	1	1	1	1	39	39
5	46	2	4	8	16	92	184
Total	171	0	10	0	34	53	351

The normal equations for estimating a, b, and c are,

$\Sigma Y = na + b\ \Sigma t + c\ \Sigma t^2$; $\Sigma tY = a\ \Sigma t + b\ \Sigma t^2 + c\ \Sigma t^3$; $\Sigma t^2Y = a\ \Sigma t^2 + b\ \Sigma t^3 + c\ \Sigma t^4$,

or, 171 = 5a + + 10c; 53 = 10b 351 = 10a + 34c

or a = 32.92, b = 5.3, c = 0.64.

Substituting the values of a, b, c in (i), we get $Y = 32.92 + 5.3t + 0.64t^2$, where t = X-3.

Hence the second degree trend equation of Y on X becomes: $Y = 32.92 + 5.3\ (X-3) + 0.64\ (X-3)^2$, or $Y = 22.78 + 1.46X + 0.64X^2$.

Example 4: The sales of particular product of a company rose from Rs. 40,000 in March to Rs. 48,000 in April 1997. The company's seasonal indices for these two months are 105 and 140 respectively. The owner of the company expressed dissatisfaction with the April sales, but the Sales Manager said that he was quite pleased with the Rs. 8,000 increase. What argument should the owner of the company have used to reply to the Sales Manager?

The owner of the company was justifiably not satisfied with the rise of sales of Rs (48,000 - 40,000) = Rs. 8,000 from March to April because on the basis of the seasonal index of April, the estimated sales for the month of April should have been, Rs. (40,000/105)*140 = Rs. 53,333 and in that case the increase in the sales would have been Rs. 13333..

Thus the actual sales amounting to Rs, 48,000 for the month of April is much below the expected sales and hence dissatisfaction of the owner is justified.

Example 5) The sales of a reputed organization rose from Rs, 1,26,000 in the month of August 1993 to Rs. 1,38,000 in the month of September 1993. The seasonal indices for the two months were 105 and 140. The General Manager was not at all satisfied with rise of the sales in the month of September 1993 by Rs. 12,000. He expected much more because of seasonal index of the month. What was his expectation of sales for the month of September 1993?

The owner of the company was justifiably not satisfied with the rise of sales of Rs (1,38,000 -1,26,000) = Rs.12000 from August to September because on the basis of the seasonal index of September, the estimated sales for the month of September should have been Rs. (126000/105)*140 = Rs.1,68,000.

Example 6): The trend equation for the yearly sales of a commodity with 1st July 1991 as origin is

$$Y_e = 96 + 28.8X + 4X^2 \text{ (where unit of X = 1 year):}$$

i) Determine the monthly trend equation with January 1992 as origin.

ii) Compute the trend values for August 1991 and March 1992.

i) The given annual trend equation, when reduced to monthly values (with 1st July, 1991 as origin), becomes:

$$Y_e = 96/12 + [28.8/(12*12)]*X + [4/(12*12*12)]*X^2,$$

or, $Y_e = 8 + 0.2\ X + 0.0023\ X^2$.......i), (with origin at 1st July, 1991)

Putting X = 0, 1, 2, 3,......., we get the trend values for 1st July, 1st August, 1st September, 1st October,..., 1991 respectively.

Now we want to shift the origin to January 1992 (viz., middle of January i.e., 15th January 1992), or more specifically, we want to shift the origin 6 ½ months forward and thus the required equation is obtained on changing x to x + 6.5 in equation i). Hence the required trend equation (January 15, 1992 as origin) becomes, $Y_e = 8 + 0.2\ (X+6.5) + 0.0023\ (X+6.5)^2 = 8 + 0.2\ X + 1.3 + 0.0023(X^2 + 13X + 42.25) = 9.3972 + 0.2299\ X + 0.0023\ X^2$.........ii)

Putting X = 0, 1, 2, 3, ..., we get the trend values for Jan 15, Feb 15, Mar 15, Apr 15,...,1992 respectively. Since Aug 15, 1991 is 5 month backward from the origin, Jan 15, 1992, thus putting x = - 5 in (ii) we get the trend value for Aug 15, 1991.

ii) The trend value for August 1991 (i.e., 15th August 1991) and for March 1992 (i.e., March 15, 19992) are obtained by substituting X = - 5 and X = 2 in equation ii respectively), thus

$Y'_{Aug\ 1991} = 9.3972 + 0.2299*(-5) + 0.00023*(-5)^2 = 8.25345$

$Y'_{March\ 1992} = 9.3972 + 0.2299*2 + 0.00023*2^2 = 9.8662.$

Example 7) Deseasonalise the following data with the help of the seasonal data given below: '

Month	Cash Balance(Y)	Seasonal Index
January	360	120
February	400	80
March	550	110
April	360	90
May	350	70
June	550	100

The deseasonalised values are obtained on dividing the given time series values (Y) by the seasonal effect- assuming that the given series follows multiplicative model of decomposition. We have:

Seasonal effect = Seasonal Index/100 = S.I./100

Hence, using multiplicative model: Y = T * S * C * I

Deseasonalised value = Y/ Seasonal effect = Y/ Seasonal effect = (Y/ S.I.) * 100

Table showing the computation of computation of deseasonalised values

Month	Cash Balance	Seasonal Index	Desaeasonalised values = (Y/ S.I.) * 100
January	360	120	300
February	400	80	500
March	550	110	500
April	360	90	400
May	350	70	500
June	550	100	550

Additive Model

If we use additive model of the time series, then the model of the moving averages for computing seasonal indices involves the following steps.

i) Obtain 12- month moving average values. These will contain trend and cyclic components, i.e., they will represent (T+C)

ii) Trend eliminated values are obtained on subtracting these moving average values from the given time series to give :

y - M.A. values = (T+S+C+I) - (T+C) = S+I

iii) Irregular Component is eliminated on averaging these (S+I) values for each month over different years and we get the preliminary indices for each month = S

iii) Sum of these indices over years should be zero. In case, it is not so the preliminary indices obtained in step iii) are adjusted to a total of zero by subtracting from each of them a constant factor.

K= 1/12 [Sum of the monthly seasonal indices]

Multiple Regression Equation: The simple regression equation of y on x describes a relationship between y and x, where, y and x are the dependent and independent variables respectively. So linear regression is a case when there is only one independent variable, x. When there are more than one independent variables but only one dependent variable and the relationship is linear, the said equation is called a multiple regression equation of y on $x_1, x_2, x_{3, \ldots} x_p$, when there are p independent variables. Thus, the multiple regression rquation can be written as,

$Y = a_0 + a_1 x_1 + a_2 x_2 + a_3 x_3 + \ldots\ldots + a_p x_p$, ...(i), and y = Y + error, the coefficients, $a_0, a_1, a_2, a_{3,} \ldots a_p$ are to be obtained by applying the least squares technique (l.s.t.). Solving the normal equations (p + 1 in number) so obtained by the method of l.s.t., we get the estimated values of the coefficients, $a_0, a_1, a_2, a_{3,} \ldots a_p$. Putting the estimated values of the coefficients in the equation (1), the estimated multiple regression equation of y on y on $x_1, x_2, x_{3,\ldots} x_p$ is obtained. The estimated multiple regression equation can then be obtained to find out the predicted values of y, given the values of $x_1, x_2, x_{3,\ldots} \ldots x_p$.

ADDENDUM – QUESTION PAPER SOLUTION

Engineering & Management Examination, December-2006

STATISTICS-I

SEMESTER-1

Time: 3 hours] **GROUP - A** [Full Marks 70

(Objective Type Questions)

1. Answer *any ten* questions of the following: 10 x 1 = 10

A. Choose the correct alternatives:

i) The median of the following data:

12, 5, 7, 10, 4, 9, 15, 14, 2 is:

a) 9 b) 10

c) 12 d) 4

ii) The highest point of the frequency curve is

a) Mean b) Median

c) Mode d) None of these

iii) Standard deviation is dependent on

a) origin only b) scale only

c) both a) and b) d) none of these

iv) Correlation coefficient lies between

a) -1 to 1 b) 0 to 1

c) 1 to 2 d) none of these

v) A.M. of 1,2,3,4,………,m is

a) m/2 b) (m+1)/2

b) 2m/2 c) none of these

vi) The mode of the following data:

2, 5, 7, 2, 4, 9, 7, 4, 2 is

a) 9 b) 7

c) 2 d) 4

vii) The line of regression concerning to the variables x and y are given by y = 32 - x and x = 13 - 0.25y. The values of the means are

a) 6.7 and 25.3 b) 4.2 and 9.7

c) 7.9 and 24.8 d) none of these

viii) Find out first and third quartile from the following data:

2, 5, 7, 10, 4, 9, 14

a) 5,12 b) 7,9

c) 5,10 d) none of these

ix) The G.M. of 3,12, and 48 is

a) 12 b) 9

c) 6 d) none of these

x) Two regression lines are $2x + 3y - 4 = 0$ and $x + 2y + 6 = 0$.

Their correlation coefficient between x an y is

a) -3/4 b) ¾

c) 4/3 d) -4/3

xi) Distinguish between attribute and variable.

ANSWERS

A) i) (a), ii) c), iii) (b), iv) (a), v) (b), vi) (c)

vii) Calculation: Here from the regression line of y on x (first one), mean of x + mean of y = 32, and from the regession line of x on y (second line), mean of x + .25*mean of y = 13. Hence, values of mean of y and mean of y are, 6.7 and 25.3 respectively (obtained by solving the above two equations). Ans: (a)

viii) (c), ix) (a), x) (a)

GROUP – B

(Short Answer Questions)

(Answer any three questions)

3 x 5 = 15

2. The weighted arithmetic mean of the four numbers 26, 28, 12, and 4 is 10.4. The weights of the first three numbers are 1, 3 and 4 respectively, find the weight of the 4th number.

Ans) Let 'w' be the weight of the fourth number, 4.

SERIAL NUMBER	x_i	f_i	$x_i f_i$
1.	26	1	26
2.	28	3	84
3.	12	4	48
4.	4	w	4w
Total		8+w=N	158+4w

Here, N = 8 + w, and we know, Mean of x = x' = $\Sigma f_i x_i /N$, or, 10.4 = (158 + 4w)/(8 + w)

or, 83.2 + 10.4w = 158 + 4w, or, 10.4w - 4w = 158 - 83.2, or, 6.4 w = 74.8, or, w = 74.8/6.4 = 11.69

Therefore, the weight corresponding to the fourth number is 11.69. (ans)

3. The means of two samples of size, 50 and 100 respectively are 54.4 and 50.3 and the standard deviations are 8 and 7. Obtain the mean and standard deviation of the sample of size 150 obtained by combining the two samples.

Ans) $n_1 = 50$; $n_2 = 100$; $x_1' = 54.4$; $x_2' = 50.3$; $\sigma_1 = 8$; $\sigma_2 = 7$; N = 150

By the problem,

$N x' = n_1 x_1' + n_2 x_2'$ or, $x' = (n_1 x_1' + n_2 x_2')/N$ = (50*54.4+ 100*50.3)/150, or, $x' = 51.6$.

By the problem, $d_1 = x_1' - x' = 54.4 - 51.6 = 2.8$, $d_2 = x_2' - x' = - 1.3$;

Now, $\sigma^2 = [(n_1\sigma_1^2 + n_2\sigma_2^2) + (n_1d_1^2 + n_2d_2^2)]/N = [(n_1\sigma_1^2 + n_2\sigma_2^2) + (n_1d_1^2 + n_2d_2^2)]$ $/N = [(50*8^2 + 100*7^2) + (50*2.8^2 + 100*(-1.3)^2)]/150 = (8100 + 561)/150 = 57.74$, or.

$\sigma = 7.59$. Therefore, the pooled standard deviation is 7.59 (Ans)

4. Calculate Mean and Median of the following frequency distribution

Class- Interval	50-59	60-69	70-79	80-89	90-99	100-109
Frequency	5	20	40	50	30	6

Ans) Table Showing the calculation of Median

Class-limit	Class Interval	Frequency (f)	Cumulative Frequency	Mid Values (x)	Y = x - 74.5/10	fy
50-59	49.5 – 59.5	5	5	54.5	-2	-10
60-69	59.5 – 69.5	20	25	64.5	-1	-20
70-79	69.5 – 79.5	40	65	74.5	0	0
80-89	79.5 – 89.5	50	115	84.5	1	50
90-99	89.5 – 99.5	30	145	94.5	2	60
100-109	99.5 – 109.5	6	151	104.5	3	18
Total	N = 151					98

Mean = 74.5 + 10 * (98/151) = 80.9, Now, N/2= 151/2= 75.5, and the median class is 79.5 – 89.5, Therefore, Median = 80 + (75.5 – 65)/ 50 = 80.21

5. Calculate S.D. from the following table:

Marks	0-10	10-20	20-30	30-40	40-50
No. of students	5	8	15	16	6

Ans) <u>Calculation of S.D.</u>

Marks	No. of Students(f)	Mid-value (x)	y = x - 25/10	y^2	fy	fy^2
0-10	5	5	-2	4	-10	20
10-20	8	15	-1	1	-8	8
20-30	15	25	0	0	0	0
30-40	16	35	1	1	16	16
40-50	6	45	2	4	12	24
Total	N = 50				10	68

$\sigma^2 = \Sigma fy^2/N - (\Sigma fy/N)^2 = 68/50 - (10/50)^2 = 1.36 - 0.04 = 1.32$; $\sigma = 1.14$. or, s.d = 1.14

6. Construct a Pie- chart from the following data:

Countries	USA	India	Egypt	Brazil	Argentina
Cotton exported	536	300	160	72	92

Table showing the calculation for construction of Pie chart

Countries	Exports	Percent of total	Angle (degree) at the Centre of pie chart
USA	536	46.21	166.36
India	300	25.87	93.13
Egypt	160	13.79	49.64
Brazil	72	6.20	22.32
Argentina	92	7.93	28.55
Total	1160	100	360

Draw a circle of any width and partition the circle in five parts (each part represents an angle made at the centre of the circle).

GROUP-C

(Answer any three questions) $3\times15=45$

1. Given the heights of 30 Indian adults males:

169	164.5	154.2	163	171.6	157.5
166.7	166.7	161.8	163.1	167	160.5
159.9	165	156.5	157.5	169.5	170.3
168.8	172.8	169	167.7	170	160.6
167.7	183.5	168	159.5	181.5	170.1

7 (a) Construct a frequency distribution of the given data with 6 class intervals and draw frequency polygon of the data.

Ans) Here, Highest value = 183.5; Lowest value = 154.2; Range = 183.5 - 154.2 = 29.3 Therefore, Class width = 29.3/6 = 4.89 = 4.9 (approx)

<u>Calculation concerning the construction of the Frequency Distribution</u>

Class Limit	Class boundary	Tally mark	Frequency
154-158.9	153.95 – 158.95	IIII	4
159-163.9	158.95 – 163.95	IIIIIII	7
164-168.9	163.95 – 168.95	IIIIIIIIII	10
169-173.9	168.95 – 173.95	IIIIIIII	8
174-178.9	173.95 – 178.95		0
179-183.9	178.95 – 183.95	II	2

7 (b) Find the median and the median class of the given data:

Class Boundary	15-25	25-35	35-45	45-55	55-65
Frequency	4	11	19	14	2

N.B. See the next Example.

Example: Calculate the median and Median class from the first and third columns of the following table: Also calculate the standard deviation.

Ans Table showing the calculation of Median

Class Limit	Class Boundary	Frequency	Cummulative Frequency
20-29	19.5 – 29.5	5	5
30-39	29.5 – 39.5	12	17
40-49	39.5 – 49.5	15	32
50-59	49.5 – 59.5	20	52
60-69	59.5 – 69.5	18	70
70-79	69.5 – 79.5	10	80
80-89	79.5 – 89.5	6	86
90-99	89.5 – 99.5	4	90
Total		N=90	

Here, N= 90, Median is the N/2 th Observation = 45th item i.e., the median class is 49.5 -59.5, and Median = 49.5 + (45 – 32)/20 * 10 = 56.

8 (b) Problem on computation of Index Number: Base year = 1982, Current year = 1985, Number of commodities = 4; Use the formulae for Laspeyres' and Paasche's Index Numbers (See the Chapter on Index Number).

9. What is time series? Obtain the trend values for the following series by fitting a second degree polynomial. Represent the trend value and the data in a suitable diagram.

Year	60	61	62	63	64	65	66
Goods carried	1478	1576	1618	1800	1922	1950	2041

See the chapter on Time Series and Construct a Bar chart showing "Goods Carried" on Y-axis.

10. Following table has some frequencies missing. Given that total frequency =1000 and that median = 413.11. Find the missing frequencies:

Values	300-325	325-350	350-355	355-360	365-370	370-375	375-380	380-385
Frequency	5	17	80	?	326	?	88	9

Ans Table showing the calculation for obtaining missing frequencies

Values	Frequency	Cummulative frequency
300-325	5	5
325-350	17	22
350-375	80	102
375-400	? f_4	$102+f_4$
400-425	326	$428+f_4$
425-450	? f_6	$428+f_4+f_6$
450-475	88	$516+f_4+f_6$
475-500	9	$525+f_4+f_6$
	$525+f_4+f_6$	

Here, $525+f_4+f_6 = 1000$, $f_4+f_6 = 475$,

$413.11 = 400 + [(500 - 102 - f_4) / 26] * 25$

$13.11 = [(398 - f_4)/26] * 25$, or, $40.86 = 9950 - 25 f_4$, or, $25 f_4 = 9909.14$

or, $f_4 = 363 = 363$. Now, $f_4+f_6 = 475$, hence, $f_6 = 112$..

Therefore, the missing frequencies are 363 and 112 respectively (ans).

11 (a) Calculate correlation coefficient from the following data:

X	10	12	13	16	17	20	25
Y	19	22	24	27	29	33	37

11 (b) Write a short note on Scatter diagram.

Ans) a) Table showing the computation concerning correlation co-efficient

X	Y	X-X'=i.e. X-16	(X-X')^2	Y-Y'=Y-27	(Y-Y')^2	(X-X')(Y-Y')
10	19	-6	36	-8	64	48
12	22	-4	16	-5	25	20
13	24	-3	9	-3	9	9
16	27	0	0	0	0	0
17	29	1	1	2	4	2
20	33	4	16	6	36	24
25	37	9	81	10	100	90
113	191		159		238	193

$X' = \Sigma X/N = 113/7 = 16.14$; $Y' = \Sigma Y/N = 191/7 = 27.2$

$r = [\Sigma(X-X')(Y-Y')]/[\sqrt{\Sigma(X-X')^2}\ \sqrt{\Sigma(Y-Y')^2}] = 193/[\sqrt{159} * \sqrt{238}] = 0.99$ (ans)

Hence the value of the correlation coefficient is 0.99.

11 (b) See the Chapter on "Statistical Treatment of Bivariate Data"

Engineering & Management Exa mination, December-2007

STATISTICS-I

SEMESTER-1

Time: 3 hours] **GROUP - A** [Full Marks 70

(Objective Type Questions)

1. Answer *any ten* questions of the following: 10×1 = 10

A. Choose the correct alternatives:

i) The relation between mean, median , mode is

a) Mean – mode = 3(mean- median) b) Mean-mode= median- mean

c) Mean- median=3 (mean- mode) d) None of these

ii) Variance is dependent on

a origin only b) scale only

c) both a) and b) d) none of these

iii) Correlation coefficient is dependent on

a) origin only b) scale only

c) both a) and b) d) none of these

iv) The median of the following data 13,6,8,11,5,10,16,15,3 is

a) 13 b) 10

c) 3 d) 15

v) The H.M. of 6,12,24 is

a) 72/7 b) 12

c) 14 d) none of these

vi) The mode of the following data, 5,3,9,11,6,2,9,6 is

a) 6 b) 9

c) 11 d) none of these

vii) Which of the following method will satisfy both time reversal and factor reversal tests?

a) Laspeyre's Method b) Paasche's Method

c) Fisher's Method d) Marshall Edgeworth Method

viii) The mean of the following data 36,40,44,48 is

a) 40.70 b) 42.00

c) 46.00 d) 45.20

ix) A graphical representation of a cumulative frequency distribution is called

a) Ogive b) Cummulative Frequency Polygon

c) Both a) and b) d) None of these

x) The correlation coefficient 'r' lies between

a) 1 to 2 b) -1 to +1

c) 0 to 1 d) none

xi) The G.M and H.M. for the numbers 2, 4, 6, 8, 10 are

a) 2.235,5.235 b) 5.211,4.379

c) 4.379,5.112 d) none of these

xii) The most common form of diagrammatic representation of a grouped frequency distribution is

a) Histogram b) Frequency polygon

c) Pictogram d) Bar diagram

xiii) The various components of time series are

a) secular trend, seasonal variation, cyclical fluctuation and irregular or random movement

b) seasonal movement random variation

c) secular variation, cyclical movement, seasonal trends, random fluctuations

d) none of these

Answers

i) (a), ii) b), iii) c), iv) b, v) a), vi) d), vii) c), viii) b, ix) a), x) b), xi) c), xii) a), xiii) a)

Group B

(Answer *any three* questions)

2.The average marks obtained in an examination by two groups of students were found to be 75 and 85 respectively. Determine the ratio of the students in the two groups if the average mark for all students was 80.

Ans) We know, $x' = (n_1x_1' + n_2x_2')/(n_1 + n_2)$; Given, $x_1' = 75$, $x_2' = 85$, and $x' = 80$

Now, from the condition given in the problem, $80n_1 + 80n_2 = 75n_1 + n_2 85$, or, $5n_1 = 5n_2$, or, $n_1/n_2 = 1:1$. Therefore the ratio of the students is equal to 1:1.

3.Find the median and the median class of the data given data below:

Class boundaries	15-25	25-35	35-45	45-55	55-65	65-75
Frequencies	4	11	19	14	0	2

Ans) Table showing the calculation for median

Class boundaries	Frequencies	Cummulative frequencies
15-25	4	4
25-35	11	15
35-45	19	34
45-55	14	48
55-65	0	48
65-75	2	50
	50	

Therefore N=50 and N/2 = 50/2. So Median is the 25th term which lies in the class 35-45, which is the median class. Therefore, Median = l + [(N/2 – C)/f *] c (the terms having usual significance) = 35 + [(25- 15)/ 19] * 10 = 45.26 (ans).

4.a) What is a pie-chart?

b) Draw a pie chart to represent the following data relating to the production cost of a manufacturer:

Production Cost	Rs.
Cost of material	9,600
Cost of labour	7,680
Direct expenses of manufacturer	2,880
Factory overhead expenses	3,840

Ans) Table showing the calculation for construction of Pie-Chart

Production Cost	Rs	Percentage	Angle
Cost of material	9,600	40	144
Cost of labour	7,680	32	115.2
Direct expenses of manufacturer	2,880	12	43.2
Factory overhead expenses	3,840	16	57.6
Total	24000	100	360

Draw a circle of any chosen radius and partition the circle into four parts, the parts being obtained in commensurate with the four angles.

5.Find index numbers by i) method of aggregrates and ii) method of relatives (using arithmetic mean) from the following:

Commodity	Base Price	Current Price
Rice	35	42
Wheat	30	35
Pulse	40	38
Fish	107	120

Ans. Table showing the calculation of Index numbers

Commodity	Base Price (p_0)	Current Price (p_1)	Price Relative $[p_1/p_0]*100$
Rice	35	42	120
Wheat	30	35	116.6
Pulse	40	38	95
Fish	107	120	112.14
	212	235	443.74

The required index number by aggregate method is

$= [\Sigma pi/ \Sigma p_0]*100 = [235/212]*100 = 110.85$

and, the required index number by method of relatives is

$= [(1/N)(\Sigma p_1/p_0)]*100 = 443.74/4 = 110.94$ (ans)

6. Calculate the mean of the following frequency distribution:

Height(x cm)	158-161	162-165	166-169	170-173	174-177	178-181
No. of men (f)	11	23	31	18	12	5

Ans. Table showing the calculation of the mean

Height (x cm) Class Limit	Height (cm) Class Interval	No. of men (f_i)	Class mark (x_i)	$f_i x_i$
158-161	157.5 – 161.5	11	159.5	1754.5
162-165	161.5 – 165.5	23	163.5	3760.5
166-169	165.5 – 169.5	31	167.5	5192.5
170-173	169.5 – 173.5	18	171.5	3087
174-177	173.5 – 177.5	12	175.5	2106
178-181	177.5 – 181.5	5	179.5	897.5
		100		16798

Therefore, Mean = $\Sigma f_i x_i / \Sigma f_i$ = 16798/ 100 = 167.98 cms (ans)

7.What do you mean by time series? Explain the different components of such a series.

[see the chapter on time series]

Group C

(Answer *any three* questions) 3 × 15 = 45

8a) Define Primary data and Secondary Data.

(See Chapter -2)

8b)The weights in pounds of 45 persons are given below :

120,174,179,176,159,175,160,97,138,159,176,151,115,105,72

170,128,112,101,123,117,93,117,99,90,113,119,128134,178

105,107,147,157,155,95,177,98,175,135,97,168,160,144,175

Arrange the data in the form of a frequency distribution table in 10 classes of equal length. Form the cumulative frequency table. Find the relative frequency of the last class – [See 2006 solved answers].

[Hints: Find out the maximum and minimum values and obtain the value of the Range. Divide the range by the number of classes (here 10) – the quotient will give the length of the class interval. Prepare the column containing Tally Marks and hence find the column representing frequencies. Calculate the columns for cumulative and relative frequencies].

9.From the data below find i) the two regression equations, ii)The coefficient of correlation between marks in Economics and Statistics, iii) the most likely marks in Statistics when the marks in Economics is 30.

Marks in Economics (x)	25	28	35	32	31	36	29	38
Marks in Statistics 9y)	43	46	49	41	36	32	31	30

Ans) Table showing the computation of components of correlation coefficient

Marks in Economics(X)	Marks in Statistics (Y)	(X-X')	$(X-X')^2$	(Y-Y')	$(Y-Y')^2$	(X-X')(Y-Y')
25	43	-6	36	5	25	-30
28	46	-3	9	8	64	-24
35	49	4	16	11	121	44
32	41	1	1	3	9	3
31	36	0	0	-2	4	0

36	32	5	25	-6	36	-30
29	31	-2	2	-7	49	14
38	30	7	49	-8	64	-56
254	308		138		372	-79

X' = 254/8 = 31.75, Y' = 308/8 = 38.5

$r = \Sigma(X-X')(Y-Y')/[\sqrt{\Sigma(X-X')^2} \times \sqrt{\Sigma(Y-Y')^2}] = (-79)/\sqrt{[138 * 372]} = (-79)/\sqrt{51336} =$ –79/226.57 = –.3487. Therefore the value of the correlation coefficient between marks in Economics and Statistics is equal to, –.3487.

Regression equation of y on x is: 'y = Y' + b_{yx} (X – X'), where

$b_{yx} = \Sigma(X-X')(Y-Y')/\Sigma(X-X')^2$, (all component parts are already calculated)

Regression equation of x on x is: x = X' + b_{xy} (Y –Y'), where

$b_{xy} = \Sigma(X-X')(Y-Y')/\Sigma(Y-Y')^2$, (all component parts are already calculated)

Thus the regression equations can be calculated and the most likely marks in Statistics (i.e., Y) when the marks in Economics is 30 (i.e., X) can be obtained on putting the value of X as 30 in the regression equation of y on x (and then finding out the corresponding value of Y).

10 (a) The number of runs scored by the cricketers A and B during the test series for each of 10 innings is shown below :

Cricketer A 34	36	45	75	12	61	40	58	82	11
Cricheter B 17	38	52	42	36	54	48	34	50	54

Ans: The means and variances of the two scores are to be computed. The higher mean is an indication of higher score but lesser variance indicates consistency which is also desired and it reflects the consistent nature of the score of the cricketer. In case any cricketer has a record of higher score and lesser variance than those of the other cricketer, then it can be said straight-way that the said cricketer is definitely better than the other cricketer in all respects.

11a).Draw ogives or calculate frequency curve from the following frequency distribution:

wages (Rs)	10-20	20-30	30-40	40-50	50-60	60-70	70-80
No. of workers	5	10	12	36	3	5	4

Calculate the median ogive.

Ans) Table showing the calculation of Median

Wages (Rs)	No. of workers	Cummulative frequencies (less than type)	Cummulative frequencies (greater than type)
10-20	5	5	75
20-30	10	15	70
30-40	12	27	60
40-50	36	63	48
50-60	3	66	12
60-70	5	71	9
70-80	4	75	4
	75		

Here N =75, N/2 = 75/2= 37.5. So Median is the 37.5th term in the series and it lies in the class interval, 40-50, which is the median class.

Therefore, Median = l + N/2 – C /f * c = 40 + [(37.5-27)/36] * 10 = 42.92 Rs

The two Ogives are to be drawn and the point of intersection of the two ogives gives the median ogive. The median ogive may be compared (greater or lesser) with the median to find out the amount of departure of the median ogive from the median computed by the formula. Ogives are drawn by hand and thus though theoretically the two medians should be same but, in effect, these two differ. .

b) Show that the arithmetic mean is never less than geometric mean, again geometric mean is never less than the harmonic mean.

Ans) Table showing the calculation of Cumulative frequency

wages (Rs)	No. of workers	L.C.F.	G.C.F.
10-20	5	0	75
20-30	10	5	70
30-40	12	15	60
40-50	36	27	48
50-60	3	63	12
60-70	5	66	9
70-80	4	71	4
		75	0

a)a) Find the missing frequencies of the following frequency distribution when the mean is 11.09.

C.I.	9.3-9.7	9.8-10.2	10.3-10.7	10.8-11.2	11.3-11.7	11.8-12.2	12.3-12.7	12.8-13.2	Total
Frequency	2	5	f_3	f_4	14	6	3	1	60

Ans Table showing the calculation of

C.I.	Frequency	Class mark	fx
9.3-9.7	2	9.5	19
9.8-10.2	5	10	50
10.3-10.7	f_3	10.5	10.5* f_3
10.8-11.2	f_4	11	11* f_4
11.3-11.7	14	11.5	161
11.8-12.2	6	12	72
12.3-12.7	3	12.5	37.5
12.8-13.2	1	13	13
	60		352.5+10.5 f_3+11 f_4

Now, total frequency = 2 + 5 + f_3 + f_4 + 14 + 6 +3 +1 = 31 + f_3+ f_4. or, 31 + f_3+ f_4 = 60 (given), or, f_3+ f_4 = 29, or, 10.5 (f_3+ f_4) = 10.5*29, or, 10.5 f_3+ 10.5 f_4= 304.5 ……….(1), Again, Mean = 11.09 (given) = (352.5+10.5 f_3+11 f_4) /60, or, 10.5 f_3+11 f4 = 312.9………(2), Recalling (1), 10.5 f_3+ 10.5 f_4 = 304.5, and using (1) and (2) simultaneously, we get, 0.5 f_4 = 8, or, f_4 =16, and finally, f_3 = 13 (as f_3+ f_4 = 29) (ans).

Index